W0050911

A.B. Sawaoka (Ed.)

Shock Waves in Materials Science

With 169 Illustrations

Springer-Verlag
Tokyo Berlin Heidelberg
New York London Paris
Hong Kong Barcelona Budapest

Akira B. Sawaoka, Ph. D.
Center for Ceramics Research, Tokyo Institute of Technology, 4259 Nagatsuta-cho,
Midori-ku, Yokohama, 227 Japan

ISBN-13: 978-4-431-68242-4   e-ISBN-13: 978-4-431-68240-0
DOI: 10.1007/978-4-431-68240-0

Printed on acid-free paper

© Springer-Verlag Tokyo 1993
Softcover reprint of the hardcover 1st edition 1993

This work is subject to copyright. All rights are reserved, whether the whole or part of the material is
concerned, specifically the rights of translation, reprinting, reuse of illustrations, recitation, broadcasting,
reproduction on microfilms or in other ways, and storage in data banks.
The use of registered names, trademarks, etc. in this publication does not imply, even in the absence of a
specific statement, that such names are exempt from the relevant protective laws and regulations and therefore free for general use.

Preface

This book is written to notice in public how attractive the shock compression technology is as a means of studying materials science. The editor believes that this book is entirely useful for scientists and engineers in the fields of materials science, chemistry, mechanical engineering, and many other areas, and not only in the limited area of shock wave phenomena.

Systematic study of shock compression started in the USA. and Soviet Union in the 1940s, in connection with the elaboration of nuclear weapons. The technology was kept secret for ten years until it was disclosed in 1950 as a means to research the equation of state of condensed matters. Measurements were made to obtain shock compression curves of mainly metals up to several hundred GPa, with the use of explosives. The major reason why metals were chosen is that poreless specimens are easier to get with metals.

It was already known that if pores are present in a solid specimen, the shock temperature at that spot will become abnormally high, and that shock compression of the solid will exhibit remarkably complicated behavior. Non-porous specimen of a brittle solid like ceramics were particularly difficult to obtain, and therefore, systematic study of shock compression of inorganic materials was not performed till the 1980s.

In the 1970s, study on new material synthesis actually started. In particular, forerunning research of dynamic compaction of powder materials was made in Soviet Union. In the 1980s, systematic study was performed in the USA., Germany, and Japan. Through this study, inhomogeneous temperature change during shock compression of porous materials was effectively applied for materials study. Furthermore, quantitative study on the dynamic compaction of mixtures accompanied with chemical reaction has also begun.

To investigate the feasibility of dynamic compression as means for materials science research and for industrial production, the editor held workshops for three times. The first workshop was opened at the Yokohama Campus of Tokyo Institute of Technology in 1986, and the second workshop at the Tokyo Campus of the same Institute in 1988, both sponsored by Tokyo Institute of Technology.

The third workshop was opened in the Zao Mountain in 1991, sponsored by Grant-in Aid for scientific Research, Priority Area "Shock Waves" given by Ministry of Education, Science and Culture, Japan. Free presentation and discussion were the main object of the workshops. Accordingly, no proceedings of the workshops were published. However, so many topics and information stated there were considered to be necessary to report in public, and this book was planned to publish the latest results of study in addition to nine topics in the workshops.

The editor would like to express his hearty gratitude to the fifteen scientists who wrote for the book. He is also thankful for the encouragement by Prof. K. Takayama of Tohoku University, who was the organizer of the research project, "Shock Waves". The research project was the sponsor of the Workshop at Zao Mountain, and the Workshop motivated the author to compile this book.

Readers of this book will possibly understand that the shock compression technology of powder and porous materials has substantial possibility as a means of new material synthesis. Shock compression phenomena of powder materials, however, have not yet been sufficiently elucidated.

Research for understanding the shock compression process of powder or porous materials are being made by analyzing the microstructure of samples recovered from shock compression. The recovered sample is strongly affected by the release process from shock compression state and by residual temperature, and thus, elucidation of the compression process is not easy. Recently, for this purpose, the estimation of temperature and pressure change in powder materials has started with computer simulation. But, the result is not very accurate.

In-situ measurement of temperature and pressure during shock compression, and comparison of the results from the observation of recovered samples and from computer simulation should be performed. Then, they may successfully illustrate the shock compression phenomena of powder materials. This is considered to be the assignment of our future research work.

January 31, 1993
Akira B. SAWAOKA

Acknowledgments

Acknowledgments expressed originally by each author are collected together and rearranged here, chapter by chapter.

Chapter 1: A. B. Sawaoka, the author, expresses his hearty gratitude, indicating that the results of study mentioned in this chapter have been obtained in cooperation with the scientists as follows: Kenichi Kondo, Hideki Tamura, and Koji Dan of Tokyo Institute of Technology, Tamotsu Akashi of Sumitomo Coal Mining Co., Masatada Araki of Nippon Oil & Fats Co., Hiroshi Kunishige of Defense Agency of Japan, and Yasuyuki Horie of North Carolina State University. The work is partly supported by Grant-in-Aid for Scientific Research, Priority Area "Shock Waves", given by Ministry of Education, Science and Culture, Japan.

Chapter 4: Y. Horie, the author, expresses his gratitude as below. The work described in Chapter 4 is the result of true collaboration of many people at three institutions: North Carolina State University, Sandia National Laboratories, and Tokyo Institute of Technology. He owes very special thanks to I. K. Simonsen at NCSU for her selfless, ever-ready involvement, interest, and superb microscopy work, to R. A. Graham and B. Morosin at SNL for friendship, visionary enlightenment, and support they have given to sustain the program at NCSU, and to A. B. Sawaoka at TITech for fellowship and generous support that made his sabbatical leave possible at TITech to muse about the subject of shock chemistry without interruptions. The author is also blessed with colleagues who have given generous support and valuable information . They are F. Y. Sorrell, K. Iyer, and J. K. Whitfiled at NCSU, M. Kipp, M. Bear, P. Taylor, M. Carr, and W. F. Hammetter at SNL, H. Tamura and Y. Oya at TITech, T. Akashi at Sumitomo Coal Mining Co, T. Taniguchi and M. Akaishi at NIRIM, N. Thadhani at Georgia Institue of Technology, and D. P. Dandekar at U. S. Army Material Technology Laboratory, and R. D. Young of Southwest Research Institute. Also, He wishes to acknowledge the students (or former students) who performed much of the work described in this article. They are M-D Hwang, S-K You, D. E. P. Hoy, and L. Bennettare at NCSU, and H. Kunishige, Y. Fukuyama, S. Watanabe at TITech.

Chapter 5: Y. Syono and M. Kikuchi, the authors, express their gratitude as follows. The article is based on the work collaborated with H. Takei and H. Takeya, Institute for Solid State Physics, University of Tokyo, S. Nakajima, CASIO Computer, M. Nagoshi, NKK Corporation, W. J. Nellis and S. T. Weir, Lawrence Livermore National Laboratory, N. Kobayashi, K. Kusaba, T. Atou, A. Tokiwa, Y. Sakaguchi, T. Oku, E. Aoyagi, T. Oh-ishi and K. Fukuoka, IMR, Tohoku University, to whom the authors are deeply indebted. The work is partly supported by Grant-in-Aid for Scientific Research, Priority Area "Shock Waves", given by Ministry of Education, Science and Culture, Japan.

Chapter 6: T. Mashimo, the author, wishes to thank D. E. Grady of sandia National Laboratories and Y. Syono of Tohoku University for their encouragements and supports in carrying out the shock compression research of ceramics. He also wishes to acknowledge the students of his laboratory, A. Nakamura and others. for their invaluable contributions, and to acknowledge the Sumitomo Electric Industries Ltd. for their experimental supports. The work is partly supported by Grant-in-Aid for Scientific Research, Priority Area "Shock Waves", given by Ministry of Education, Science and Culture, Japan.

Chapter 7: M. A. Meyers, S. S. Shanga, and K. Hokamoto, the authors, express their gratitude as follows. The research described herein has been supported, over the 1988-1992 period, by National Science Foundation Grants DMR 8713258 and DMR 91-5835R1. They also thank McDonnell-Douglas, General Electric, the State of New Mexico (through the Center for Explosives Technology Research, in Socorro, New Mexico), and United Technologies Government Products Division for support prior to that period. The help of N. N. Thadhani, Georgia Institute of Technology, has been invaluable. The work described here has been carried out in collaboration with N. N. Thadhani, A. Szecket, L. H. Yu, A. Ferreira, S. L. Wang, and S. N. Chang. The authors benefited immensely from numerous discussions, over the past eight years, with A. B. Sawaoka, Tokyo Institute of Technology, and R. A. Graham, Sandia National Laboratories. The research described here and in the papers cited in the references would not have been possible without the dedicated support of engineers and technicians of the TERA (Terminal Effects Research and Analysis) and CETR (Center for Explosives Technology Research) groups at New Mexico Tech.

Chapter 9: K. Nagayama, the author, wishes to thank T. Murakami of Kobe Design University for his encouragement and continued interest during the course of this work. The author is deeply indebted to Mr. Y. Mori of their staff member for his collaboration for the project and experimental works, especially for the Mach reflection and conical convergence observation. He wishes to thank T. Mashimo of Kumamoto University for his valuable discussions and collaboration and friendship when the author belonged to Kumamoto University. Ultrasonic measurements of polyethylene has been made by a research associate, H. Okabe of their Department, to whom the author wishes to thank for his kind cooperation. The author also wishes to thank S. Ozaki of Try Engineering for the construction of gas gun and camera body. The work is partly supported by Grant-in-Aid for Scientific Research, Priority Area "Shock Waves", given by Ministry of Education, Science and Culture, Japan.

Table of Contents

List of Contributors

BREUSOV, Oleg N., Institute of Structural Macrokinetics, Russian Academy of Sciences, Chernogolovka, Moscow Region 142432, Russia

DREMIN, Anatoly N., Institute of Chemical Physics, Chernogolovka, Moscow Region 142432, Russia

GORDOPOLOV, Yury, Institute of Structural Macrokinetics, Russian Academy of Sciences, Chernogolovka, Moscow Region 142432, Russia

GRAHAM, Robert A., Sandia National Laboratories, Albuquerque, New Mexico 87185, USA

HOKAMOTO, K., Department of Mechanical Engineering, Kumamoto University, Kumamoto 860, Japan

HORIE, Yasuyuki., Department of Materials Sciences, North Carolina State University, Raleigh, North Carolina 27695-7908, USA.

KIKUCHI, Masae, Institute for Materials Research, Tohoku University, Aoba, Sendai 980, Japan.

MASHOMO, Tutomu, High Energy Rate Laboratory, Kumamoto University, Kumamoto 860, Japan.

MERZHANOV, Alexander, Institute of Structural Macrokinetics, Russian Academy of Sciences, Chernogolovka, Moscow Region 142432, Russia

MEYERS, Marc A., Department of Applied Mechanics and Engineering Sciences, University of California at San Diego, La Jolla, California 92093, USA

NAGAYAMA, Kunihito, Department of Applied Science, Kyushu University, Higashiku, Fukuoka 812, Japan

SAWAOKA, Akira B., Center for Ceramics Research, Tokyo Institute of Technology, Midori, Yokohama 227, Japan

SHANG, Shi-Shyan, Department of Applied Mechanics and Engineering Sciences, University of California at San Diego, La Jolla, California92093, USA

SYONO, Yasuhiko, Institute for Materials Research, Tohoku University, Aoba,Sendai980, Japan

THADHANI, Naresh N., School of Materials Science and Engineering, Georgia Institute of Technology, Atlanta, Georgia 30332-0245, USA

Chapter 1
Heterogeneous Distribution of Temperatures and Pressures in the Shock Recovery Fixtures and its Utilization to Materials Science Study

A. B. SAWAOKA

1. Introduction

It is well known that passage of strong shock waves in a solid will cause defects or disorders in the disposition of atoms which constitute the solid. For this reason, chemical reactivity and sinterability of shock-treated bulk or powder materials become larger than before the shock treatment. A typical example of this fact is found in the remarkable increase in catalytic activity of titanium oxide powder after shock treatment [1].

Phase transition to a high dense form is induced by shock compression of a solid, and this also is commonly known. Presence of coesite and stishovite, both high dense forms of silicon oxide, is reported to have been found in a crater, which was produced by the impact of a meteorite on the Earth [2]. These dense forms of silicon oxide were yielded with shock high pressure generated by the impact of a meteorite on the Earth.

Since early 1970s production of micro-polycrystalline diamond powder have been made by using shock waves generated with impact of a cylindrical driver tube at high speed, at E.I. DuPont Company in USA [3]. Recently, two Japanese companies have started production of similar kind of diamond in Hokkaido. In Russia, carbon powders mixed in explosives have been shock-treated for the fabrication of ultra very fine diamond powder [4].

Commercial production of wurtzite type BN by shock compression has been performed by Nippon Oil & Fats Co., Ltd. since 1983. The powder product is used as a raw material for cutting tools.

Thermodynamically, shock compression of materials is a non equilibrium process. Materials with unique microstructure and physical properties can be obtained with shock compression of bulk or powder materials. Very often shock treated materials will provide valuable information for the study of materials science. When such interesting material is gained with shock compression, we must acquire the information on the history of stress and heat to which the material is subjected. However, estimation of the straining and heating process through characterization of recovered samples is very difficult, because of extreme complexity of shock compression process of solids.

Evaluation of temperature in powder during shock compression is one of the difficult subjects of study. Regarding powder material, it is reported that during shock compression, temperature on the powder surface is remarkably higher than inside of the powder [5-7]. A compact of high density can be obtained by shock compression of powder while its surface is being melted. Metallurgical bonding between grains of diamond powder is said to be the most difficult process. Shock compression can achieve strong and dense compact of diamond powder [8,9]. Application of this technique to commercial fabrication, however, involves a number of difficult problems to overcome.

1

Ejection of metal jet from the inner wall of the metal capsule into porous specimen is an unavoidable phenomenon. The present author has developed a synthesizing method with fewer contamination, in which interference of shock wave is utilized in the specimen.

In this chapter, shock treatment of solids and related recovery technique are discussed as tools for materials science study, and not for commercial production.

2. Reasonable Size of Recovery Fixture

To the investigation of mechanical and thermal properties of compacts recovered after shock compression, crackfree samples of at least 5 mm is essential. Cracking of a compact is a serious problem in dynamic compaction processing of powder materials.

Recovered samples are bulk compacts in general. But, in the compacts, amount of cracks will increase as shock pressure increases. Even when only small segments of cracked compacts are obtained, many kinds of characterization can be performed by using the segments. The amount of samples necessary for electron microscopic observation is only 1 mg. X-ray powder diffraction requires samples of several tens of milligrams. Generally speaking, sample mass of several tens of mg is necessary for the measurements of electrical or magnetic properties, except some specific measurements as of critical current in a superconductor.

Metallic capsules are used in the shock recovery experiments of such brittle samples as powders or ceramics. Shock impedance of these samples is normally smaller than that of the capsules. During the passage of plane shock waves through the metal of capsules, temperature or pressure in the samples is not homogeneous, and differs substantially according to each point in the capsules. Accordingly, recovered samples should first be cut into small pieces of less than several mm, and then characterization is made using each piece individually.

To reduce the difference of pressure or temperature distribution at each point in the metallic capsule during shock compression, the aspect ratio, that is, the ratio of diameter to thickness of the sample, should be made as large as possible. For example, when a sample having 1 mm thickness is required, the diameter of the sample chamber is preferably more than 5 mm. The capsule diameter must be at least 10 mm for a sample chamber of 5 mm in diameter. In the shock compression experiment with a gun, where a flyer plate is impinged against a capsule, the necessary diameter of the plate is 12 mm at the minimum.

Since 1974, the author has performed so many number of shock recovery experiments using stainless steel capsules of 24 mm in outer diameter and 40 mm thick with aspect ratio of 2.4 [10-27]. In these sample chambers, pressure value at each point differs more than twice at the maximum. Many kinds of information could be obtained at the same time from each of the samples through precise characterization at different points of one sample. In this chapter, pressure distribution in the capsule obtained by computer simulation and by experiment is compared and discussed.

As a means of generating shock very high pressure having extremely steep pressure gradient in a sample, a technique utilizing Mach stem by shock wave interference can be applied.

3. Shock Wave Reflection in Solids

During shock compression of a solid, if the pressure is sufficiently high, the solid will behave like a fluid. The critical pressure at which a solid apparently changes into a fluid is called Hugoniot elastic limit (HEL). In the pressure range higher than HEL, pressure or volume of a specimen during shock compression can be obtained by applying the preservation law, that is, mass and momentum of the specimen before shock compression remain unchanged during compression.

In practice, pressure or volume during shock compression is estimable from measurement of shock wave velocity and particle velocity in the specimen. This calculation is performed based on the two kinds of preservation law given above.

Product of shock wave velocity and density of a material is named shock impedance. Two plates of different shock impedance are joined together, and when shock waves pass the plates from lower shock impedance to higher, shock waves reflect back at the boundary surface. On the contrary, passage of waves from higher shock impedance to lower will cause the shock waves to pass through the boundary. To confirm this fact, the author measured particle velocity and pressure of quartz glass which is sandwiched between two sapphire plates with higher shock impedance [28].

Figure 1 shows sample assembly for the shock experiment. _ The principle of electromagnetic induction is applied for the measurement of particle velocity. The particle velocity gauge consists of a copper foil circuit, 10 mm thick, and is backed up with 12.5 mm thick polyimide film. The flyer plate, driver plate, delay plate, and reflection plate are made of sapphire. The particle velocity gauge is placed between two quartz glass disks of 0.44 mm thick. The quartz glass sample is sandwiched between the delay plate and the reflection plate. These plates and the gauge are connected with epoxy resin. The width of the binder is less than 15 mm.

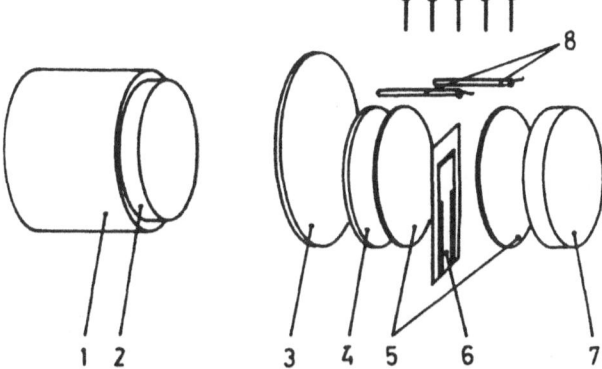

Fig. 1 Target assembly for measuring particle-velocity changes due to shock reverberation using electromagnetic induction. 1-plastic sabot, 2-flyer plate, 3-driver plate, 4-delay plate, 5-fused quartz plate, 6-particle-velocity gauge, 7-reflector plate, 8-self-shorting pins for oscilloscope triggering. H-magnetic field. 2-4 and 7 are made of sapphire crystal (after Kondo et al [28]).

The flyer plate impacts at a velocity of 2.87±0.03 mm/μs against the sample assembly shown in Fig. 1. The flyer plate is accelerated by the two stage light gas gun. The sample assembly is fixed in the magnetic field of 646 gauss. A trace of gauge output signal recorded on the oscilloscope is shown in Fig. 2. The variation of voltage corresponds with particle velocity. In Fig. 2, variation of particle velocity is shown, which is measured by the impedance matching method. The results of measurement and calculation agree in good order. The value of shock pressure can be estimated from the particle velocity.

Pressure in the quartz glass can be measured by replacing particle velocity gauge with manganin gauge. Pressure measurement is performed based on the fact that electric resistance of a manganin foil varies in proportion to pressure. No magnetic field is necessary for the manganin gauge. Change of electric resistance in manganin gauge is shown in Fig. 3. Stepwise pressure increase in quartz glass is shown, as a result of shock wave reflection. In Table 1, pressure change obtained from measurement and from calculation is shown. Both results are seen in good agreement with each other. The pressure attainable in the sample can approach the pressure value in the high impedance material by reducing the width of the sample which is sandwiched between the high impedance materials.

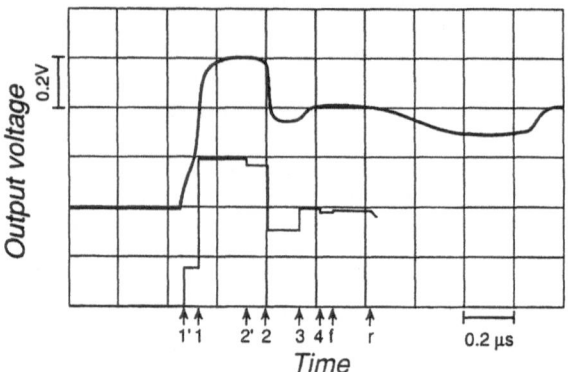

Fig. 2 Particle-velocity changes due to multiple shock reverberations. The upper curve is a trace from a photograph recorded on the oscilloscope in the particle-velocity experiment, and the lower is the calculated curve using the impedance matching method (after Kondo et al [28]).

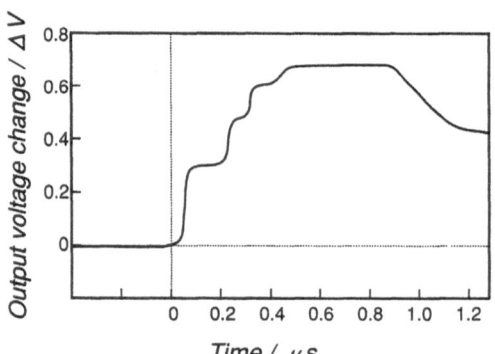

Fig. 3 Pressure changes due to multiple shock reverberations as shown by the manganin gauge profile. Pressure is nearly proportional to the resistance ratio $\Delta R_1/R_0$, obtained from the voltage ratio $\Delta V_1/V_0$ (after Kondo et al [28])

Table 1 Comparison of the observed shock pressure values with the calculated one

State	Pressure (GPa)		
	Calculated	Observed	Difference
1	14.2	14.3	+0.1
2	24.5	24.0	-0.5
3	31.5	29.3	-2.2
4	35.0	...	...
f	40.5	34.1	-6.4

4. Recovery Assembly for a Very Thin Specimen, Sandwiched between High Impedance Materials

Pressure generated in a low impedance sample, which is sandwiched between high impedance metals, is almost on the similar level with the pressure in the high impedance metal, when the thickness of the sample is very small. This is because shock waves, that enter low impedance sample from high impedance material, continue to reflect between the metal plates on both sides, and the pressure in the sample draws nearer and nearer to the pressure in the metal. Thus, the pressure in the sample increases in stepwise manner as shown in Fig. 4. The area shown with hatching in the volume-pressure diagram in Fig. 4 is equivalent to the internal energy increment due to shock compression.

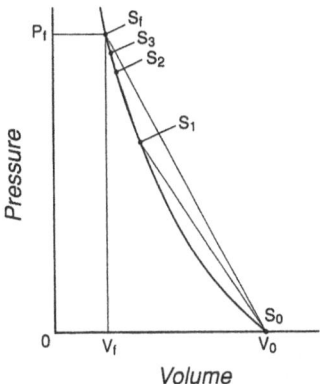

Fig. 4 Pressure-volume diagram comparing the specific energy behind the shock wave front.

The internal energy increase due to simple shock wave compression without wave reflection corresponds to the area $\Delta V_0 S_f V_f$ in the figure. It is quite evident that the internal energy increase in case of stepwise pressure growth is smaller. This means that shock temperature is smaller for stepwise pressure increase.

Nellis et al. developed a technique to generate shock high pressure using a thin sample sandwiched between materials of high shock impedance, and made shock synthesis experiment of Nb3Si [29]. The shock recovery assembly developed by Nellis et al. is shown in Fig. 5. Flyer plate of 22 mm in diameter, made of Ta, is accelerated in the two stage light gas gun, and is impinged against shock assembly.

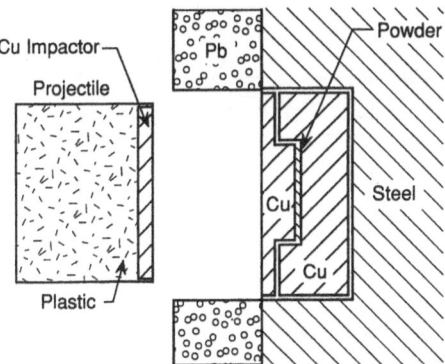

Fig. 5 Schematic of sample recovery fixture and incident projectile (after Nellis et al [30]).

In case of powder sample, volume reduction due to shock compression is comparatively large, and high shock temperature is reached in spite of the compression accompanied with shock wave reflection. The temperature directly after the passage of shock waves is called residual temperature. The residual temperature gained with shock compression of powder material is much higher than the temperature of metallic plates on both sides of the material.

Heat in the hot sample is transmitted to the metallic plates by thermal conduction, and is rapidly cooled. Nellis et al reported on the estimation of melting and rapid solidification due to shock compression, where Cu-Zr powder sample was contained in a copper capsule with a sample chamber of 500 μm in width [30].

Kondo et al reported that a new type of diamond was gained by impacting flyer plate against shock assembly of 50 μm thick carbon film which was sandwiched between copper plates [31].

5. Recovery Fixture Having Thick Specimen Chamber

Recovery of shocked sample is not difficult, if the thickness of the sample is very small. But, as the thickness becomes larger, deformation of recovery fixture due to shock compression becomes larger, which makes recovery work more difficult.

5.1 Gun Recovery Experiment

When a projectile consisted of flyer plate and plastic sabot impacts against a solid body at high speed, a deep crater is formed on the target. The sequence of the crater formation is schematically shown in Fig. 6, where a flyer plate is impacted against a semi-infinite metal target. When the metal plate impacts the target sample, the shock waves propagate in a tapering manner into a point through the sample. The compressed volume forms the shape of a reducing circular cone. The boundary surface receives a large discontinuous shearing force, and one must consider this factor in the design of impact recovery fixtures.

When a billiard ball is struck, the kinetic energy is transferred to the struck ball, and the striking ball stops while the struck ball alone starts to move. This principle is called the momentum trap, and with this method, it has become possible to recover samples without causing significant damage.

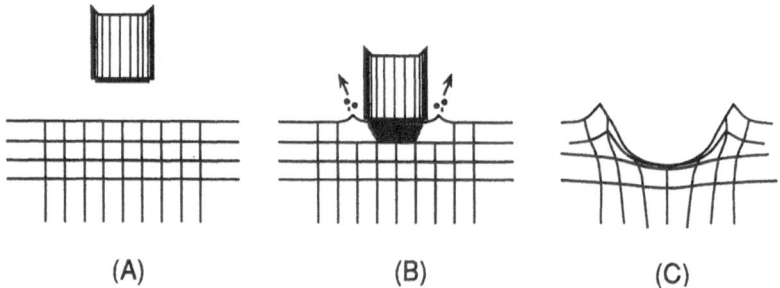

(A)   (B)   (C)

Fig. 6 Crater formation: (A)-before impact, (B)-during impact, (C)-after impact.

After many trial and error runs, an impact recovery device was designed as shown in Fig. 7. The sample material is contained in a recovery capsule of stainless steel and the capsule is placed in a mild steel cylinder. A thick stainless steel disk of the same diameter as the cylinder is placed behind the capsule, and both the cylinder and the disk are placed in a mild steel container. When the projectile impacts the recovery capsule and the shock waves pass through the capsule and reach the thick disk in the rear, the disk begins to move to the rear through the mild steel container. The steel cylinder spreads outward around the capsule. The impact energy is transmitted as much as possible from the sample capsule to surrounding objects to prevent damage to the sample.

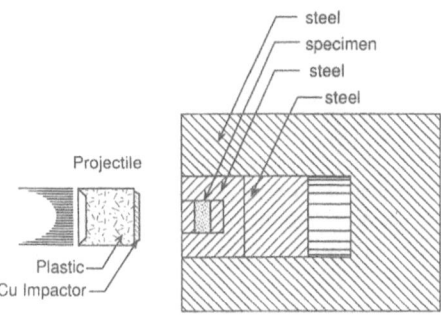

Fig. 7 Cross sectional view of recovery assembly for hyper velocity impact.

If the diameter of flyer plate is large compared to the size of recovery capsule, recovery is less difficult. The present authors have experimented on the operation of 20 mm and 40 mm powder guns whose maximum projectile velocity is 2 km/s.

In the case of projectile of 20 mm diameter, the largest allowable diameter of the capsule and the sample chamber are 18 mm and 10 mm, respectively, and proper thickness of the sample is 4 mm. In the case of 40 mm projectile, other parameters are almost twice as large.

5.2 Explosive Recovery Experiments

Two kinds of recovery fixture are shown in Fig. 8. The recovery capsules containing specimen can be shock-treated by hypervelocity impact of metal flyer plate. Fixture consists of a plane wave generator and a momentum trap type holder. Two kinds of plane wave generator were developed; the mouse trap type [32] and the punched triangle sheet explosive type [33]. The former has been used at Tokyo Institute of Technology and the latter used at New Mexico Tech, USA.

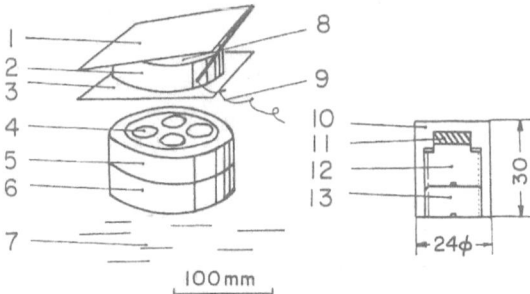

Fig. 8 Mouse trap type shock recovery fixture. 1-sheet explosive and aluminum plate, 2-plastic ring, 3-flyer plate, 4-capsule, 5-steel ring, 6-steel disk, 7-water, 8-main explosive, 9-detonator, 10-capsule, 11- sample, 12, 13-plug.

A steel flyer plate with 3.2 mm or 4.3 mm thickness is accelerated downwards by the detonation of a main explosive charge, which is initiated by impacting aluminum plate of the mouse trap type plane wave generator or glass plate of other generator. The flyer plate is accelerated over a 10-25 mm standoff distance and impacts the end face of the capsules embedded in the cavity of momentum trap recovery fixture.

A numerical simulation of pressure and temperature history in the recovery capsule has demonstrated using two dimensional CSQ code [34,35] and PISCES CODE [36,37] that inhomogeneous distribution of pressures and temperatures existed in a steel capsule containing ceramics powders. Studies [8,9] have shown that highly dense polycrystalline diamond can be obtained by using the momentum trap recovery fixture. However, crack in recovered compacts have been a serious hindrance for industrial applications. The control of residual temperature as well as shock temperature is thought to be essential in elimination of the cracks. For instance, utilization of exothermal chemical reaction is known to be an effective technique for residual temperature control [38]. It was also demonstrated that diamond and silicon system may serve as a model system and that 7.2 vol% silicon is the optimum content [39].

6. Numerical Simulation of Shock Compression in the Recovery Capsule

The purpose of this study is to show insight into the shock compression processing of diamond powders through use of the two dimensional PISCES code with special focus on sample thickness [37].

Fig. 9 shows a schematic of the shock recovery assembly. The radial boundary is defined to be rigid along the perimeter of the capsule. This restriction will have no effects on the loading conditions of powder samples for the time duration of our interest. The flyer plate is made of mild steel and has the thickness of 4.3 mm. The diameter of the powder chamber is 12 mm. Shock processing conditions are as follows: impact velocity, 2.5 km/s; initial powder density, 65% of theoretical; powder samples, diamond without additives and a mixture of diamond and 7.2 vol% silicon.

The PISCES code performs integration of the continuum differential equations of the balance of mass, momentum, and energy using a finite difference scheme. The constitutive model of powder mixtures used was discussed in [36].

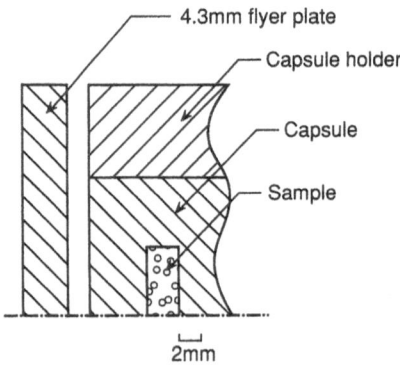

Fig. 9 Schematic layout of the shock recovery assembly used for numerical simulation.

Calculations were performed for both diamond without additives and diamond with silicon. The sample thickness of the pure diamond was 5 mm. For the mixed powder, we investigating of the three thicknesses of 5, 3, and 2 mm were made. In the case of the latter, the model calculations have included the effect of the exothermal chemical reaction between diamond and silicon. However, the effect of the exothermic heat on pressure is found to be negligibly small.

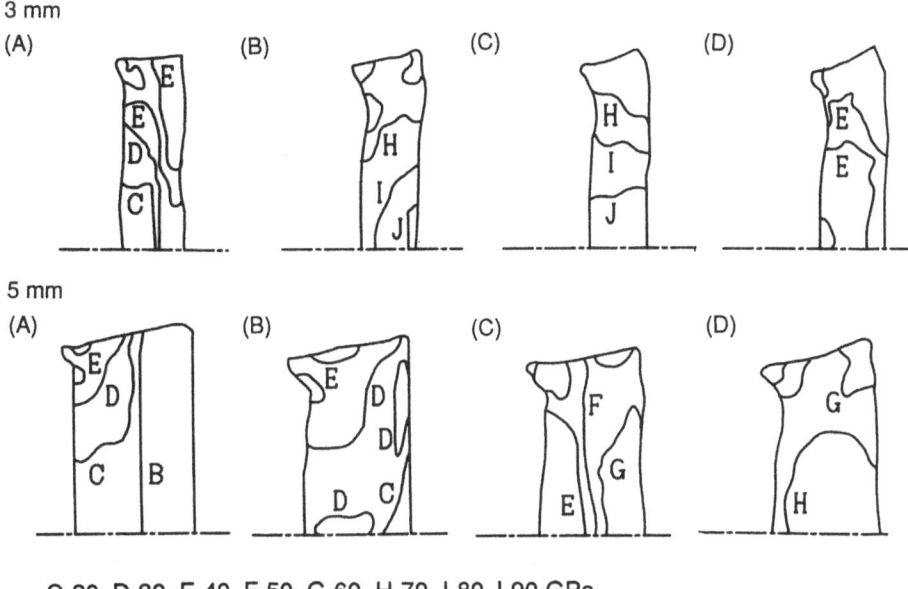

C-20, D-30, E-40, F-50, G-60, H-70, I-80, J-90 GPa

Fig. 10 Pressure contours in the sample with 3 and 5 mm initial thickness. (A)-1.5 μs after impact, (B)-1.8 μs, (C)-2.0 μs, (D)-2.4 μs (after Sawaoka et al [37]).

Selected results of the calculation are shown in Figs. 10. These figures show pressure contours in the mixed powder sample at (A) 1.5 μs, (B) 1.8 μs, (C) 2.0 μs and (D) 2.4 μs after impact. The sample thickness was 3 and 5 mm, respectively. Fig. 11 shows pressure histories at two locations on the center axis; middle of the sample and near the rear surface. They will be referred to as point-1 and point-2 in discussion.

There are several characteristic features that one can discern from the examination of Figs. 10 and 11.

(1) The development of pressure contours in time is strongly influenced by the magnitude of the sample thickness. The pattern is very complex due to the interactions among incident shock, residual wave emanating from the front edge of the sample, and reflected waves from the rear surface. However, on the central axis, essential features of the pressure histories are very similar. Obviously, the thinner the sample, the smaller the difference in pressure history between the two locations.

(2) The maximum pressure on the center axis (90 GPa) is found with the 3 mm thick sample, indicating the existence of optimum shock wave interactions to magnify pressure.

(3) In the 5 mm sample the region of the maximum pressure (region H) did not cover the front surface area.

(4) The magnitude of the initial shock at point-1 remains constant at about 15 GPa, but the corresponding magnitude at point-2 is influenced by the sample thickness. The magnitude increased from 35 to 50 GPa as the thickness decreased from 5 mm to 2 mm.

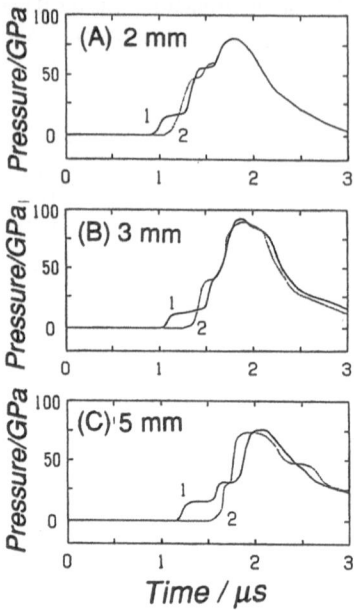

Fig. 11 Pressure histories in the powder sample with initial thickness: (A) 2, (B) 3, and (C) 5 mm: 1- middle point, 2- point near the rear surface on the center axis (after Sawaoka et al [37]).

In the multiple shock processing of powder materials, a major portion of internal energy increase is caused by the first shock compression. For instance, in the case of 5 mm sample, the first shock at p-1 is 14 GPa and compresses the powder from 65% to 90% of theoretical. Subsequent shocks of 32 and 74 GPa bring the powder to the near theoretical value. On the other hand at point-2, the final compression is achieved by two jumps consisting of 32 and 72 GPa shock. The change of internal energy at point-2 is much larger than that at p-1. Accordingly, the same observation can be made of shock temperature in spite of the comparable peak pressures. Table 2 illustrates a similar trend for the 2 and 3 mm samples.

Table 2 Step and maximum pressure values at middle point-1 and point near the rear surface on the center axis -2 in the mixture of diamond and 7.2 volume % silicon (after Sawaoka et al [37])

Initial sample thickness (mm)	Point-1			Point-2		
	1st step	2nd step	max	1st step	2nd step	max
	(GPa)			(GPa)		
5	14	32	74	32	-	72
3	16	42	94	42	-	90
2	16	56	80	45	60	80

Probably, the most significant observations one can make from the calculation are 1) that even with the thinnest sample (2 mm), two dimensional effects cannot be ignored and 2) that the thinner the sample, the closer the overall loading becomes to that of adiabatic compression.

7. Shock Compression of a Solid by Means of Converging Shock waves

Al'tshuler et al. reported on the converging technique of shock waves as a means of production of very high shock pressure. When two kinds of shock waves come into oblique collision, Mach stem is formed, which is an ultra high pressure field of relatively uniform pressure distribution. Extremely steep pressure gradient exists in the outskirts of Mach stem. It is possible to expect unique microstructure or metastable new phase induced in a solid by the steep pressure gradient.

Syono et al. assembled a rod of Nb3Si alloy with low shock wave velocity inside a Ta cylinder with high shock wave velocity. They produced an ultra high pressure range, which is called Mach disk, in the rod, by impinging flyer plate against the assembly [37]. The object of their study was synthesis of Nb3Si high pressure phase. In this section, as rather a new application of this technique, discussion is made on the application of the steep pressure gradient generated around Mach disk.

7.1 Simulation of Conically Converging Shock Wave in the Rod-in-Cylinder Structure [40]

The configuration used in the simulation is shown in a sectioned drawing in Fig. 12. A rod of 5 mm in diameter and 60 mm long is inserted in a cylinder of 2024 aluminum with 24 mm outer diameter. The simulation calculation was performed for a collision of copper disk flyer plate of 3 mm thick against recovery assembly at a velocity of 2 km/s. The computer code applied was PISCES as described in the preceding section. Material constants shown in Table 3 are used for the calculation.

The results of the simulation are shown in Fig. 13 and 14. From Fig. 13, it was found that on the central axis of the copper rod an ultra high pressure region of over 100 GPa was built up after 2.4 µs following the collision of flyer the plate. This ultra high pressure region was generated due to the Mach effect. Stress gradient of extreme steepness was found on the periphery of the Mach disk.

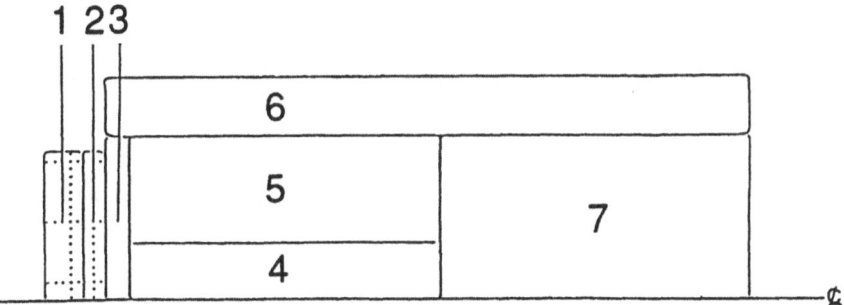

Fig. 12 Rod in cylinder assembly for conically convergent shock wave simulation. 1-Sabot, 2-Flyer plate, 3-Driver plate, 4-Inner rod, 5-Outer cylinder, 6-Holder, 7-Momentum trap.

Table 3 Some physical properties of target materials

Material	Initial density ρ_0 (g/cm^3)	C_0^* (km/s)	S^*	Grüneisen coefficient
Copper	8.92	3.91	1.51	1.99
2024 alminium alloy	2.78	5.37	1.29	2.00

*C_0 and S were obtained from the relation $Us = C_0 + S \cdot U_p$

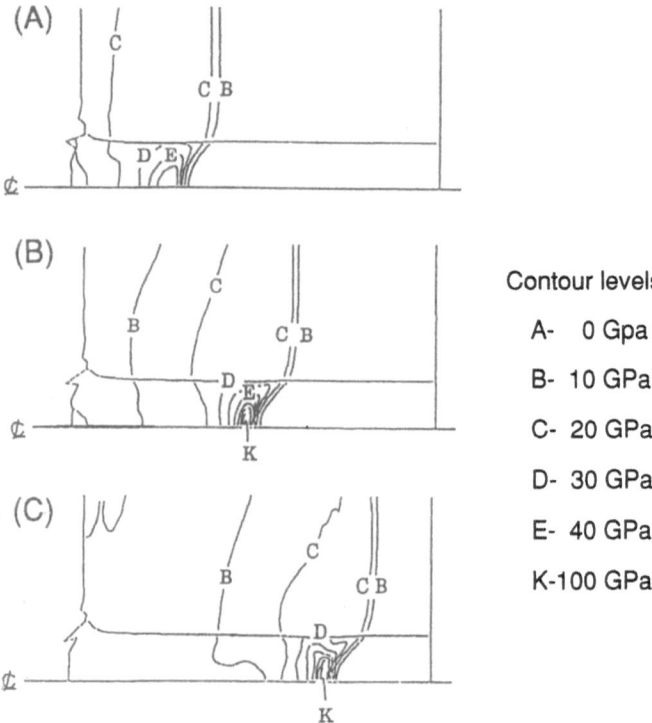

Contour levels

A-   0 Gpa

B- 10 GPa

C- 20 GPa

D- 30 GPa

E- 40 GPa

K-100 GPa

Fig. 13 Pressure contours in 5 mm central rod of copper and surrounding cylinder of 2024 aluminum: (A)-2.0, (B)-2.6, (C)-3.2 μs after impact (after Kunishige et al [40]).

In Fig. 14, pressure change in the course of time is shown at several points along the central axis of the rod. The peak pressure increases with the progress of shock waves, as shown in the figure, and the pressure saturates at about 100 GPa after a progress of 20 mm from the surface of collision. A contour of the maximum pressure in the rod is shown in Fig. 15.

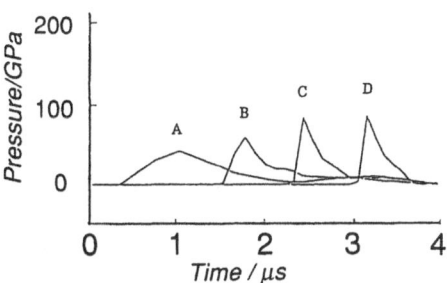

Fig. 14 Pressure-time histories to the several positions on the central axis in 5 mm copper rod. A-0.75, B-5.75, C-10.75, D- 15.75 mm from the front surface (after Kunishige et al [40]).

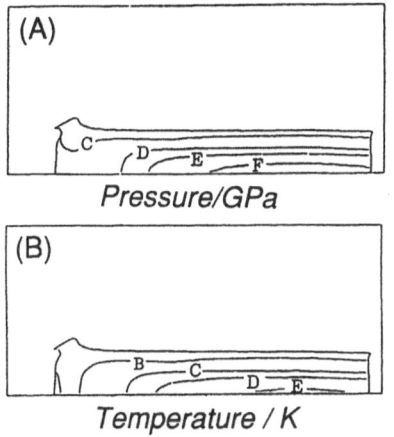

Contour levels

(A) Pressure   (B) Temperature

(A) Pressure	(B) Temperature
A- 0 GPa	A- 0 K
B- 20 GPa	B-1000 K
C- 40 GPa	C-2000 K
D- 60 GPa	D-3000 K
E- 80 GPa	E-4000 K
F-100 GPa	

Fig. 15 Maximum pressure and temperature contours in 5 mm rod (after Kunishige et al [40]).

7.2 Shock Compression of Iron by Using the Conically Converging Technique [41]

An iron rod of 99.999% purity was shock-treated using the shock recover fixture shown in Fig. 12. The iron rod was 5 mm in diameter, 60 mm long. It was placed inside a cylinder of 2024 Al with the outer diameter of 24 mm. An iron plate of 3.2 mm thick was impinged against a momentum trap, which contained a capsule, at velocities of 1.6, 2.2, and 2.3 km/s.

In Fig. 16, optical micrographs of the sample section are shown before shock treatment and after 2.2 km/s impact. It is estimable that convergence of shock waves and formation of Mach disk as seen in Fig. 17 took place due to the impact at 2.2 km/s. The micrograph in Fig. 16 indicates a microstructure corresponding to the behavior of shock waves.

Iron is known to make phase transition from α-phase to ϵ-phase at 13 GPa. Also it transforms to γ-phase at a shock high temperature. Both ϵ-phase and γ-phase are not kept frozen, but go back to α-phase under the atmospheric pressure and temperature. However, microstructure of iron changes substantially due to its experience of either ϵ- or γ-phase. In particular, shear stress gives considerable influence on the phase change. To our regret, scientific knowledge at present cannot quantitatively explain the relation between the microstructure of converging shock compressed iron and shock compression history. The authors are making more quantitative study with samples of single crystal Copper.

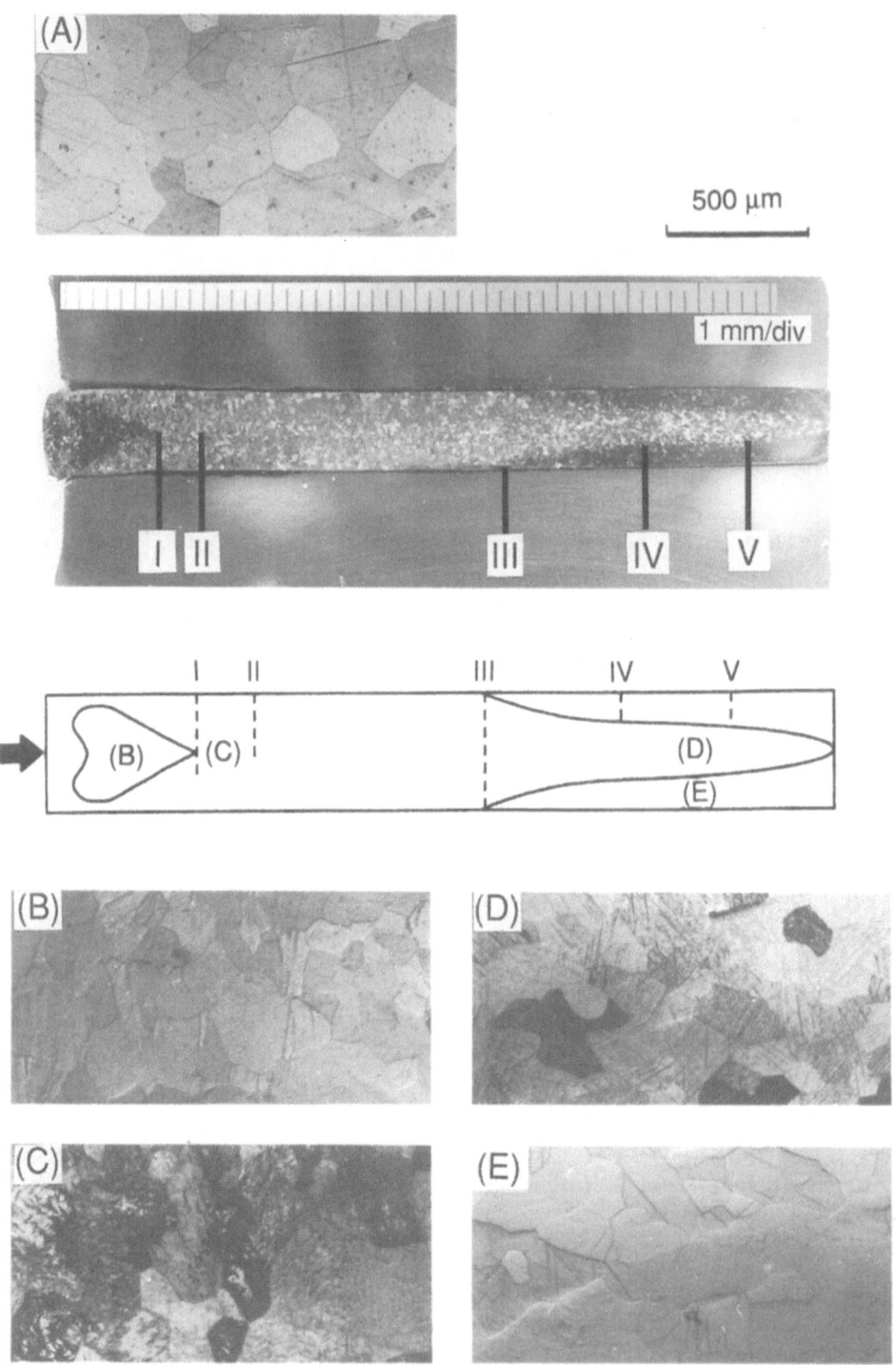

Fig. 16 Cross sectional view of the recovered iron rod. Impact velocity of the flyer plate 2.2 km/s, (A)-before and (B)~(E)-after shock treatment (after Dan et al [41]).

14

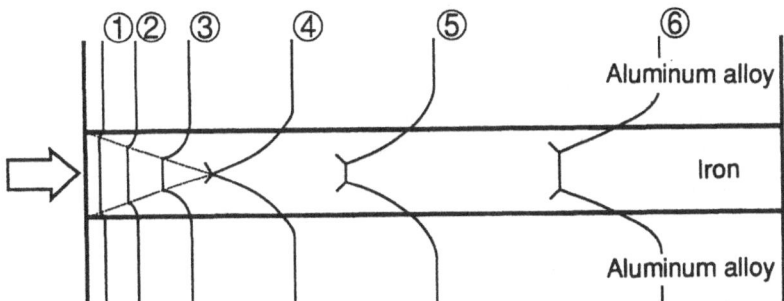

Fig. 17 Formation and propagation of a conically converging shock wave in an iron rod.

8. Conclusions

Observation of microstructure and measurement of physical property made on a shock-compressed solids can disclose the effect of shock compression on the sample material. If the sample is a metal with sufficient mechanical strength, recovery of shocked sample is easy, but if it is brittle like ceramics, its recovery is difficult. A metal capsule is usually used for recovery fixture. This is the case also for powder sample or porous ones.

The shock impedance of powder or porous samples is much less than that of the capsule metal. Shock waves, consequently, propagate faster in the capsule material than in the sample, going into the sample through the inner sides of the capsule. Thus, the history of pressure and temperature in each part of the sample becomes very complicated, and is different at each spot.

The distribution of pressure and temperature in the metal capsule is estimable at present only with computer simulation. The reliability of calculated result is not very high. Development of a reliable simulation method will be one of the important subjects for study in the future. If the history in detail of pressure and temperature is known in each part of the sample, plenty of information can be simultaneously obtained through observation of small parts in one capsule and with analysis thereof.

Ejection of metal jet from the inner wall of the metal capsule into the powder sample is an unavoidable phenomenon, as a result, causing contamination in the specimen. Thickness of a specimen in the capsule for synthesis experiment should desirably be as large as possible, and characterization of the synthesized sample should be made by using specimen taken from a part of the sample sufficiently apart from the capsule walls.

The present author has developed a synthesizing method with fewer contamination, in which interference of shock waves is utilized in a capsule.

References

1. Graham RA (1981) *Bulletin of the American Physical Society 25*: 495.
2. Chao ECT (1967) *Science 156*: 192.
3. Balchan AS, Cowan GR (1972) US Patent 3, 667, 911.
4. Dremin AN, Breusov ON (1993) In: this book 17.
5. Raybould D (1981) *J. Mater. Sci. 16*: 589.
6. Morris DG (1981) *Metal. Science 54*: 116.

7. Kondo K, Soga S, Sawaoka A, Araki M. (1985) *J. Mater. Sci. 20*: 1033.
8. Akashi T, Sawaoka AB (1987) *J. Mater. Sci. 22*: 3276.
9. Kondo K, Sawai S (1990) *J. Am. Ceram. Soc. 73*: 1983.
10. Swaoka A, Soma T, Saito S (1974) *Japan, J. Appl. Phys. 13*: 891.
11. Soma T, Sawaoka A, Saito S (1974) *Mat. Res. Bull. 9*: 755.
12. Soma T, Sawaoka A, Saito S (1975) *Proc. 4th International Conf. on High Pressure-1974, 446.*
13. Akashi T, Sawaoka A, Saito S, Araki M (1976) *Japan. J. Appl. Phys. 15*: 891.
14. Mashimo T, Nishii K, Soma T, Sawaoka A (1980) *Phys. Chem. Minerals 5*: 367.
15. Soga S, Kondo K, Sawaoka A, Tanaka Y (1983) *J. Mat. Sci. Letters 2*: 673.
16. Akashi T, Sawaoka A (1984) *Mater. Letters 3:* 11.
17. Kondo K, Soga S, Sawaoka A, Araki M (1985) *J. Mat. Sci. 20*: 1033.
18. Akashi T, Lotrich V, Sawaoka A, Beauchamp EK (1985) *J. Am. Ceram. Soc. 68*: C-322.
19. Akashi T, Sawaoka AB (1986) *J. Am. Ceram .Soc. 69*: C-78.
20. Kondo K, Soga S, Rapoport E, Sawaoka A (1986) *J. Mat. Sci. 21*: 1579.
21. Akashi T, Pak H-R, Sawaoka AB (1986) *J. Mat. Sci. 21*: 4060.
22. Akashi T, Sawaoka AB (1987) *J. Mat. Sci. 22*: 1031.
23. Akashi T, Sawaoka AB (1987) *J. Mat. Sci. 22*: 1127.
24. Akashi T, Sawaoka AB (1987) *J. Mat. Sci. 22*: 3276.
25. Akashi T, Sawaoka AB (1988) *Advanced Ceramics Mat. 3:* 288.
26. Kunishige H, Sawaoka AB, Akashi T, Horie Y (1991) *Shock Waves 1*: 165.
27. Simonsen I, Horie Y, Akashi T, Sawaoka AB (1992) *J . Mat. Sci. 27*: 1735.
28. Kondo K, Yasumoto Y, Sugiura H, Sawaoka A (1981) *J. Appl. Phys. 52*: 772.
29. Nellis WJ, Radousky HB, Geballe TH, Hammod RH, Koch R, Hull GH (1986) *Applied Physics Letters 49*: 413.
30. Nellis WJ, Gourdin WH, Maple MB (1988) *Shock Waves in Cnodensed Matter-1987*: edited by Schmidt BC, Holmes NC, Eleesevier Science Publishers B. V., New York, 407.
31. Hirai H, Kondo K (1991) *Science 253*: 772.
32. Akashi T, Sawaoka A (1976) *Japan J. Appl. Phys. 15*: 891.
33. Akashi T, Lotrich V, Sawaoka AB, Beauchamp EK (1985) *J. Am. Ceram. Soc. 68*: C-322.
34. Norwood FR, Graham RA, Sawaoka AB (1986) Shock Waves in Condensed Matter -1985: Gupta YM, Plenum Press, New York, 837.
35. Norwood FR, Graham RA (1992) *Shock-Wave and High-Strain-Rate Phenomena in Materials* : edited by Meyers MA, Murr LE, Staudhammer KP, Dekker. Inc, New York, 989.
36. Kunishige H, Horie Y, Sawaoka AB (1990) *Shock Compression of Condensed Matter-1989* : edited by Schmidt SC, Johnson JN, Davison LW, Elesevier Science Publishers B.V., New York, 527.
37. Sawaoka AB, Kunishige H, Tamura H, Horie Y (1992) Shock Compression of Condensed Matter-1991: edited by Schmidt SC, Dick RD, Forbes JW, Tasker DG, Eleesevier Science Publishers B. V., New York, 621.
38. Akashi T, Sawaoka AB (1988) *Shock Waves in Condensed Matter-1987* : edited by Schmidt BC, Holmes NC, Eleesevier Science Publishers B.V., New York, 423.
39. Akashi T, Sawaoka AB (1987) United States Patent 4, 695, 321.
40. Kunishige H, Horie Y, Sawaoka AB (1990) *Proceedings of the 1989 National Symposium on Shock Wave Phenomena* : Tohoku University, Sendai, 181.
41. Dan k, Tamura H, Kunishige H, Sawaoka AB, Mori T (1992) *Proceedings of Shock Wave Symposium, Sendai-1991* : edited by Takayama K, Springer-Verlag Tokyo, 181.

Chapter 2
Dynamic Synthesis of Superhard Materials

A. N. Dremin and O. N. Breusov

1. Introduction

Some ineffective attempts to use explosion energy for diamond synthesis had already been made at the end of the previous century [1,2]. However, the opinion confirmed gradually that it was difficult to get considerable structural change of any substance due to shock compression of short duration. Bancroft, Peterson and Minshall had been the first who declared, on the contrary to the generally accepted opinion, the possibility of polymorphic transformation by shock compression. In 1956, they published shock compression data for iron [3]. According to this data the compressibility of iron increased sharply at a shock pressure of approximately 13 GPa. The authors associated the change in compressibility with the polymorphic transformation of iron into some new modification.

Bancroft et al. had investigated metal. In 1959, Dremin and Adadurov published their shock compression data for marble [4]. Marble is a sedimentary rock, with calcium carbonate as the main component. However, it had been discovered that the compressibility of marble, as of iron at 13 GPa, increased sharply at 14.6 GPa shock pressure. In 1961 Alder and Chiristian reported an analogous behavior of carbon at about 20 GPa [5]. It should be noted that the unusual change in compressibility for marble and carbon had been implied by the authors to be also governed by their polymorphic transformations. However, since the change in shock compression curves can be governed by various processes some doubt about the possibility of substance structure transformation under shock compression shorter than 1 μs remained until 1961. Then, DeCarli and Jamieson [6] announced that they had achieved diamonds in some carbon samples recovered after shock compression. Since diamond modification of carbon is a high-pressure modification the finding could be considered as the first direct experimental verification for the structural transformation of a crystalline lattice into another by shock compression.

It should be mentioned that Bancroft's and his coauthors's assumption that the sharp change of the shock compression curve of iron is governed by its polymorphic transformation was also substantiated later [7-9]. It turned out to be a γ-Fe transformation into a new previously unknown iron modification; it was named ε-Fe.

As a result of above-mentioned verifications shock compression has become firmly established as a new method of polymorphic modification synthesis. Undoubtedly, the most important achievements in business, from the practical point of view, had been the synthesis of diamonds from non-diamond forms of carbon [10-11] and diamond-like boron nitride modification from its graphite-like phase [12-14].

2. Synthesis Studies of Diamond and High Dense Forms of Boron Nitride at Chernogolovka

Our investigations on the problem of diamonds synthesis with the help of explosion energy began in Chernogolovka in 1959. At that time there had been some doubts about the possibility of substance structure transformation by shock compression. Therefore we decided to perform two kinds of investigations simultaneously - to determine substance shock compression and to study the samples recovered after shock compression. Disashed graphite (Taiga's deposit) had been used in the experiments. Soon after the start "shock" diamonds had been found in the recovered graphite samples. However the diamond yield after the shock of 40-60 GPa had been rather low (3-5%). At the same time one could see from the graphite shock adiabat that at pressures higher than 40 GPa the full transformation of graphite into its dense modification took place*.

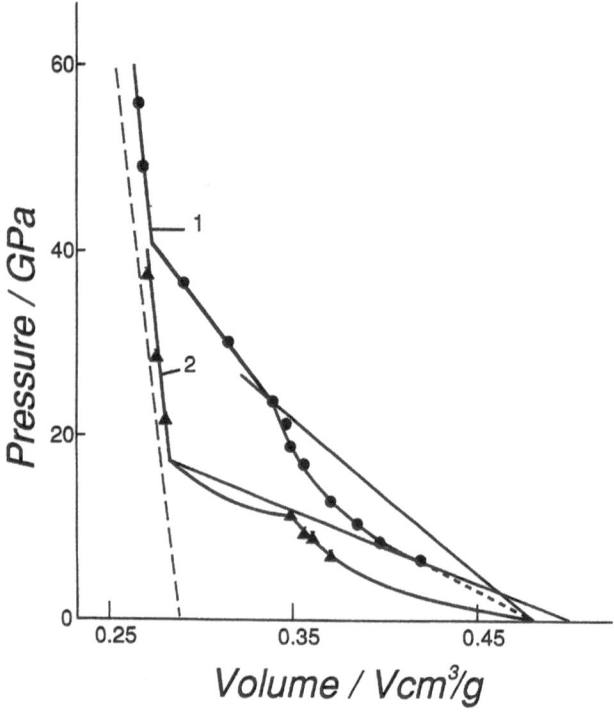

Fig. 1 Shock compression curves of 1-graphite [19] and 2-boron nitride [12].

---

* It should be noted that the Hugoniots for the various graphites are different. Thus, shock experiments at 50 GPa on high initial density pyrolytic graphite performed by Doran [15] and Coleburn [16] showed no evidence of shock compression anomalies that could be related to a transformation into the diamond structure. But McQueen's and Marsh's data for pyrolytic graphite had also testified to its transformation into diamond. The shock parameters for the start of the transformation are depending on the initial density of the sample [17]. According to Gust's data the parameters for natural, synthetic and vitreous graphites are also different [18].

The shock compression data for graphite [19] and boron nitride [12] are shown in Fig. 1. Taking the graphite shock compression data into account, it had been assumed that the diamond originated under shock compression has been transformed back into graphite under the shock residual temperature effect.

Because of the low yield of diamond in the experiments it had been decided to continue subsequent work with boron nitride which is crystallochemical analog to carbon. In particular, boron nitride has polymorphic modifications similar to these of carbon [20-22]: hexagonal   (g-BN) and rhombohedral of low density (2.25-2.29 g/cm^3), cubic structure of intermediate density (2.8 g/cm^3), hexagonal wurtzite-like (w-BN) and cubic (c-BN) modifications of high density(3.48-3.51 g/cm^3). C-BN modification had been synthesized for the first time by Wentorf [21] in 1957. In 1963 Bandy and Wentorof also first announced w-BN modification synthesis [22]. Both c-BN and w-BN modifications had been synthesized at static conditions.

High-density modifications of carbon and boron nitride are metastable at normal pressure. The shock pressure of low-to-high density transformation of g-BN turned out to be approximately two times lower than that of the corresponding shock pressure for graphite. One can see from Fig. 1 that graphite transformation into diamond starts at shock pressures of about 20 GPa. Its complete transformation takes place at a shock pressure of about 40 GPa while these parameters for g-BN are 12 and 20 GPa. It follows from the data that BN residual temperature after the shock necessary for the low-to-high density transformation is lower than in the case of carbon. Obviously, lower shock pressure, necessary for the transformation, had conditioned rather substantial (~50%) high-density modification yield already in the very first recovering experiments. Obtained BN high-density modification turned out to be w-BN [12], what had been revealed by x-ray and infra-red analyses in 1965.

G-BN adiabat at shock pressures higher than 20 GPa testified in favour of its full transformation into w-BN. Nevertheless the high density modification had not been found within the recovered samples subjected to a single shock of some 20 GPa. In the experiment the intensity of the initial shock wave of 20 GPa decreased during its motion along the sample. It had been about to determine the high-density modification yield dependence on single shock intensity by measuring the intensity and the yield changes along the sample. However high-density modifications had not been detected in any sample. At the same time the infra-red spectrum of g-BN taken from the recovered sample (Fig. 2(B)) turned out to be different from that of the initial g-BN (A) and it coincided with the infrared spectrum (C) of g-BN obtained by annealing of w-BN modifications. The finding meant that the high density modification particles originated under single shock compression had been smaller than that of the critical one [23] at the shock residual temperature. Therefore they had been easily transformed back into low-density modification. Naturally the double recrystallization had an influence on the recovered sample infra-red spectrum.

In ordinary experiments the initial thin layer had been dynamically compressed between two walls of high shock impedance steel which the recovering device consisted of. It is well known that the dynamically compressed material temperature as well as its residual temperature are smaller in comparison to these of a single shock with the same final intensity.

It followed from the above mentioned that the decrease of the residual temperature would promote the new modification yield. To verify the assumption some special investigations had been performed. In the experiments various substances with shock impedance different from the initial modification (g-BN) had been mixed with the modification. In this case g-BN particles were compressed dynamically in the course of reverberating shock waves. Liquid additive substances turned out to be better for their capability to occupy all free volume between g-BN particles. Water testified to be best for achieving additional advantages, namely a high heat capacity. Adding water w-BN yield as large as 95-98% had been obtained in some laboratory experiments; w-BN yield average is about 70-80% in the material industrial production.

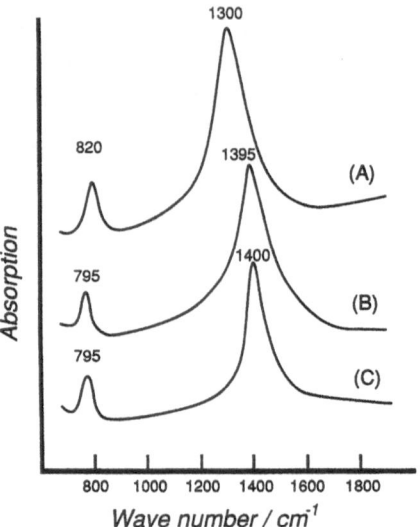

Fig. 2   Infra-red spectra of g-BN modification samples: (A)-initial, (B)-recovered after 20 GPa shock compression, and (C)-obtained by annealing of w-BN modification.

It should be noted that the additive influence on w-BN yield is probably governed not only by the residual temperature of the sample but also by the relative displacements of the w-BN particles. The displacements can promote w-BN small particles strong sticking together and the sticking results in the origin of particles with size larger than the critical one at the samples residual temperature. w-BN polycrystal particles mean size had been approximately 0.5 μm and the average of their coherent scattering regions 15 nm [24]. The material has been named "Chernobor" in honor of Chernogolovka scientific city where it had been produced for the first time [12]. Chernobor has found wide application mostly for the production of cubic boron nitride under high static pressure and temperature combined effect of the impact-and-thermic-strength polycrystals. It appeared that cutting tools made of the polycrystals could treat interrupted surfaces of hardened steels, cast irons and hard-facing alloys. The tool efficiency increased when operating with cooling liquid [25].

After w-BN modification "explosion" production had been arranged the investigation on the "explosion diamond" problem has recommenced. At this occasion a new fundamental approach to the problem has been elaborated. The detonation method for the production of super hard materials has been developed [26-32]. The method is very effective and obviously it will be of long time usage for substance explosion treatment. It has some principal differences to the shock wave methods. Means of shock wave generation and substances under treatment are separated in shock wave methods. In the case of the detonation method explosive and substance under treatment are mixed. The explosive detonation high pressure and temperature effects on the particles of the substances under investigation are the substantial points of the detonation method. At detonation the additive particle substance (for instance graphite) is transformed into a high-density form (in our case diamond). Fast cooling of diamond particles takes place at the process of the explosion product expansion in the rarefaction wave behind the detonation front. The cooling does not allow diamond to transform into graphite and favours its rather high yield.

To achieve these results many investigations had to be performed. The focus had been on the diamond yield dependence from carbons material nature, the particle size of carbon and content in the mixed explosive charge, the mixed charge size, shape and density, the kind of explosive in the mixed charge, the quality of the explosive and carbon material mixing, the nature of gaseous media and their pressure inside the explosion chamber, the mixed charge weight and the explosion chamber volume, as well as on some other characteristics of the method. Highly crystalline graphite of natural and artificial origin, soot, oil coke and other carbon-containing materials had been used for the investigation. Naturally the gas dynamic investigations of the explosive mixture detonation processes had been also undertaken to obtain optimal conditions for the diamond synthesis. However, in spite of the above mentioned large-scale investigations performed it should be stressed out that the problem needs even some further investigations.

For practical realization of the detonation method special installations had been designed, made and put into practice. Some mechanization and automatization of the detonation synthesis technological process had been employed in the installation. It consumes a little more than ten kg of explosives to produce one kg of diamond powder. The size of the diamond particles ranges from 0.005 to 10 μm.

The diamond obtained by the detonation method turned out to be good for many applications, such as drill bit and cutting tool production, cutting, grinding and polishing of brilliants, wear proof coatings etc. The Russian product name of the detonation diamond is "Dalan". This abbreviation stands for: D - detonation, al - first two letters of the Russian name for diamond "Almaz" and an - Academy of Sciences Russian name "Academy Nauk".

3. Considerations on Super Hard Materials Dynamic Synthesis Mechanism

It is known that shock wave methods of super hard materials production are in wide use in many countries. Undoubtedly the detonation method will be widely employed in the nearest future. However, in spite of the apparent practical progress in the explosion synthesis it still remains an unexplained process. It is still unclear what mechanism governs shock wave rate of structural transformation of one crystalline lattice into another. Some considerations on the mechanism of super hard materials shock compression synthesis are proposed in this paper. The considerations are based on the shock compression data and on the data of physic-chemical investigations of samples recovered after shock compression.

Shock compression data of graphite and graphite-like boron nitride testified to their complete transformation into high-density modifications to some extent for each substance shock pressure; ~40 GPa for graphite [19] and ~20 GPa for graphite-like boron nitride [12]. That means the transformations take place inside the shock wave front. At the same time the distribution for the particle size of the high-density modification turns out to be dependent on the duration of shock compression. But it has been revealed that particles which can be single and polycrystal both consist of coherent scattering regions (CSRs) of about 10 nm mean size [24].

So the findings testified to the high-density CSRs origin during the process of initial substance compression inside the shock wave front. Since the CSRs appear inside the front their size can not exceed the front width value. Obviously, the CSRs mean size is similar to that of the shock wave front width. According to the theory [33,34] the width is governed by mass- and heat-transfer processes, which are substantial characteristics. The fact that the process takes place inside the shock wave front means that the initial substance changes completely into an enormous number ($\sim 10^{18}$/cm^3) of high-density modification CSRs of finest size (~ 10 nm). One can evaluate the transformation time being in the range of some ps by dividing the shock wave front width by the shock wave velocity.

Naturally the question arises: what is the mechanism of the CSR fast origin inside the shock wave front ? One can introduce the following consideration on the problem. It is known that shock compression in solids is not hydrostatic. From macroscopic point of view it is one-dimensional. However, at shock pressure higher than that corresponding to the Hugoniot elastic limit (P_{HEL}, see Fig. 3) a transition from one-dimensional to three-dimensional compression takes place. During the transition process materials inside the shock wave front can change into fragments, the fragments relative displacements are occurring in the transversal direction of the shock motion. It is a well established fact for brittle materials [35] and it is not improbable for any condensed substance owing to a tremendous rate of its loading inside the shock wave front [34]. It is obvious that due to fragments motion the substance will be transformed into single or polycrystal particles of the substance initial modifications if the shock intensity is lower than that necessary for polymorphic transformation (P_t in Fig. 3). However, if the shock intensity is higher than P_t the substance is transformed into new modification.

It should be mentioned that from the hydrodynamics point of view, the transition from one-to-three dimensional compression is analogous to polymorphic transformation with an increase of the density. The only difference is that a new crystalline lattice originates at the process of substance polymorphic transformation and in the case of the material's loss of strength one has some crystalline lattice but with smaller parameters. In both cases two-shock configurations appear within some proper pressure ranges; Fig. 3 (P_b-P_{HEL}) shows the pressure interval in which two-shock configurations exist due to transition from one-two-three dimensional compression and (P_c-P_t) - due to the initial transformation of the material into some new modification. What is not obvious but has been implied a few times is the fact that the second shock wave front and the process transition time sometimes are of the same duration. The matter is that the second shock wave is always weak and broadens during its motion [33].

As far as the mechanism of the CSRs fast origin inside the shock wave front is concerned it can be both martensitic and reconstructive. It should be mentioned that it has been originally implied that only martensitic transformations can proceed in the range of a μs. Therefore an exceptional importance has been attached to initial substance crystalline lattice peculiarities and perfection. It has been implied that with shock compression diamond only originates from rombohedral graphite [6]. The ineffective attempt to synthesize boron-nitride diamond-like modifications from turbostatic graphite-like one has been exclusively associated to the high grade of disorderation of the initial substance structure [36]. It has been implied also that shock compression of carbon and boron nitride high-density cubic modifications proceeds inevitably through the stage of the origin of the substances high-density hexagonal modifications - lonsdaleite and wurtzite-like boron nitride. Again, these synthesis of high-density modifications is supposed to proceed through the martensitic mechanism and their transformation into cubic modifications - through the reconstructive mechanism [37, 38]. However, it has been elucidated that the assumption is not quite true. The matter is that cubic diamond has been synthesized by shock compression of amorphous carbon [39], and high-density cubic boron nitride has been synthesized both from highly crystallized [12] and turbostatic boron nitride [40]. However, cubic boron-nitride yield at detonation synthesis from graphite-like boron nitride turned out to be approximately two times larger than that

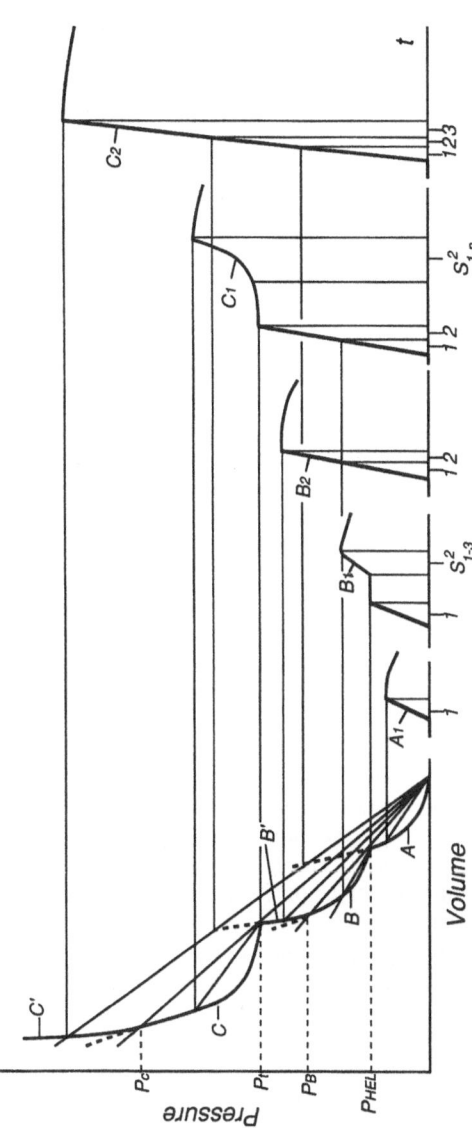

Fig. 3   Pressure P-Volume V shock compression curves and shock waves profiles over time for some materials which displace an elastic behavior up to the HEL pressure P_{HEL} and start to transform into a new modification at proper pressure P_t. A - the material elastic one-dimensional single shock compression curve. B - the material two shock three-dimensional compression curve; B' - the material single shock three-dimensional compression curve. C - new modification two shock compression curve; C' - new modification single shock compression curve. $(P_b - P_{HEL})$ - pressure interval in which two shock configurations exist due to the one-to-three dimensional compression transition. $(P_c - P_t)$- pressure interval in which two shock configurations exist due to the initial material transformation into some new modification. A_1 - elastic single shock pressure profile; 1 - the shock wave front, one-dimensional elastic compression takes place inside the front. B_1 - two shock configuration originating due to the material loss of strength at the P_{HEL}; 1 - the first elastic shock wave front; initial material prepares to be broken into fragments inside the front, the fragments relative displacement is taking place inside the second shock wave front S^2_{1-3} and results in the three-dimensional compression of the material. The second shock is weak [33] and its front broadens during the shock motion; therefore the second wave front width is not constant and has to be larger than the one-to-three dimensional transition time. B_2 single shock three-dimensional compression pressure profile; 2 - one-to-three dimensional transition zone. C_1 - two shock configuration originating due to this material transformation into a new modification. The transformation takes place inside the shock wave front S^2_{3-n}. Since the shock is weak it broadens during its motion; due to this reason the second wave front width as in the case of B_1 for the one-two-three dimensional compression, transformation is not constant and has to be larger than the transformation time. C_2 - single shock, new modification compression pressure profile; 3 - initial-to-new modification transformation zone. The total width of the shock front is about 10 nm.

from wurtzite-like modification [41]. It is even more surprising that diamond can be synthesized with dynamic effect from some carbon chemical compounds [42-50]. All afore-mentioned data testify to the possibility of the reconstructive mechanism of the super hard materials shock synthesis.

So, at present both martensitic and reconstructive mechanisms are believed to be possible. However, it should be mentioned that a martensitic mechanism works at shock synthesis apparently only from highly-crystallized low-density carbon and boron-nitride modifications, whereas a reconstructive one can work at the synthesis from both amorphous as well as highly-crystallized materials. For highly-crystallized low-density modifications a martensitic transformation takes place probably just during the process of the initial substance fragments transversal displacements inside the shock front (see Fig. 3: C_1 and C_2 shock wave pressure profiles). In the case of the materials reconstructive and diffusive transformation the defect originated at the very beginning of the compression process is probably the new modification nuclei [51], the fragments relative displacements promoting the nuclei growth and the transformation by such a way the initial substance fragments into the new modification CSRs. Obviously, in both cases the new modification CSRs sizes do not exceed the sizes of the initial modification fragments originated within the shock front.

Amorphous carbon and boron nitride are transformed into their high-density modifications most likely through a reconstructive mechanism. However, it is not improbable that only the materials low-density modifications of crystalline state (hexagonal or rhombohedral graphite as well as graphite-like boron nitride) originate through a reconstructive mechanism at the very beginning of the transformation process. As far as the materials high-density modifications are concerned they could come into being during the final stages of the transformation process through a martensitic mechanism [52]. At any case the initial material is transformed completely into the high-density modification CSRs. Since the transformation takes place inside the shock wave front the CSRs sizes can not exceed the size of the shock wave front width which is the materials characteristic. The CSRs mean size of the super-hard materials produced by the shock synthesis from amorphous carbon and boron nitride turns out to be similar to that produced from carbon and boron nitride crystallines.

Thus for shock pressure higher than P_C (see Fig. 3) the entire substance inside the shock wave front is transformed into the new modification CSRs. It should be noted that the substance behind the front will be in a monocrystalline state if the initial sample is a monocrystal. The transformation of the initial monocrystal into the new one is possible since the new modification CSRs crystallographic axes are collinear. Therefore, the CSRs stick easily together as soon as their relative motion ends, the CSRs interfaces being the monocrystal dislocations.

During shock compression the samples are usually compacts of powder of either low-density modifications substances alone (mono- or polycrystalline) or more often a mixture with some cooling additives and at detonation synthesis a mixture with explosives. In all these cases each single low-density modification particle is transformed into the high-density modification particle. The new modification particles at favorable orientation can stick together with polycrystalline material formation. However both monocrystalline as well as polycrystalline materials are chaotically broken into some particles in the rarefaction wave. The particle size depends on the material strength and the stress gradient behind the shock front. The smaller the gradient, the larger the particle size. It is obvious that during the destruction process some relative displacements of particles arises. The displacements result in the origin of liquid melted layers of substance between the particles. Naturally, the liquid layers crystalline very rapidly into the new modification as soon as the displacement is over.

It is obvious that the grinding process, the liquid melted layers origin and their fast crystallization into the high-density modification repeats unless and until the substance state corresponds to states higher than the reverse transformation hysteresis line 2 (Fig. 4). The crystallization results again in a strong sticking together of the particles. The high-density modification polycrystal particles will not be broken more as soon as the substance state corresponds to states lower than the hysteresis line 2. The material further fragmentation will proceed in the rarefaction wave over along the low-density modification layers originating due to the liquid melted layers crystallization into the low-density modification at these states.

As the stress gradient behind the shock front is very large it is not possible to produce new modification polycrystal particles of big size. Usually the shock pressures higher than 20-40 GPa are used for the production of super hard materials, at which the total time of the shock compression ranges from some μs to some ten μs. Thus, for example, charges in the range of some ten kilograms are used for the production of wurtzite-like boron nitride modification [12-14]. The maximum of the distribution curve for the modified particles is corresponding to a size of about 1 μm. Balchan's and Cowan's explosion method of diamond production employs charges up to some tons resulting in diamond particles of some ten μm [53].

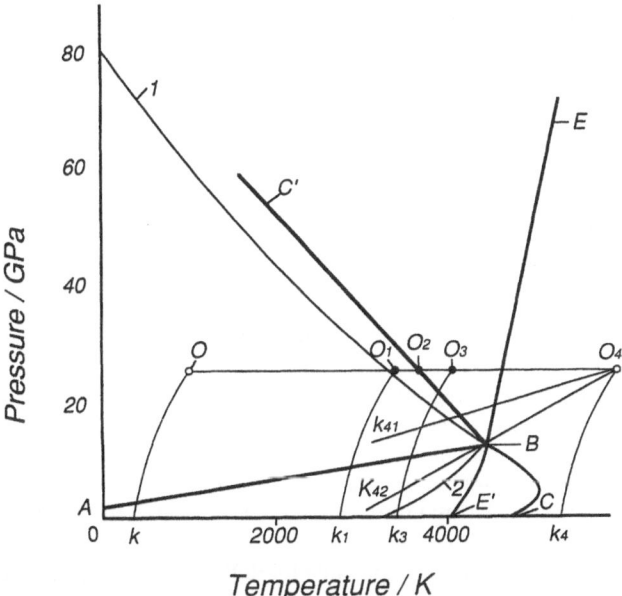

Fig. 4 P-T diagram for carbon according to van Thiel and Ree [57]. AB - carbon-diamond equilibrium line, BE - diamond melting curve (BE' - the curves metastable part), BC - carbon melting curve (BC' - the curves metastable part), 1 and 2 - hysteresis curves of carbon - to - diamond and diamond - to - carbon transformations at shock compression; OK, O_1K_1,...., O_4K_{42}, O_4K_4 - rarefaction curves from different points of some isobar OO_4.

It is well known that carbon and boron nitride high-density modifications are thermodynamically unstable at temperature and normal pressure. They are not transformed into their low-density modifications due to the existence of a large energetic barrier between the high- and low-density modifications. Owing to the barrier the high-density modification particles can be cooled adiabatically in the rarefaction wave and recovered after the shock compression. However, it is also well known that the high-density modification yield is not always complete.

Thus, on the one hand, according to available shock compression data, the low-density modifications are transformed completely into their high-density modifications inside the shock wave front of proper intensity and, on the other hand, the samples recovered after the same shock compression testify in favour of incomplete transformation. Moreover, the high-density modification (w-BN) yield dependence on the low-density modification (g-BN) sample initial temperature turns out to be extreme [54]. Obviously to interpret the data of the shock-induced substance transformation process one has to take the shock-compressed substance temperature as well as the pressure and pressure gradient into account since it is well known that both martensitic and reconstructive transformations are thermoactivated processes [55, 56].

It is advisable to take the shock-compressed substance temperature by comparison of shock compression data with corresponding pressure-temperature state diagrams into account. It is possible to compare the P, T-state diagrams with the states along the Hugoniots of the materials. However, at shock compression with material sample recovering, the maximum pressure in the sample is reached through some shock reverberations. In this case the P, T-states can differ considerably from that of a single shock under maximum pressure. Besides that the P, T-states are some times changed in wide ranges by shock compression of porous as well as preliminary cooled or heated samples. Therefore it is probably convenient to compare the P, T-states along some isobar line. At the same time it is necessary to take into consideration that first order phase transformations are characterized by appreciable hysteresis. The phenomenon is especially expressed for super fast transformations by shock compression.

A carbon P-T diagram according to van Thiel and Ree [57] is presented in Fig. 4. It is: AB - graphite-diamond equilibrium line, BE - diamond melting curve (BE' - the curves metastable part), BC - graphite melting Curve (BC's - the curves metastable part), 1 and 2 -hysteresis curves of carbon-to-diamond and diamond-to-carbon transformations at shock compression. The hysteresis curves are not thermodynamic but kinetic lines; many factors including such as sensitivity of methods to detect the beginning of the transformations, the chemical and physical purity of the initial substance, its crystallochemical perfection, dispersity and certainly the change rate of the thermodynamic parameters which are governing the curves position. Therefore the lines position can differ considerably for the different kinds of graphite as well as for different conditions of dynamic compression. For example it can be different for single and multi-stepped shock compressions. In Fig. 4 a martensitic transformation is presented, carbon-to-diamond transformation hysteresis curve crossing the pressure axis [55,56]. That shows the transformation can be forced at any temperature only by increasing the pressure. This is impossible for diffusive transformation at low temperature [55,56].

The hysteresis positions in Fig. 4 are arbitrary. However it is of little importance for the qualitative analyses of the various states of carbon. Nevertheless it should be noted that the carbon-to-diamond transformation hysteresis curve either coincides or is close to the carbon metastable melting curve (BC' in Fig. 4) [58].

Let's examine the behavior of carbon samples at single shock compression to P-T states along the isobar line 00_4 (Fig. 4). According to afore-said diamond can not originate at the state corresponding to point 0 since it is the left side from the hysteresis curve 1. The state pressure is larger than the P_{HEL} (see Fig. 3), that is the initial graphite crystals are already broken into fragments of 10 nm mean size and the graphite compression is already three-dimensional. However, the pressure is smaller than P_t; that is carbon-to-diamond transformation can not proceed since the shock compression temperature is insufficient to overcome some thermoactivating barrier. The shock-compressed graphite rarefaction proceeds along some isentropy OK, only graphite structure change (crystal perfection, dispersity) and rhombohedral graphite domains can be observed in the recovered samples. It is known that the domains originate at the process of hexagonal graphite mechanical grinding [59].

The OO_4 isobar lines crossing point with the hysteresis curve 1 corresponds to carbon Hugoniot break at the P_t pressure (see Fig. 3). Therefore diamond originates at the state of point O_1 locating the right side of the hysteresis curve 1. The transformation scale at the states probably depends on the temperature, but is independent from the transformation mechanism way. There is some experimental verification of the notion. In the experiments with g-BN samples quasiisentropic shock compression, which does not lead to substance high shock heating, showed no evidence of high-density modifications of boron nitride in the samples with an initial temperature ranging from that of liquid nitrogen to room temperature. As the temperature further increased the transformation had taken place [60]. However, if the transformation is of martensitic nature it can be complete at states just at the right side of the hysteresis curve 1.

So diamond origin process is facilitated and accelerated by an increase of the temperature. However, sample residual temperature increases simultaneously. The thermal annealing scale of the high-density modification increases also. Obviously this is the reason of the experimentally revealed extreme dependence of the high-density modification yield on the low-density modifications initial sample temperature [54].

Point O_2 of the OO_4 isobar line is a particular one; it lies on graphites metastable melting curve BC'. This means that graphite can not exist in crystalline state at the right side of the point since it is impossible to overheat any substance over its melting temperature irrespective whether the melting is ordinary or metastable. Crystalline graphite transformation into metastable melt (for example, at the O_3 point state) sharply increases the mobility of the carbon atoms and facilitates diamond origin as carbon stable modification at the P-T state. Bundy and Kasper have connected graphite-to-diamond direct transformation line in carbon P-T diagram just with graphite metastable melting [58]. This state region is especially favourable for super hard material dynamic synthesis since the initial modification crystals complete melting hampers the high-density modification subsequent annealing. It is known that the process needs some time for the low-density modification nuclei origin at the other side of the reverse transformation hysteresis curve 2 (Fig. 4).

The K_3 point is the right side of the reverse transformation hysteresis curve 2. It means that the O_3K_3 isoisentrope rarefaction leads to diamond kinetic instability regions. In order to prevent diamonds complete graphitization rapid cooling is necessary after the rarefaction. Mixed samples of carbon materials and some additives which have lower residual temperatures in comparison to that of diamond are usually employed for this purpose [53]. Detonation of explosive and carbon material mixed charges [26-29] are other ways to accelerate the cooling. Inert gases and detonation products adiabatic scattering results in fast cooling of gases, detonation products and diamond separate particles.

A peculiarity of the detonation method [26] for substance treatment is its capability to reach high temperature of explosive detonation gaseous products (5,000 - 6,000K [61]) at pressures of 20-30 GPa (O_4 point, Fig. 4). Composition of samples recovered after rarefactions from these states depends on the rarefaction rate. The CSRs size of diamonds crystallizing from liquid carbon will not associated with the shock wave front width and diamond crystals can be more perfect in comparison with that of produced through solid-to-solid mechanism if the rarefaction curve O_4K_{41} intersects the diamond melting curve BE. No nuclei of both graphite and diamond will originate from overcooled freezing liquid and carbon will remain in glassy state if the rarefaction curve O_4K_{42} enter the solid stability region near B point and does not go out the hysteresis zone between the curves 1 and 2. In fact, at some experiments performed in the department glassy carbon has been obtained. The chemical properties of the material (stability to boiling chloric acid) do not correspond to ordinary vitreous carbon but to diamond; its density is about 2.7 g/cm^3 [62].

So, the above-presented consideration on the P-T diagram of carbon with due regard for the position of the hysteresis curves of direct and reverse transformations, for substance fragmentation at shock pressures higher the HEL pressure and for residual temperature effects has given the possibility to realize the following: why does the size of high-density modifications particles depend on the total time of shock compression? why is the high-density modification yield dependence on shock amplitude of extreme nature? why is the yield zero at relatively low temperatures and quasiisentropic compression although the transformation takes place at the same effect on preliminary heated sample and as the temperature further increases the yield decreases again? how do polycrystal particles originate from single crystal graphite? why do the Hugoniots of hexagonal graphites of different crystal structure perfection differ? what are lonsdaleite origin conditions? The possibility of ordinary and diamond-like vitreous carbon production is also substantiated.

One can repeat the consideration presented for crystalline graphite also for amorphous carbon (soot, vitreous carbon and so on); it is possible to assume that amorphous carbon densification will take place without its state change, but only due to the atoms coordinate number increase from 3 to 4. However, in the case of crystalline material origin their formation will be always governed by a diffusive stage irrespective of the transformation process way: whether the high-density modification will originate at once or the low-density modification will crystallize at the beginning and then will be transformed into high-density modification through a martensitic mechanism.

The considerations on dynamic synthesis of diamond and high-density modifications of boron nitride will not be complete if available experimental data on the synthesis of hexagonal modifications-lonsdaleite and wurtzite-like boron nitride are left aside. The modifications are not believed to have thermodynamic stability regions in the P, T diagrams since they have not been synthesized yet from carbon and nitride high-density cubic modifications. Moreover it is also noted that the modifications are transformed into cubic high-density modifications at high pressure and temperatures (about 2,000-3,000 K) and at these temperatures and normal pressure they are transformed into graphite and graphite-like boron nitride [53].

The high-density hexagonal modifications origin possibility is governed by hexagonal graphite and graphite-like boron nitride martensitic transformation at temperatures where the high-density hexagonal modifications are still kinetically stable. In the case of carbon the process becomes complicated. Hexagonal graphite partial transformation into rhombohedral one at any mechanical treatment (including shock compression) is the reason for the complication. And rhombohedral graphite is capable of martensitic transformation into cubic diamond [6]. Therefore, some mixture of cubic diamond and lonsdaleite or some structures with interchange of diamond and lonsdaleite fragments originate at hexagonal graphite shock compression at low temperatures (up to 2,000 K) [63]. However, it should be noted that lonsdaleite has been obtained also from some amorphous carbon [62]. The extent of the material transformation into lonsdaleite has been revealed to be even larger than into cubic diamond. One would think that lonsdaleite can originate also through a reconstructive mechanism. However, the opposite way is possible: hexagonal graphite originates from amorphous carbon through a reconstructive mechanism and is then transformed into lonsdaleite through a martensitic mechanism.

The above-presented considerations on super hard materials shock synthesis mechanisms had been founded on the assumption of direct solid-solid transformation of carbon and boron nitride low-density modifications just into their high-density modifications. The synthesis features have been interpreted on the base of this assumption. However, the features can be interpreted on the very other base, namely, on the assumption that the transformation high rate is governed by liquid phase origin, that is on the assumption of a solid-liquid-solid transformation way. The matter is that a sharp increase of the density can be associated not only with high-density crystalline particles origin but also with liquid densification. It means the densification of liquid layers between the initial material fragments. It is obvious that the liquid part of the whole material increases with the increase of the shock pressure. At shock pressures $P_t > P > P_{HEL}$ (see Fig. 3) the layers temperature is larger than that of the fragments and smaller than the materials melting temperature. At pressures $P > P_t$ the layer temperature exceeds the melting temperature, with the atoms coordinate number increasing from 3 to 4. The increase of the coordinate number is accompanied by volume decrement at the material Hugoniot. In this case the hysteresis curve 1 has to coincide with carbon metastable melting curve BC' (see Fig. 4). However, it should be noticed that the O_2 point (graphite melting point) is deep in the diamond stability region. It means that just as the molten graphite is transformed into diamond it will be highly overcooled and it would be true to consider the molten as a diamond-like glass. It can be investigated if one arranges some favourable conditions for quenching to yield the material. It has been said above that diamond-like glass has been obtained at some dynamic experiments [62].

The temperature increase (change to O_3 point and further) promotes origin and growth of diamond crystals from the melt. The growth can continue in rarefaction waves, lasting some μs. For 10 nm single crystal size it corresponds to ~ 1 cm/s growth rate.

For the isobar points the right side of diamond melting curve (for example O_4) the only possible transformation is the solid-liquid-solid way.

It follows from the above-presented considerations on shock compression mechanisms that they are valid for the detonation method as well. The fact is that the considerations are applicable to each of the separate low-density modification particles since their size is always considerably larger than the shock front width. Therefore one can use the regularities of the shock wave motion in an infinite space for the interpretation of the shock motion effect within the low-density modification separate particles.

However, it should be noted that the considerations are not applicable for the interpretation of the detonation mechanism of ultra-dispersive (3-5 nm particle size) diamond synthesis [42-49]. Obviously the ultra-dispersive diamond formation mechanism differs in essence from that of the shock and ordinary detonation methods. With these methods carbon and boron nitride low-density modifications are transformed into their high-density modifications and at ultra-dispersive diamond detonation synthesis diamond particles originate at detonation of some high explosive of negative oxygen balance as a chemical product of the explosive detonation process itself. Trotil (TNT) is the most often used explosive-component for the synthesis.

It has been shown that ultra-dispersive diamonds originate inside the explosives detonation waves chemical reaction zone. That follows from some experimental findings. First, the diamond particles mean size turns out to be independent from the total time of the explosives detonation wave effect [45]. Second, it had been found out that the explosive detonation velocity dependence on its initial density changes sharply at some proper density characteristic of the explosive [64-67], and the electroconductivity of the explosive detonation products are changing considerably at the same proper density [67-69].

It has been revealed by transmission electron microscopy [46] and x-ray analyses [70, 71] that ultra-dispersive diamonds consist of separate particles and aggregates. The two-stages process (both of a diffusive nature) of diamonds formation has been introduced to interpret the result [45, 70, 71]. According to [45] small (~2 nm size) compact clusters of free carbon atoms originate during the first stage, the compacts ten in number can merge together during the second stage with the formation of diamond clusters of 5 nm in size. And according to [70, 71] the compacts of ~4 nm in size originate at once during the first stage and during the second stage they combine into some aggregates (clusters) of fractal structure (up to 20 nm in size).

It should be noted that the assumption is questionable, whether the first stage is governed by free carbon atoms clustering. It looks more probable that the diamonds originate in the course of the explosives molecules fragments joining with simultaneous separation of unnecessary atoms or groups of atoms. As far as the fragments are concerned they can appear due to the nonequilibrium process of explosive molecules destruction inside the shock front of the detonation wave [72].

4. Conclusion

Phase compositions of carbon and boron nitride samples recovered after their dynamic treatment, high-density modifications yield as well as the modifications particle size are at least three subjects of a discussion of possible mechanisms of carbon and boron nitride polymorphic transformations under dynamic effects. Obviously the transformations can proceed through both martensitic and reconstructive ways.

The products of martensitic transformations can be lonsdaleite (from hexagonal graphite), cubic diamond (from rhombohedral graphite) and w-BN (from graphite-like hexagonal BN). High-density hexagonal modifications (lonsdaleit and w-BN), according to generally-accepted notion are originating only through a martensitic way.

The reconstructive transformation of any form of carbon and boron nitride results in the substances cubic high-density modifications origin. Crystalline graphitic structures can appear as some intermediate products of amorphous carbon transformation.

Reconstructive transformations activation barriers are very high (200 kcal/mol for g-BN to c-BN and for w-BN to c-BN transformations [73]). Therefore they proceed at P, T - states considerably different from the equilibrium states. Experimental data testify to activation energy high value also for martensitic transformations. It means that the transformations P, T - conditions at dynamic compression are governed by the corresponding position of the hysteresis lines; the positions of the equilibrium lines do not play any role. It is known that the initial temperature of the graphite samples, contrary to boron nitride, has an influence on the transformation shock wave pressure [74, 75] although the P, T-diagrams of both substances are similar. However, the findings testify to different positions of the substances transformation hysteresis lines. In the case of boron nitride the positions turns out to be obviously parallel to the temperature axis.

Micro stresses appearing during martensitic transformation processes usually hamper complete transformation [76]. However, the transformations proceed at pressures considerably higher than that corresponding to the Hugoniot elastic limit, at hydrostatic conditions. There are no micro stresses which hamper the transformations at these conditions and therefore the transformations origin and progress are governed only by the shock compression temperature.

Low-density single-crystal fragments originated at pressure higher then the HEL are collinear [77]. Therefore, when the fragments are transformed martensitically into the high-density modifications CSRs, the CSRs will be collinear as well. The fragments interface layer temperature is higher than the fragments one. At the fragments relative displacements the temperature increases up to the substance melting temperature. As soon as displacements are over the layers fast epytaxial crystallization takes place so that the substance behind the shock front becomes a single-crystal with defects of large number. The single-crystal CSRs size corresponds to that of the initial single-crystal fragments and does not exceed the shock wave front width (~10 nm). The high-density modification single-crystals are thermodynamically unstable because of the defects. Therefore the substance recrystallization can begin even during the rarefaction process resulting in the transformation of single-crystals into polycrystals.

At high defect rates martensitic transformations are impossible. In this case high-density modifications can originate only through a reconstructive way which needs higher temperatures and consequently higher shock pressures. Obviously, because of that reason high-grade pyrolytic graphite transformation into diamond begins at 21 GPa where as low-grade one at 24, 31, and 42 GPa (for three different shots) [78].

For reconstructive transformation the most favourable sites of new modification nuclei origin are initial substance fragments interfaces. In this case the new modifications CSRs will not exceed the fragment size, that is the shock wave front width (~10 nm). The substance behind the shock front will be polycrystal.

The higher the dynamic effect temperature, the larger the extent of both the martensitic and reconstructive transformation process. The possibility of diamond-like glass origin should be taken into account if the temperature is higher than the initial substance metastable melting temperature. The glass can be recovered if the sample cooling rate during and after the rarefaction is fast. In the other case it can be transformed into ordinary glass or crystallize into both high- and low-pressure modifications. The CSRs size will not be conditioned by the shock wave front width if the diamond-like glass crystallizes.

The rate of the samples temperature decrease (exactly speaking the P, T rarefaction line), the sample residual temperature value as well as the sample cooling rate after the rarefaction are the main factors for dynamic effects which govern the possibility of high-pressure modifications recovering. It should be noted that the larger and more perfect high-pressure modification particles are the easier they can be recovered. Therefore, according to available experimental data the effective shock compression time has to be increased. The rapid cooling rate of the high-pressure modifications is provided either by shock compression of mixed samples of low-pressure modifications and some additives, which have lower residual temperature in comparison to that of high-pressure modifications, or by usage of the detonation method characterized by extremely fast cooling.

References

1. Majorana M (1897) *Chem Zentrallblatt 2*: 887; (1898) 1: 1065.
2. Crookes W (1905) *Chem. Zentrallblatt 2*: 1153.
3. Bancroft D, Peterson E, Minshall S (1956) *J. Appl. Phys. 27*: 291.
4. Dremin AN, Adadurov GA (1959) *Dokl. Acad. Nauk SSSR 128*: 261 (in Russian).
5. Alder J, Christian RH (1961) *Phys. Rev. Letters 7*: 367.
6. DeCarli PS, Jamieson JC (1961) *Science 133*: 1821.
7. Jamieson JC, Lowson AW (1962) *J. Appl. Phys. 33*: 776.
8. Clendenen RL, Drickamer HA (1964) *J. Phys. Chem. Solids 25*: 865.
9. Takahashi T, Bassett WA (1964) *Science 145*: 483.
10. DeCarli PS (1966) U.S. Patent No.3238019.
11. Cowan YR, Dannington BW, Wood P, Chester W, Pa., Holtzman AH (1968) U.S. Patent No.3401019.
12. Adadurov GA, Aliev ZG, Atovmyan LO et al. (1967) *Dokl. Acad. Nauk SSSR 172*: 1066 (in Russian).

13. Dremin AN, Breusov ON, Bavina TV, Pershin SV (1977) U.S. Patent No.4014979.
14. Adadurov GA, Bavin TV, Ananjin AV, Drobishev VN, Dubovitsky FI, Pershin SV, Tatsy VF, Dremin AN, Rogacheva AI, Messinev MJu, Apollonov VN, Zemlyakova LG, Doronin VN (1982) G.B.Patent No.2090239.
15. Doran DG (1964) *J. Appl. Phys. 34*: 844.
16. Coleburn NL (1964) *J. Chem. Phys. 40*: 71.
17. McQueen RG, Marsh SP (1968) *Symposium on High Dynamic Pressure*-1987, Paris: Gordon and Breach, New York.
18. Gust WH (1980) *Phys. Rev. 22 B*: 4744.
19. Dremin AN, Perchin SV (1968) *Fiz. Goreniya i Vzriva 1*: 112 (in Russian).
20. Pease RS (1952) *Acta Crystallogr. 5*: 356.
21. Wentorf RH (1957) *J. Chem. Phys. 26*: 956.
22. Bandy FP, Wentorf RH (1963) *J. Chem. Phys. 38*: 1144.
23. Danning V (1961) In : Chemistry of Solid States, Moscow, Foreign Literature: 213 Moscow (in Russian).
24. Kurdumov AV, Pilyankevitch AN, Franzevich JN (1973) *Poroshkovaya Metallugiya 10*: 57 (in Russian).
25. Bogorodsky ES, Goryatshev NS, Aparnikov GL (1978) *Sinteticheskie Almazi 5*: 21 (in Russian).
26. Adadurov GA, Ananjin AV, Bavina TV, Breusov ON, Drobishev VN, Dubovitsky FI, Pershin SV, Tatsy VF, Dremin AN, Rogacheva AI, Messinev MJu, Apollonov VN, Zemlyakova LG, Doronin VN (1982) PCT/SU N 80/00136, publ. WO N 82/00458.
27. Drobishev VN (1983) *Fizika Goreniya i Vzriva 5*: 168 (in Russian).
28. Adadurov GA, Ananjin AV, Breusov ON, Dremin AN, Drobishev VN, Rogacheva AI, Pershin SV, Tatsy VF (1988) *The 7th Int. Symp. on Use of Explosive Energy in Manufacuturing Metallic Matireal's of New Prope-Ties, Pardubice 11*: 338.
29. Adadurov GA, Ananjin AV, Breusov ON, Dremin AN, Drobishev VN, Kurdyumov AV, Pershin SV, Rogacheva AI, Tatsy VF (1989) *Proc. X-th Int. Conf. HERF, Lublyana, Yugoslavia*: 11.
30. Adadurov GA, Breusov ON, Drobishev VN, Rogacheva AI, Tatsy VF (1979) *Trudi VNIIPHTRI. Phizica impulsnich vozdeistvi 44:* (74) 157 (in Russian).
31. Aparnikov GL, Breusov ON, Grusdov VV, Drobishev VN, Rogacheva AI, Tatsy VF (1980) *Almazi i sverchtverdie materiali 8*: 1 (in Russian).
32. Adadurov GA, Baluev AV, Breusov ON, Drobishev VN, Rogacheva AI, Sapegin AM, Tatsy VF (1977) *Neorganicheskie materiali 13*: 649 (in Russian).
33. Zeldovich YaB (1946) *The shock Wave Theory and Leading in the Gasodynamics. Moscow-Leningrad,*: Acad. Press.
34. Zeldovich YaB, Raizer YuP (1963) *Physics of shock Waves and High-Temperature Hudrodynamic Phenomena*: Moscow, Fizmatgiz.
35. Fowles GR (1969) *J. Geophys. Res.72*: 5729.
36. DeCarli P (1967) *Bull. Am. Phys. Soc. 12*: 1127.
37. Kurdumov AV (1975) *Dokl. Acad. Nauk 221*: 21 (in Russian).
38. Kurdumov AV, Pilyankevitch AN (1981) *Super hard materials, Institute of Super Hard Materials Ukrainian Academy of Sci. 1*: 44 (in Russian).
39. DeCarli P (1969) U.S. Patent No.1299614.
40. Borimtshuk NI, Zelyavski VB, Kurdjumov AV, Mel'nikova VA, Pilyankevich AN, Rogovaya IG, Yarosh VV (1989) *Dolk. Acad. Nauk SSSR 306*: 1381 (in Russian).
41. Adadurov GA, Ananjin AV, Bavina TV, Breusov ON, Dremin AN, Pershin SV, Tatsy VF (1981) *Proc. of the 2nd Meting of Explosive Working of Materials,*: Novosibirsk , USSR, 313.

42. Savvakin GI (1981) *Almazi i Sverchtverdie Materiali*: (4), 1 (in Russian).
43. Staver AM, Gubareva NV, Lyamkin AI, Petrov EA (1984) *Fizika Gorenia i Vzriva 5*: 100 (in Russian).
44. Savvakin GI, Trefilov VI, Fenochka BV (1985) *Dokl. Acad. Nauk SSSR 282*: 5 (in Russian).
45. Lyamkin AI, Petrov EA, Ershov AP, Sakovich GV, Staver AM, Titov VM (1988) *Dokl. Acad. Nauk SSSR 302*: 611 (in Russian)
46. Greiner RN, Philips DS, Johuson JD, Volk F (1988) *Nature 333*: 440.
47. Titov VM, Anisichkin VF, Mal'kov IJu (1989) *Phizika Goreniya i vzriva 25*: 117 (in Russian).
48. Johnson JD, (1989) *9th Int. Symp. on Detonation*: Portland, 417.
49. Volkov KV, Danilenko VV, Elin BV (1990) *Fizika Goreniya i Vzriva 3*: 123 (in Russian).
50. Pershin SV, Pyaternev SV, Rogatsheva AI (1985) *Fizika Goreniya i Vzriva 5*: 128 (in Russian).
51. Altshuler LV (1978) *Zh. Prikl. Mekh. i Tehn. Fiz. 4*: 93 (in Russian).
52. Kurdumov AV - Private Communication.
53. Balchan AS, Cowan GR (1972) U.S. Patent No.3667911.
54. Dulin IN, Altshuler LV, Vatschenko VJa et al. (1969) *Fizika Tverdogo Tela 11*: 1252 (in Russian).
55. Estrin EI (1973) *Problemi Metallovedeniya i Fiziki Metallov, Moscow 2*: 100 (in Russian.).
56. Estrin EI (1975) *Problemi Metallovedeniya i Fiziki Metallov, Moscow 5*: 75 (in Russian).
57. Van Thiel M, Ree FH (1989) *Int. J. Thermophsics 10*: 227.
58. Bundy FP, Kasper JS (1967) *J. Chem. Phys. 46*: 3437.
59. Cochanovska A (1955) *Acta. Techn. Acad. Scihung 13*: 43.
60. Adadurov GA, Bavina TV, Breusov ON (1981) *Fizika Goreniya i Vzriva 2*: 159 (in Russian).
61. Gibson FC et al. (1958) *J. Appl. Phys. 29*: 628.
62. to be published.
63. Cowan GR, Dunnington BW, Holtzman AH (1968) U.S. Patent No.3401019.
64. Urizar MJ, James E, Smith LC (1961) *Phys. Fluids. 4*: 262.
65. Dremin AN, Pershin SV, Pyaternev SV, Tsaplin DN (1989) *Fizika Goreniya i Vzriva 25*: 141 (in Russian).
66. Pershin SV, Tsaplin DN (1991) *Proc. of the 5th All Union Meeting on Detonation*, 191: Kransnoyarsk, 237 (in Russian).
67. Pershin SV, Tsaplin DN, Antipenko AG (1991) *Proc. of the 5th All Union Meeting on Detonation, 191*: Kransnoyarsk, 239 (in Russian).
68. Antipenko AG, Pershin SV, Tsaplin DN (1989) *Proc. IXth All union Symposium on Combustion and Explosion*:: Chernogolovka, 104.
69. Antipenko AG, Pershin SV, Tsaplin DN (1989) *HERF Xth Int.Conf.* : Ljublyana, Yugoslavia, 170.
70. Ershov AP, Kupershtoch AL, Kolomiitshuk VN (1990) J. Jech. Phys. Letters 16: 42 (in Russian).
71. Ershov AP, Kupershtoch AL (1991) *Fizika Gorenia i Vzriva 27*: 111 (Russian).
72. Dremin AN, Klimenko VJu, Davidova ON, Zoludeva TA (1989) *9th Int. Symp.* on Detonation, Portland, 724.
73. Corrigan FR, Bundy FP (1975) *J. Chem. Phys. 63*: 3812.

74. Ananjin AV, Dremin AN, Kanel GI, Pershin SV (1978) *Detonation. Crystal Phenomena,Shock Wave Physico-Chemical Changes*.: Chernogòlovka, Moscow, Nauka: 111.
75. Razorenov SV, Kanel GI, Ovchinnikov AA (1981) *Detonation-1981 The 2nd All-Union Conf.* Chernogolovka (Russia): 70.
76. Roitburd AL Ju. (1964) *Problemi metallovedenia i Phizici metallov H. Metallurgia* 8: 235 (Russian).
77. Ananjin AV, Breusov ON, Dremin AN, Pershin SV, Tatsy VF (1974) *Phizika Goreniya i vzriva 10*: 426 (Russian).
78. Ezskine DJ, Nellis WJ (1991) *Nature, 349*: 317.

Chapter 3
Solid State Reactivity of Shock-Processed Solids

R. A. Graham and N. N.Thadhani

1. Introduction

Under high pressure shock-compressive loading, materials are forced into unusual and distinctive states not achieved in other processes. The rapidity of the application of high pressures, and the stress states are controlled by the inertial responses of the solids themselves as they are plastically deformed to the high pressure states which bring them into momentary balance with the pressures of the loading sources. Because of the large plastic deformations, unusually large concentrations of defects are produced at all levels of description: atomic, mesoscopic, microscopic and macroscopic. The high pressures, large deformations, rapidity of the process, elevated temperatures, large defect concentrations, and unusual defect configurations and states, provide the potentially interesting possibility for shock processing to produce unusual metastable states or unusual net-shape-processed solids.

There has been considerable exploration of materials synthesis and processing under high pressure shock compression, but utilization of the approach has proven to be a difficult process. Little systematic study has been devoted to mechanistic aspects of the shock process.

Perhaps the most significant technological interest is with highly porous powders or powder mixtures. In such systems, shock compaction can be used to achieve fully dense, net shapes of difficult-to-sinter materials. It has, however, proven to be difficult to fabricate high quality samples due to cracking accompanying the rapid release of pressure. Other approaches to shock processing involve synergistic processes in which the shock-modified material serves as starting source for a subsequent thermal and mechanical processing. In this latter case, the enhanced solid state reactivity can significantly alter the kinetics and thermal conditions of the post-shock process, thereby yielding materials with refined microstructures. Shock processing of mixed powder systems is of interest for materials synthesis, but control of the synthesized product requires knowledge of the material in the shock-modified state. In any event, evaluation of the potential for shock processing and synthesis requires description of the shock-modified solids, particularly their solid state reactivities.

In the present chapter a summary is given of various studies which provide direct quantification of different aspects of solid state reactivity of shock-modified solids. The various investigations include manifestations of reactivity from structural, chemical, physical and deformational (mechanical) processes in single- and multiple-component systems. There are numerous studies reported in the literature from a number of materials research groups, but the present authors have worked within a systematic, highly reproducible program based on standardized shock-modification systems with controlled, quantitative loading [1,2], and that work is emphasized.

2.   Shock Modification of Shock-Processed Solids

There have been considerable efforts to subject powder compacts of low starting density to controlled shock compression loading, preserve them for post-shock examination, and develop descriptions of their structural and defect characteristics [3,4]. These studies, principally on inorganic, refractory solids, show that the shock process results in highly defective materials with ubiquitous evidence for plastic deformation to levels similar to that of cold-worked metals. The defects are characterized by substantial x-ray diffraction line broadening analysis due to residual strain and crystallite size reduction [5], high concentrations of point defects [6], and direct microstructural observations by transmission electron microscopy [7].

It is well known that defects control solid state reactivity [8,9], but only in the most ideal circumstances can direct, quantitative relationships be developed between defect configurations and solid state reactivity. It is a particularly significant observation that the shock-formed defects are in configurations (e.g. lack of crystallographic-shear features, different deformation features within a single grain [10]), concentrations and combinations (rutile [6] e.g., with saturation concentrations of both dislocations and point defects) not normally encountered in conventionally processed materials. Further, residual strains are typically significantly more anisotropic [11,12], than observed from other types of deformations. Crystallite sizes of shock-processed solids are also typically, significantly smaller than those produced due to other deformation processes. Given the defect observations, substantially enhanced solid state reactivity is to be expected from shock-processed solids, Nevertheless, it is not possible to quantify the effects or anticipate how these defects will affect various solid state processes. Accordingly, various types of direct reactivity studies have been carried out to characterize the magnitudes of the various effects of solid state processing features.

In the present chapter, structural, chemical, mechanical and physical manifestations of solid state reactivity of shock-processed solids are reviewed. The various studies include solid-solid effects, solid-liquid effects and solid-gas effects. The work includes studies of materials processed as single-component systems, as well as materials processed as powder mixtures which we can term multiple-component systems. It is to be observed that each particular manifestation of enhanced solid state reactivity has its own specific features.

3.   Single-Component Systems

3.1 Solid-Solid Interactions

3.1.1 Solid Structural Effects
Monoclinic phase ZrO_2 powder compacts were subjected to controlled shock compression from 5 to 27 GPa and preserved for post-shock analysis. The samples were studied with TEM and x-ray diffraction line broadening. The samples showed considerable line broadening due to both residual strain and crystallite size reduction. Some of the material was also found to contain small quantities of tetragonal phase material. Hammetter and coworkers [13] studied the thermally induced monoclinic to tetragonal structural transformation of the shock-processed powder with differential thermal analysis. The measurements of the onset temperature of the transformation showed that the temperature was strongly influenced by the magnitude of the shock pressure. As shown in Fig. 1, the temperature was reduced from the unshocked value of 1,172°C to about 1,160°C in the most extreme case. The decrease in transformation temperature was observed at pressures greater than about 17 GPa. This shock

pressure correlated well with that at which the crystallite size is found to be substantially reduced. With this observation, it appears that the large effect of shock-processing on monoclinic to tetragonal structural transformation temperature in ZrO_2 is due to the shock-induced reduction in crystallite size.

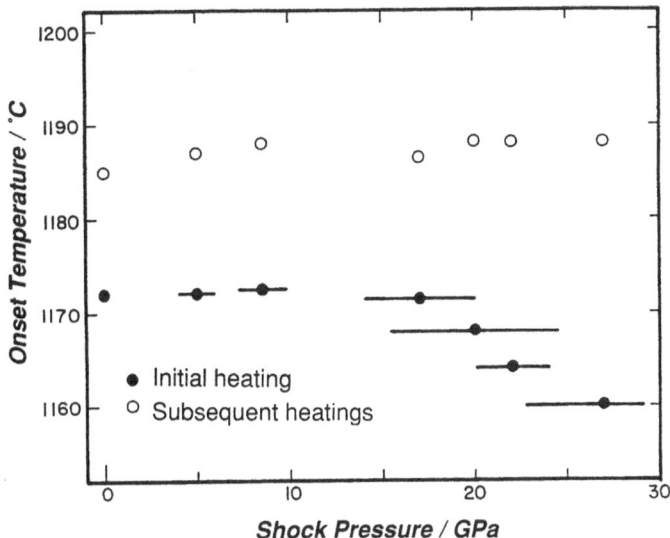

Fig. 1 Difference in measured onset temperature of the monoclinic-to-tetragonal phase transition between initial heating and subsequent heatings (after Hammetter et al [13]).

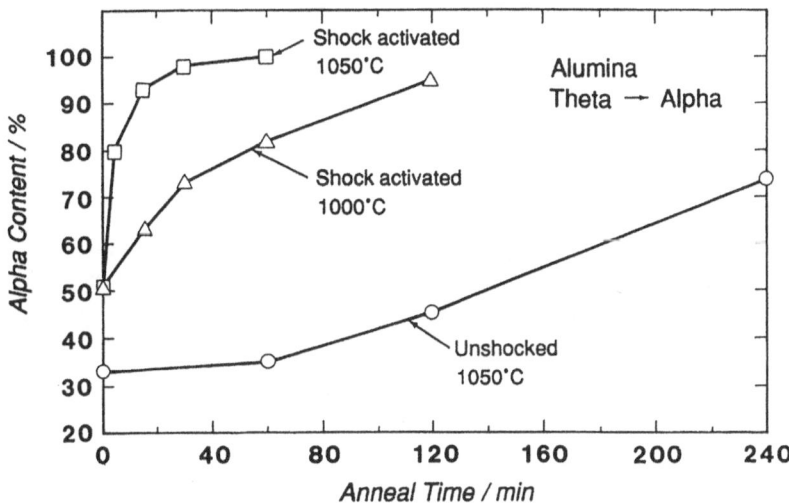

Fig. 2 α-phase content as a function of heat-treatment time for shocked and unshocked alumina (after Beauchamp and Carr [14]).

Similar observations have been carried out on an Al_2O_3 powder compact subjected to a peak shock pressure of 27 GPa in a study by Beauchamp and coworkers [14]. In this case the starting powder contained a mixture of θ-phase (70%) and α-phase (balance) material. After shock compression the powder contained about 50% α-phase. As shown in Fig. 2, the starting and shock-processed powders were annealed at temperatures near 1,050°C for times up to 240 minutes and the structural phases were identified. Whereas the untreated powder showed an incubation time of about 60 min before there was significant transformation to the α–phase material, the shock-processed powder showed very rapid conversion to the α-phase with almost complete conversion within 30 min. The effect was interpreted to be the result of large shock-formed concentrations of nuclei of the α-phase in the highly defective material. The nuclei were apparently too small for direct observation in TEM and x-ray diffraction characterizations. The growth process under annealing was found to be characterized by a linear growth process.

3.1.2 Solid Chemical Effects

The shock-processed ZrO_2 whose structural transformation characterizations were studied in the work of Hammetter and coworkers [13] was used in a solid state chemical reaction with PbO to form the compound $PbZrO_3$ in the work of Hankey and coworkers [15]. The starting material was found to exhibit a significant endothermic event in Differential Thermal Analysis (DTA) studies, and the onset temperature and "sharpness" of the reaction was found to be substantially reduced as a function of the peak shock pressure. The observations (shown in Fig. 3) reveal that the powder processed at the highest shock condition exhibited onset temperatures and reaction kinetics like that of a ZrO_2 powder with much higher specific surface. X-ray diffraction observations confirmed that the reaction product was $PbZrO_3$.

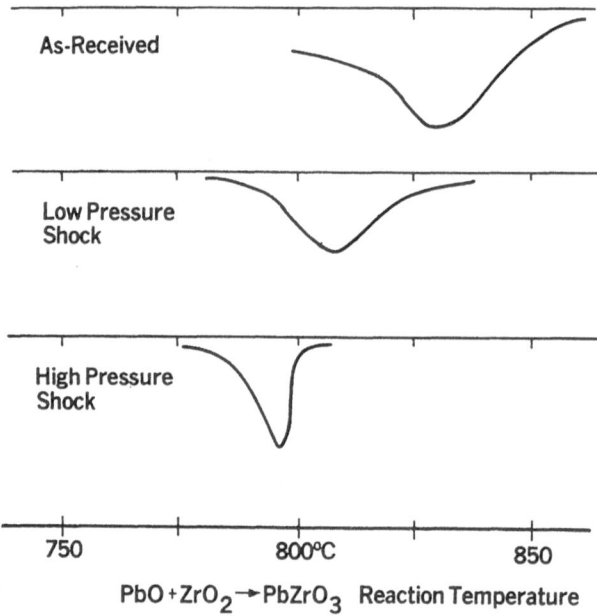

Fig. 3 Differential thermal analysis trace of the reaction of PbO and ZrO_2 in the unshocked and shocked state (after Hankey et al [15]).

3.1.3 Solid Deformational (Mechanical) Effects

The deformational (mechanical) effects produced in powders due to shock compression can also be advantageously utilized, particularly in the case of ceramics, to promote solid state mass transport kinetics, which can prove extremely important, for example, in enhancing the sinterability of otherwise difficult-to-sinter powders. Sintering of powders by thermally-induced coalescence and bonding of grains, generally occurs by diffusional transport. However, plastic deformation of powder particles by dislocation motion or by twinning can contribute significantly to sintering. While ceramics, in general, contain relatively low dislocation or twin densities in individual grains, shock treatment can introduce structural modifications and defect states that can greatly enhance the solid state reactivity of the powders, and play an important role in sintering.

Beauchamp [16] has reviewed the literature on shock-activated sintering and noted that the various shock-compression effects responsible for enhancing solid state reactivity of powders are: (a) point defects; (b) dislocations and higher order defects; (c) residual strains; (d) particle comminution; (e) reduction in grain crystallite size; (f) exposure of fresh surfaces; (g) structural phase changes; (h) interparticle bonding/agglomeration; and (i) surface reactivity.

Pressureless Sintering of Shock-Activated Powders

Based on the rationale of shock-induced, solid state reactivity, Bergmann and Barrington [17,18] at Dupont, were the first to perform an extensive study of the effect of shock activation on the sinterability of ceramic powders. They shock processed a variety of ceramic oxides (Al_2O_3, MgO, UO_2, $CaCO_3$, ZrO_2, and $BaFe_{12}O_{19}$), as well as SiC and B_4C powders, using the cylindrical implosion geometry, and investigated the residual microstructures and the physical properties of the recovered deagglomerated (milled) powders.

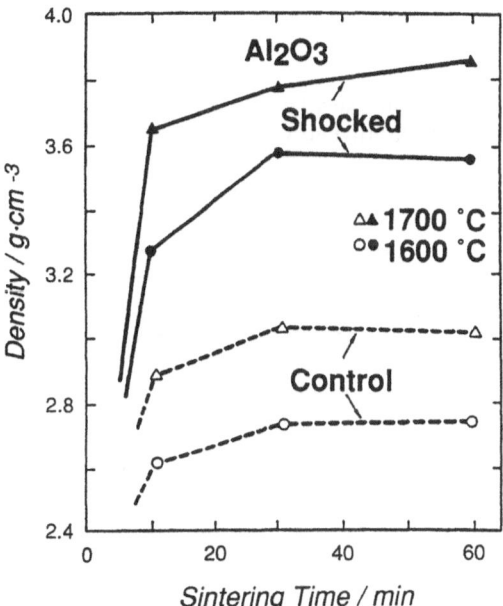

Fig. 4 Plot comparing the densification and sintering behavior of shock-activated and unshocked Al_2O_3 powders (after Bergmann and Barrington [17]).

The densification behavior of shock-processed Al_2O_3, compared with that of an unshocked "Al_2O_3 control" powder is shown in Fig. 4. It is apparent that shock-processed alumina sinters at higher rates and to higher final densities than the unshocked powders. Furthermore, the green densities (after cold pressing and prior to sintering) are also higher for the shocked powders. A similar densification behavior was observed by the authors in all the other shock-processed powders that also showed extensive x-ray line broadening. While x-ray line broadening and enhanced sinterability in SiC and B_4C was attributed to crystallite size reduction, the oxide powders showed only modest surface area increase (≈ 20 %). However, a large population of defects was observed to be introduced in the shocked oxide powders.

Heckel and Youngblood [19] have attributed x-ray line broadening in shock-activated Al_2O_3 powders to the formation of fine (< 100 nm) coherently diffracting domains and lattice microstrain. Prümmer and Ziegler [20] have also examined the effect of shock loading on microstrain and substructure in shocked Al_2O_3 powders of different particle sizes. In their investigation, fine grain size Al_2O_3 powders (≈ 4 μm) showed essentially no change at low shock pressure, but at higher pressures rapid increases in lattice distortion and decrease in sub-grain size were observed. Coarse grain size powders (≈ 280 μm) showed evidence of increasing lattice distortion and grain-size reduction starting at lower pressures and reached a saturation stage. Contrasting the effect of shock activation with ball milling of alumina powders for 157 hours, they observed that shock-processed fine powders exhibited twice as much lattice distortion and dislocation density as ball-milled fine powders. The shocked coarse powders, produced lower dislocation density and particle size reduction than ball-milled coarse powders.

The densification behavior of shock-processed Al_2O_3 powders subjected to quantitative loading has also been studied by Morosin and coworkers as summarized in the review by Beauchamp [16]. A marked increase in densification rate in the early stages of sintering was observed, followed by a rapid decrease in rate as sintering progresses. As shown in Fig. 5, sintering of the shocked compact resulted in an initial rapid increase, but a final density lower than that of unshocked sintered powders. The overall poor densification behavior of shock-treated and sintered powders was attributed to nonuniform sintering and formation of large well-densified zones (agglomerates) surrounded by porous regions.

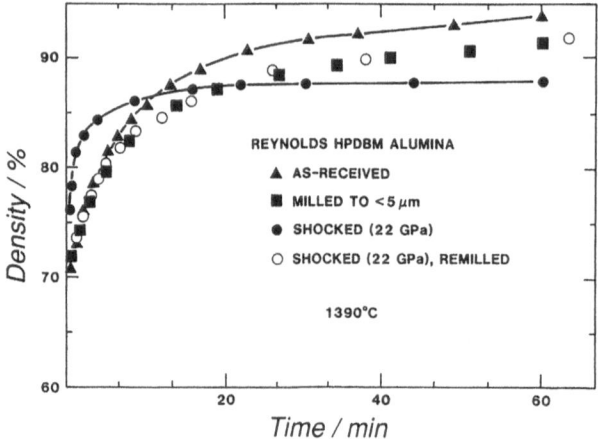

Fig. 5 Sintering density versus time for shocked compacts, shocked/reground powders and unshocked alumina powders (after Beauchamp [16]).

While powders in the agglomerates underwent rapid early-stage sintering, no bonding was achieved between the agglomerates. In contrast, no such dense agglomerates with surrounding pores were present in unshocked powders; thus, uniform densification was attained throughout the compacts. It was also observed that upon "regrinding" the shocked agglomerates, uniform sintering was attained similar to unshocked powders.

Direct evidence of dislocations and deformation twins in explosively shocked alumina powder has been obtained by Hoenig and Yost [21] using transmission electron microscopy. Misoriented dislocation bands aligned in one direction and deformation twins along basal planes were predominantly observed. Evidence of recrystallization due to presence of single crystallites larger than the original grains has also been observed.

Studies of post-shock sintering behavior of amorphous Si_3N_4 have been attempted by Hoenig and Yost [21]. The sintering was performed at 1,373 K and 1,823 K, but practically no densification occurred. However, at 1,373 K, a 43% increase in hardness was observed, while at 1,823 K, the hardness decreased markedly due to crystallization of the amorphous Si_3N_4 phase. Beauchamp et al [30] have also studied the effect of shock treatment on the sinterability of Si_3N_4 powders by indirectly exploring the rate of transformation of α-Si_3N_4 to β-Si_3N_4 at the sintering temperatures. They observed that the transformation kinetics were enhanced by a factor of two, which also significantly affected the sinterability of powders. The effect of shock compression on transformation kinetics is discussed later in Section 3.2.

It is thus generally observed that shock treatment of powders results in retention of extensive plastic deformation structures (point and line defects), grain size refinement (due to large concentrations of line defects), and modification of surface character (due to opening of fresh surfaces or simply surface cleansing) all of which can be significantly effective in enhancing powder sinterability. Alumina powder, for example, has shown maximum evidence of plastic deformation effects. However, the formation of large fragmented agglomerates, with well-bonded powder grains but very poor inter-agglomerate bonding, can actually inhibit powder sinterability and interfere with the potentially beneficial effects of shock activation. Only in the shock activation work of Bergmann and Barrington [17,18], was a consistent influence on enhanced sinterability observed. However, in their case, sintering was performed after milling and de-agglomerating the shock-treated powders.

No direct attempts have yet been made to attain a uniform degree of shock modification (by proper design of shock recovery system) and minimized agglomeration of powders during shock treatment (by varying initial packing densities). Thus, the problem of powder agglomeration has always hindered the proper understanding and utilization of the otherwise beneficial effects of shock activation.

Hot-Pressing of Shock-Processed Powders
In order to elucidate and correlate the role of individual shock-modification processes on powder sintering, researchers at Sandia National Laboratories have performed an extensive investigation on hot-pressing of shock-processed AlN, Al_2O_3, TiB_2, and TiC. Changes in densification behavior were monitored along with microstructural studies to distinguish the densification mechanisms. Much of their work has been summarized in the review by Beauchamp and Carr [22].

Characterization of the four shock-processed ceramic powders revealed that in the case of AlN, a consistent 6% increase in green density of the powder compacts was attained in contrast to unshocked powders [22,23]. Further x-ray diffraction line broadening showed an increase of internal strain from < 0.01% to $\approx$ 0.3% independent of shock pressure from 14 to 27 GPa [24]. A corresponding dislocation density increase (based on TEM analysis) from $\approx$ 10^6 /cm^2 to > 10^{11} /cm^2 was also observed due to shock treatment [25]. The Al_2O_3 powder also showed a similar increase in lattice strain, from 0.03% in the starting powder to ≈ 0.2% at 4 GPa pressure, 0.3% at 17 GPa, and 0.5% above 22 GPa [26,27]. In contrast, the shock treated TiB_2 and TiC powders [24] showed much lower residual strains: at 27 GPa only 0.03% strain was obtained in TiB_2 (and dislocation density of $10^9/cm^2$), while TiC showed 0.08% strain at 17 GPa and 0.1% strain at 22 GPa. Upon hot-pressing of the shock-modified and reground powders, Beauchamp and Carr [22,23,28] observed that the densification behavior of individual ceramics was consistent with respective lattice deformation characteristics monitored by XRD line broadening.

Shock-processed AlN powder [28] exhibited maximum increase of lattice strain, and also showed maximum enhancement in powder sinterability. Higher initial green density, initial densification rates, and final compact density were observed in the shock-processed AlN powders in contrast to unshocked powders during hot-pressing. According to the density-versus-time plots for AlN shown in Fig. 6, an increase in hot-pressing pressure from 26 MPa to 47 MPa resulted in 10% increase in overall density for both unshocked and shock-activated powders. On the other hand, an ≈ 6% increase in final compact density was obtained with shock-activated powders in contrast to unshocked powders at both hot-press pressures.

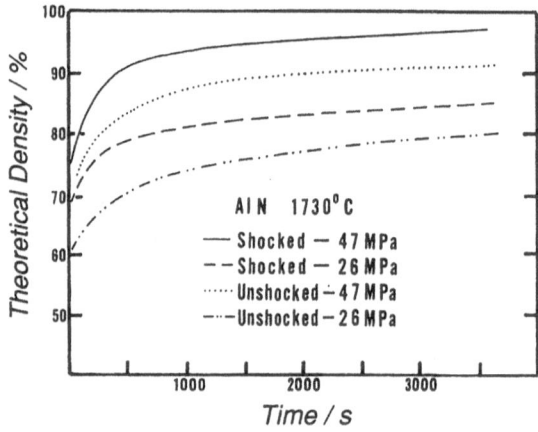

Fig. 6 Densification behavior of shock-processed and unshocked AlN powder upon hot-pressing at 26 and 47 MPa pressure (after Beauchamp et al [28]).

The most significant effect of shock-processing on hot-pressing of AlN powder was revealed by monitoring the densification rate (at 80% of theoretical density) as a function of ram pressure, as shown in Fig. 7. In the case of unshocked AlN powder, the densification rate for hot-pressing at 1,730°C showed an initial gradual increase with increasing ram pressure, and then a rapid transition to a much higher densification rate after a specific pressure. At 1680°C, the densification rate versus pressure curve follows the same shallow slope observed at lower ram pressures at higher temperature. In contrast, the shock-processed (17 GPa) AlN powder showed a significantly increased densification rate for both temperatures (1,680°C and 1,730°C) at any given ram pressure. The densification curves parallel the slope of the higher-temperature and higher-pressure densification rate curves of unshocked powders.

Transmission electron microscopy performed by Beauchamp and Carr [25,28] on hot-pressed compacts of unshocked and shock-processed AlN powders showed different types of deformation (defect) substructures, that could be considered responsible for the different densification rates observed in Fig. 7. The hot-pressed, unshocked compacts showed: (i) fine

(0.5-0.7 μm) grains with very low dislocation density (10^6 /cm^2) and small necks between particles, but essentially the same microstructure as that of starting powders; (ii) equiaxed moderate size (1-2 μm) grains containing few dislocations either isolated or in loops (dislocation density 10^8 /cm^2) in addition to planar defects (antiphase boundaries); and (iii) few large grains (3-5 μm) assumed to be grown during hot-pressing and containing small pores, dislocations ($\approx 10^8$ /cm^2), and planar defects located around grain peripheries. The hot-pressed, shock-processed powder compacts also revealed similar inhomogeneous microstructures: (i) 0.5 to 5 μm size grains with dense (10^{10} /cm^2) dislocation structures (Fig. 8 (A)); (ii) recrystallized equiaxed 1-2 μm size dislocation-free grains (< 10^6 /cm^2), and (iii) equiaxed 1-2 μm grains exhibiting moderate dislocation densities ($10^9 - 10^{10}$ /cm^2), with dislocations occurring as loose tangles and well-developed linear or hexagonal arrays (shown in Fig. 8(B)), formed due to deformation of recrystallized grains during hot-pressing.

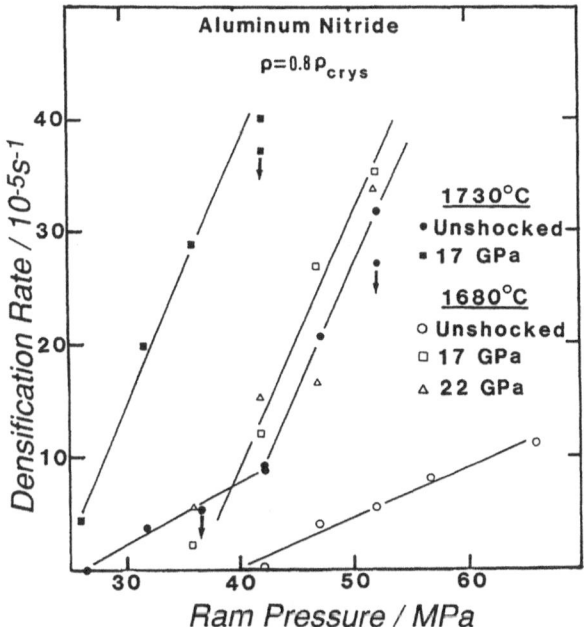

Fig. 7 Densification rate as a function of hot-press ram pressure for aluminum nitride (after Beauchamp et al [28]).

The overall microstructural characteristics revealed that in unshocked AlN powders, plastic deformation during hot-pressing occurred by dislocation motion and grain boundary transport. Initially, at low densities when grain contact areas were small, and thus high local stresses were present, the dislocation processes dominated. Subsequently, as densification proceeds, contact areas increased and local stresses decreased, thereby causing dislocation motion to stop. Subsequent densification then occurred only by diffusional processes. The density at which dislocation motion ceases to operate depends on hot-pressing pressure and temperature. Thus, the change in slope to a higher densification rate observed in Fig. 7, is due to dominance of plastic deformation by dislocation motion during hot-pressing above the threshold pressure.

Shock-activated powders on the other hand, contained significantly higher initial dislocation densities. During hot-pressing, their stored strain energy makes them microstructurally unstable. Thus, upon hot-pressing, high-stresses at contact points resulted in localized dynamic recrystallization. The softer recrystallized grains then continued to plastically deform and densify under the applied load. In essence, higher initial dislocation densities present in shock-processed AlN powders promote dynamic recrystallization during hot pressing which allows plastic flow by dislocation motion to remain active at much higher densities, thereby resulting in increasing densification.

The densification behavior of shock-activated Al_2O_3 powder [23,27], at different shock pressures, compared to that of unshocked powder is shown in the plot of densification rate versus hot-press ram pressure in Fig. 9. The most significant effect of shock-processing upon hot-pressing was observed in terms of higher initial densification rates, similar to the pressureless sintering results discussed earlier.

SEM analysis of hot-pressed, shock-processed Al_2O_3 powder confirmed the effect showing presence of very dense agglomerates surrounded by porous (less-dense) regions.

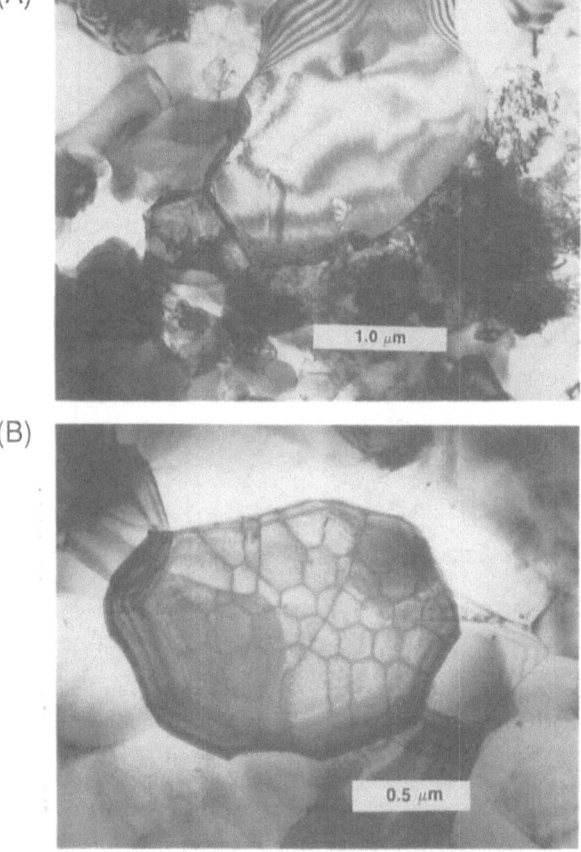

Fig. 8 TEM image of hot-pressed shock-processed AlN compact showing (A)dense dislocation structures and (B) hexagonal dislocation arrays in plastically deformed (during hot-pressing) recrystallized grains (after Carr and Beauchamp [28]).

Thus, according to the authors, the higher initial densification rates were due to rapid sintering of the agglomerates, either because of the presence of shock-induced defects or higher powder packing characteristics, or both. However, while the grains in the agglomerates continue to rapidly densify, the porous regions between agglomerates cannot sinter as rapidly and are eventually prevented from fully densifying due to the constraint of the dense zones. Thus, the hot-pressed, shock-processed powders have lower final compact densities in contrast to unshocked powders that undergo uniform densification rates.

Beauchamp and Carr [22,23,27] also observed that hot-pressing of shock-processed and reground (deagglomerated) alumina powders resulted in a uniform densification behavior similar to that of unshocked powders. Due to the similar uniform densification behavior of reground, shock-activated and unshocked alumina powders, they inferred that the enhancement in initial sinterability of shock-activated powders was not because of the presence of shock-induced internal defect structures, but due to grain-packing characteristics. It should be realized however, that x-ray line broadening analysis revealed extensive residual strain (0.5%) in the Al_2O_3 powder shocked at 27 GPa pressure.

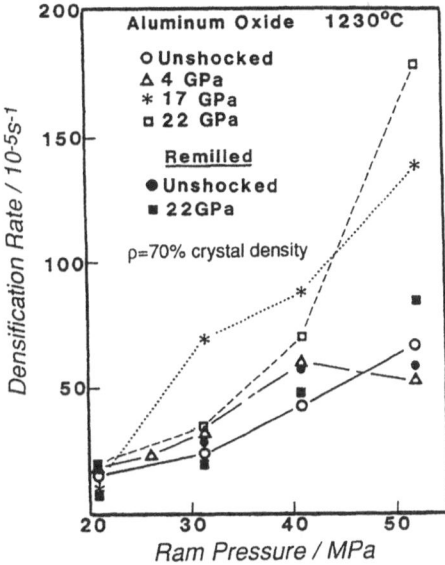

Fig. 9 Plot comparing densification rate versus hot-press ram pressure for shock-activated alumina (at different shock pressures) with unshocked powder (after Beauchamp et al [22]).

The densification behavior of shock-activated and unshocked titanium diboride and titanium carbide powders [22] is shown in the plots of Fig. 10 (A) and (B), respectively. The densification rate of TiB_2 (at 75% theoretical density) plotted as a function of hot-press ram pressure at 2,000°C, shows large scatter and poor reproducibility (Fig. 10(A)), and practically no significant difference between shock-activated and unshocked powders. SEM analysis of hot-pressed TiB_2 compacts showed uniform porosity distribution in the case of unshocked powders, while shock-processed powders showed 250 μm diameter, fairly dense zones surrounded by regions of high porosity.

The density-versus-time plot (Fig. 10 (B)) for TiC hot-pressed at 1,600°C at 52 MPa pressure, showed that the starting densities of shock-processed powders were ≈ 5 % higher than for unshocked powders. However, the densification rates decreased more rapidly for shock-activated powders reaching approximately the same final density (98%) as that of unshocked TiC powders. SEM analysis of hot-pressed TiC compacts revealed similar variations in porosity (dense 250 μm diameter zones surrounded by regions of high porosity) in shock-activated powders and uniform porosity in unshocked powders, as that observed in both hot-pressed, shock-processed and unshocked TiB$_2$ compacts.

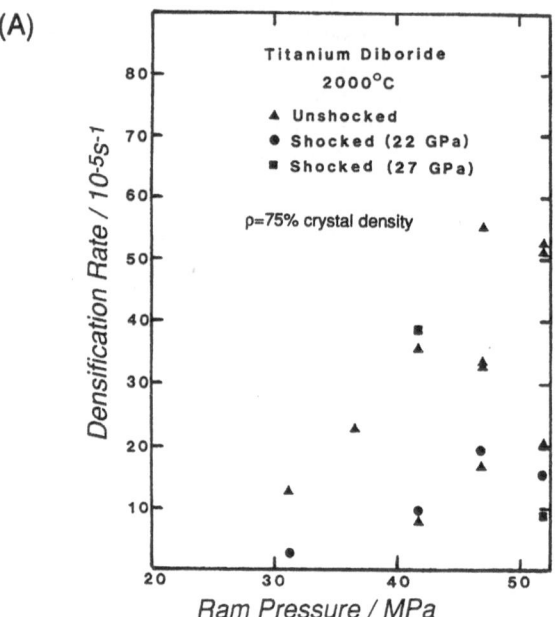

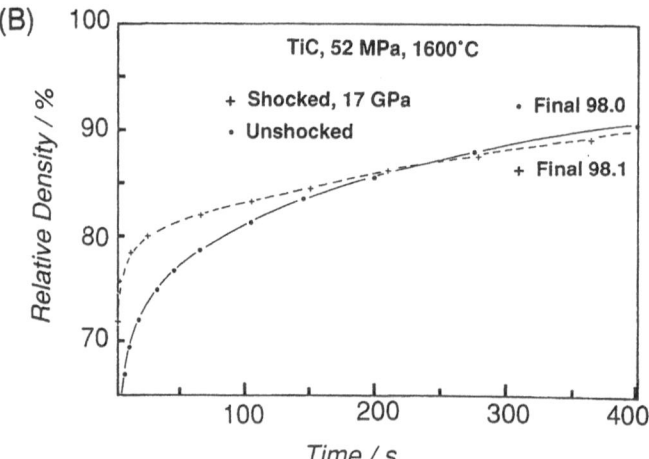

Fig. 10 (A) Densification rate as a function of hot-press ram pressure for shock-processed and unshocked TiB$_2$, (B) Densification behavior of unshocked and shock-activated TiC (after Beauchamp [22]).

In general, from Beauchamp and Carr's studies [22,23], it has been observed that only AlN showed a distinct improvement in densification behavior during the entire hot pressing cycle, due to shock-activation effects. Shock treatment of AlN (at 14-27 GPa) introduced very high concentrations of dislocations that persisted at hot-pressing temperatures. Subsequent dynamic recrystallization, and therefore continued plastic deformation, resulted in a densification process dominated by dislocation motion, which maintained high densification rates. Shock-activated Al_2O_3, and TiC to a lesser extent, showed initial higher densification rates than those in unshocked powders. However, the densification rates decreased rapidly with time, and consequently the hot-pressed, shock-processed compacts showed lower final densities than unshocked powders. Although shock treatment introduced significant residual strain, at least in Al_2O_3 (0.2 at 4 GPa, 0.3 at 17 GPa, and 0.5% at 22 GPa), the formation of high density agglomerates with surrounding low density regions, dominated the overall densification behavior. In the case of TiB_2, the densification rate results showed large scatter, but no evident difference between unshocked and shock-activated powders was observed. At the same time shock-processed TiB_2 powder [24] also showed very low residual strain (only 0.03% at 27 GPa).

It is thus evident that the maximum benefit of shock activation of ceramics is obtained if large residual strains, due to deformation-induced defects and other shock-compression effects, are introduced in the powders. At the same time, formation of large agglomerates needs to be prevented, since upon subsequent sintering (pressure or pressure-less), the grains in the agglomerates undergo rapid densification to high densities, but then inhibit further bonding between agglomerates thereby yielding low-density final compacts. In order to introduce bulk plastic deformation to utilize shock-modification effects, it is essential to uniformly shock load the powders at stress levels above their "crush-up strength." Heterogeneous shock-loading and increased deformation of particle interfaces, causes formation of dense regions which fragment into agglomerates (due to stress-wave interactions), and subsequently become impossible to sinter. The heterogeneous loading effects become more severe with increasing pressure and lower green density. While uniform shock loading and maximum bulk plastic deformation of the ceramic powders need to be maintained, formation of dense agglomerates needs to be minimized, such that after shock treatment, sintering (pressure or pressureless) can occur uniformly throughout the compact.

3.1.4 Solid Physical Effects

Hematite (α-Fe_2O_3) exhibits a magnetic spin-flip transition known as the Morin transition. The transition has been shown to be quite sensitive to the effects of crystallite size, impurities, mechanical grinding and static high pressure. Williamson and coworkers [29] have studied the Morin transition in hematite with Mössbauer and magnetization measurements to attempt to identify the nature of the influence of atomic level defects.

Hematite is an antiferromagnetic material exhibiting a weak ferromagnetism (WF) above the Morin temperature of about 260 K. The high sensitivity of the Morin transition to various defect modifications is due to the delicate balance of two competing contributions to magnetic anisotropy energy. Powder compacts of a commercially available hematite were subjected to peak shock pressures of 8, 17, and 27 GPa, and the shock-processed powders were studied with XRD, Mössbauer spectroscopy and magnetization.

The maximum in residual strain and the minimum in crystallite size as indicated by x-ray diffraction occurred at the intermediate pressure of 17 GPa. The Mössbauer measurements of the weak ferromagnetic fraction as the shock-processed powders were cooled and heated

through the transition temperature are shown in Fig. 11. It is observed that the onset
temperature for the transition upon cooling is substantially reduced with the maximum effect
observed at the intermediate pressure of 17 GPa. Furthermore, unlike the starting material, the
WF is not reduced to zero magnetization. A very substantial proportion of the material
remains in the WF state. The largest effect occurs at the intermediate pressure values. This
same sample shows a greater hysteresis effect upon warming.

Based on the x-ray diffraction, magnetization and Mössbauer measurements, the authors
[29] concluded that the reduction in Morin temperature was due to reduced crystallite size.
The absence of the Morin temperature for a large fraction of Fe sites was attributed to the
introduction of lattice defects. The modification attributed to defects was similar in nature and
magnitude to that observed when 5 mole % of Al_2O_3 is added to α-Fe_2O_3. The thermal
hysteresis was thought to be best explained by the residual strain. The shock-induced defects
were thus found to cause extreme effects not encountered in other material-modification
processes.

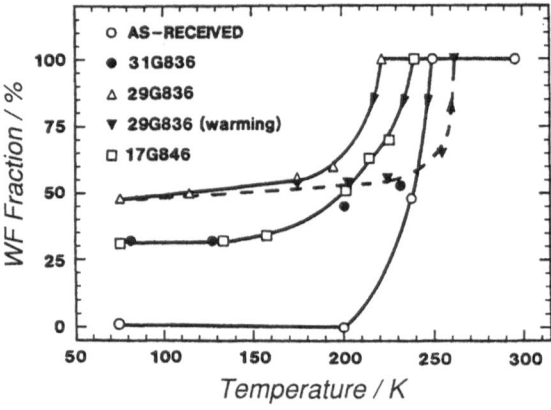

Fig. 11 Mössbauer measurements of weak ferromagnetic fraction, as shock-processed hematite powders are
cooled and heated through the transition temperature show major effects. Experiment numbers 31G836, 29G836,
and 17G846 correspond to shock pressures of 8, 17, and 27 GPa, respectively (After Williamson et al [29]).

3.2 Solid-Liquid Interactions

The influence of shock-processing on sinterability of Si_3N_4 shows the effect of dissolution of
liquid phase MgO-SiO_2 on the α-to-β transformation. Beauchamp and coworkers [30]
subjected α-phase Si_3N_4 to controlled shock compression and studied the influence of shock
processing on the dissolution behavior by adding 5 wt% MgO to the powders and carrying out
determinations of the conversion to β-phase after heating at temperatures of 1,600°C and
1,700°C for times from 0.5 to 2 hours. As shown in Fig. 12, compared to the starting material
with the same additive, the conversion from α-to-β phase was substantially enhanced.
Fig. 13 illustrates the conversion process whereby oxides on the surface of the Si_3N_4 and the
MgO additive form a silicate melt in which the α-phase material dissolves. The β-phase then
precipitates from the melt. Thus, the observed enhanced conversion rate of the shock-
processed material is a manifestation of a greatly enhanced dissolution rate.

The most carefully studied dissolution studies of shock-processed powders was reported
by Casey and coworkers [31] who quantitatively determined dissolution rates of shock-

processed rutile in hydrofluoric acid. The work included well annealed rutile, as-shocked rutile with large concentrations of point and line defects, and annealed, shock-processed rutile in which dislocation and point defect densities were varied.

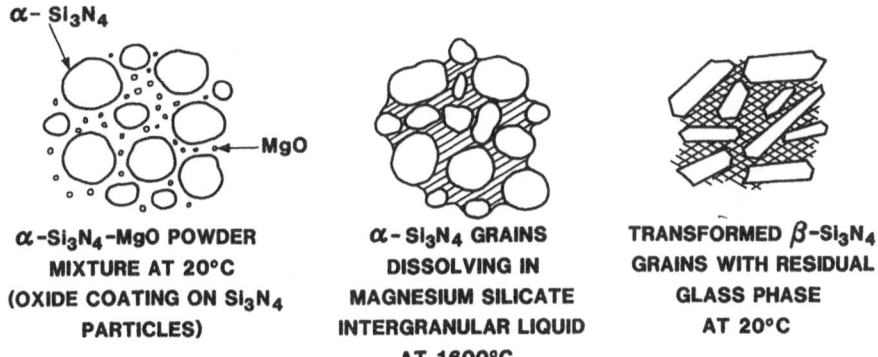

Fig. 12 When MgO is added to α-phase Si_3N_4 powder host, dissolution of impurities in the MgO at high temperature results in a melt from which β-phase Si_3N_4 is obtained.

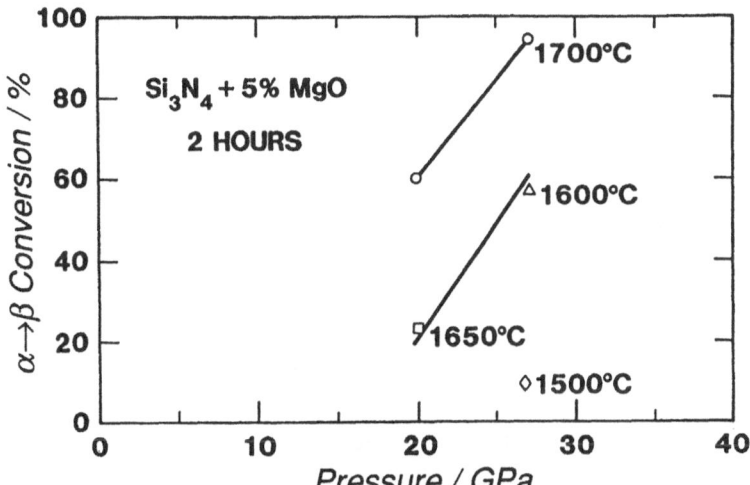

Fig. 13 The α-to-β conversion rate of shock-processed Si_3N_4 mixed with small amounts of MgO are shown to be strongly influenced by the shock process (Beauchamp et al [30])

Residual strain and point defect concentrations of the rutile samples in the as-shocked condition and after annealing are shown in Fig. 14. Note that the point defect concentrations are reduced to background levels at temperatures above $\approx 450°C$. As indicated by residual strain, the dislocation density remains high up to $\approx 1,000°C$. Other materials analysis included TEM, SEM and specific surface measurements.

Figure 15 shows the measured dissolution rates of samples annealed at the various temperatures. Although an enhanced dissolution rate is observed that is well correlated with maximum dislocation density, the effect is relatively modest. This observation suggests that the classic theory in which dislocations control dissolution rate may be in error.

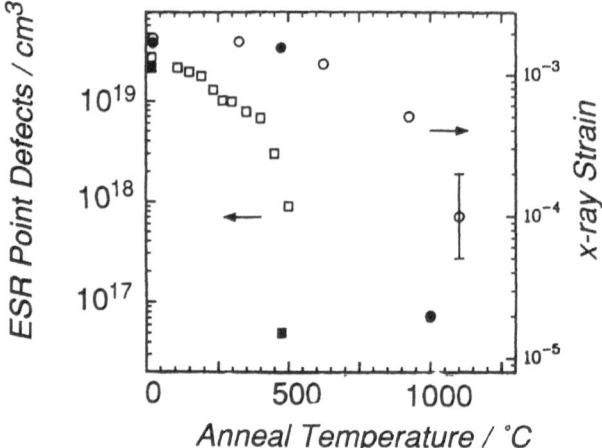

Fig. 14 Residual strain and paramagnetic defect concentrations after annealing for one hour at the indicated temperatures. Open circles and squares are for a sample subjected to a pressure of 20 GPa, and filled circles and squares are for sample exposed to 27 GPa (after Casey et al [31]).

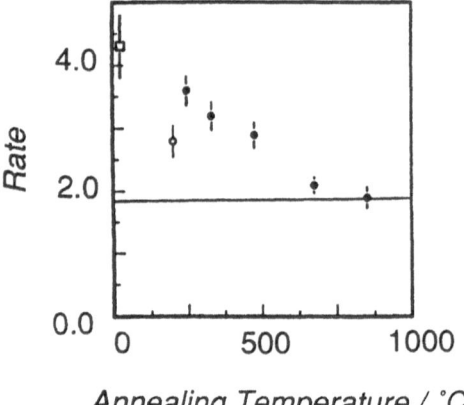

Fig. 15 Measured dissolution rates of shocked rutile samples annealed at the various temperatures (after Casey et al [31]).

3.3 Solid-Gas Interactions

The potential use of shock-processed solid powders as catalysts has prompted a number of investigations of catalytic activity which involve interactions of gas phase materials with solids. The earliest work in the United States was that of Golden and coworkers [32] who examined the oxidation of CO with shock-processed rutile. The rutile was the same material that was carefully characterized in the various studies reported above. As shown in Fig. 16, very substantial enhancements in catalytic activity were observed, and the magnitude of the effect was dependent upon the shock conditions. The effect was found to be relatively persistent.

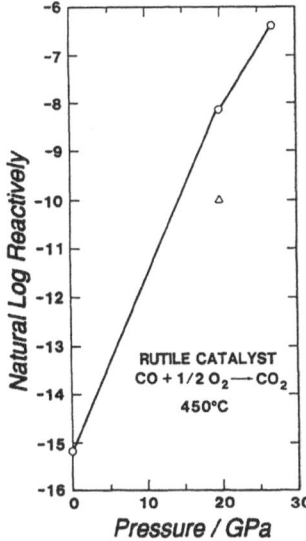

Fig. 16 Catalytic activity of shocked rutile as a function of shock conditions (after Golden et al [32]).

A similar study on enhanced catalytic activity of shock-processed ZnO in CO oxidation and methanol synthesis was reported by Williams and coworkers [33]. In this case the question of change in the selectivity of the shock-processed material was examined. As shown in Fig. 17, significant enhancements in reaction rates were found as a function of shock pressure, but the selectivity of conversion for methanol synthesis was not altered by shock modification.

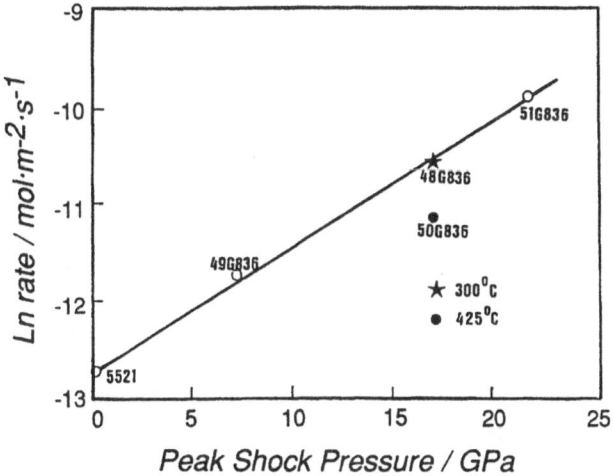

Fig. 17 Enhancements in reaction rates of oxidation of CO with rutile as a function of shock pressure. Sample numbers and corresponding shock pressures are: No. 48G836 - 7.5 GPa; No. 48G836 - 16.0 GPa; No. 50G836 - 16.0 GPa; and No. 51G836 - 22.0 GPa (After Williams and coworkers [33]).

The effect of shock processing on the oxidation of CO in ZnO was also studied, and, as shown in Fig. 18, the oxidation rate was increased a factor of 17, with the largest increase occurring at the peak shock pressure. These two studies by Williams and coworkers [33] have shown greatly increased surface activity of shock-processed oxides, and provide a basis for consideration of shock processing for production of unique catalysts.

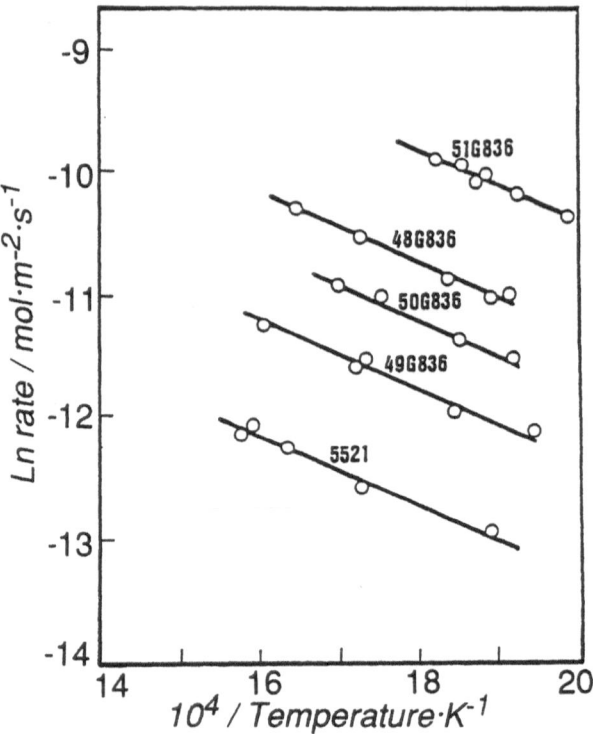

Fig. 18 Arrhenius plots for CO oxidation rates over ZnO powders, with stoichiometric CO/CO_2 feed at 1 atm. (after Williams et al [33]).

4. Multiple-Component Systems

Shock-compression of powders of multiple components produces a unique configuration of extensively plastically deformed and intimately mixed powder components, as well as dense packing characteristics. Such a configuration can significantly promote solid-state chemical reactivity of the powders leading to either post-shock chemical reactions occurring because of defect-enhanced, solid-solid diffusional flow at otherwise unusually lower temperatures, or even chemical reactions occurring in the shock-compression state due to solid-state mechanochemical processes. In this section, we will limit the discussion to the effects of shock-processing on post-shock chemical reactions. The shock-processing effects will be compared with conventional processing approaches involving diffusion reactions occurring during combustion of multiple components resulting in compound formation.

4.1 Conventional Reaction Processing

Self-sustaining chemical reactions, between multiple components in a powder mixture, have been utilized to synthesize a variety of intermetallic compounds, ceramics, as well as composite materials [34,35]. The chemical reactions are initiated by either the "combustion" mode (in systems with higher adiabatic reaction temperature) or by the "thermal explosion / reactive sintering" mode (in systems with lower adiabatic temperature) [35]. In the "combustion" mode the reaction is ignited by heating a green powder mixture compact at one end with an electric spark (or hot wire or even a laser or ion beam), which results in a reaction wave propagation along the compact axis. In general, the reaction is initiated at the melting temperature of the low-melting point constituent and involves dissolution of the higher melting temperature solid into the melt, followed by solidification of the compound, and subsequent solid-state diffusion of the remaining component into the intermediate compound. Reaction-wave propagation rates for such combustion-type processes range between 0.5 to 0.25 cm/s. In the "thermal explosion" mode, the powder mix is adiabatically heated in a furnace until the onset of chemical reaction at a temperature close to the melting point of one of the components.

Reactions occurring with the onset of liquid phase formation in the "thermal explosion" mode, can be inhibited if enough time is allowed for solid-state thermally activated diffusion to occur, for example, during slow heating. In recent differential scanning calorimetry (DSC) experiments on Pd-Sn colaminated composites, Bordeaux and Yavari [36] demonstrated partial occurrence of a solid-state diffusion reaction (with peak at ≈393 K) followed by the main exothermic reaction occurring with the melting of Sn (peak at ≈510 K). Fig. 19 shows their DSC thermograms obtained at different heating rates. It is seen that with increasing heating rates, the onset and peak of the solid-state reaction are shifted to higher temperatures. At heating rates greater than 160 K/min, the solid-state peak is totally inhibited, and only the main exothermic reaction occurs with the melting of Sn at ≈ 520 K. Alternatively, at lower heating rates, solid-state diffusion dominates the reaction process, and almost complete reaction can occur in the solid-state, thereby leaving no heat available to be otherwise released subsequently via a liquid-state reaction.

In an alternative process, with calorimetric studies of kinetics of intermetallic nucleation during reactions in Al/Ni thin films, Ma et al [37] showed evidence that interdiffusion of Al and Ni led to initial formation of metastable solid solutions. The Al_3Ni intermetallic subsequently nucleated in the interdiffused regions, only at certain preferred intergranular defect sites. The final growth of the intermetallic phase then occurred by a diffusion-limited process.

Liquid state reactions can also be inhibited if solid-solid diffusivities of powder mixture components are accelerated due to presence of internal defects. Additionally, reduction of diffusion distances (e.g., in ball-milling) and generation of fresh/cleansed contacts between surfaces of intimately mixed powder components (e.g., in shock-processing) can also significantly promote solid-state diffusion. Ball-milling as well as shock treatment processes not only enhance solid-state diffusivities, but also result in mechanochemical activation of powder mixture components leading to enhanced chemical reactivity. Hida and Lin [28] performed high-speed planetary milling of SiO_2-Al thermite mixtures, as a pretreatment for attaining mechanochemical activation. Fig. 20 shows DTA patterns of silica-quartz sand and Al powder mixtures milled for different time durations. While unmilled SiO_2+Al powder mixtures react at approximately 600°C, the reaction onset temperature in milled powder mixtures decreases with increasing milling time. The enhanced reactivity in milled SiO_2-Al powder mixtures, has been attributed to a "mechanochemical activation" phenomena, which strongly influences the solid-state diffusion rates and causes reactions to occur at lower temperatures.

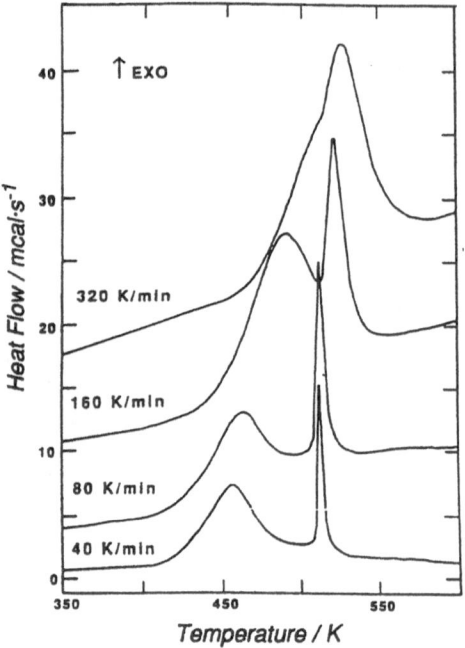

Fig. 19 DSC thermograms obtained at different heating rates for a Pd-Sn colaminated composite (after Bordeaux and Yavari [36]).

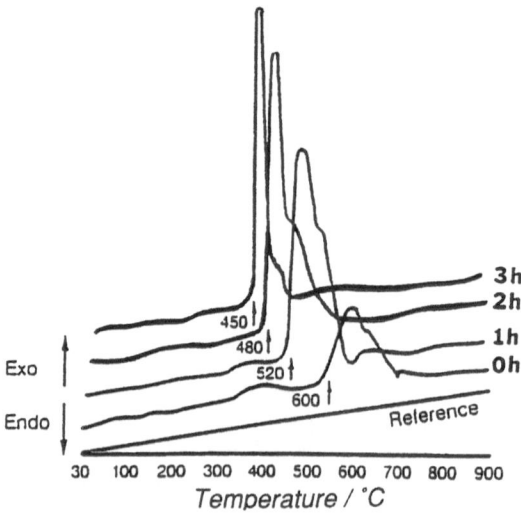

Fig. 20 DTA traces of ground SiO_2 and Al powder mixtures showing reaction onset temperature variation as function of milling time (after Hida and Lin [38]).

Similar effects have also been observed in ball-milled Ni-Al powder mixtures [39]. While intermetallic formation in Ni-Al starts to occur within 2 hours of ball milling, complete alloying occurs within 4 hours. DTA of powder mixtures, ball-milled for various times up to 4 hours, revealed different extent of reaction indicated by energy released. As shown in the plot in Fig. 21, the total energy released (heat of reaction) decreased with increasing milling time, and practically no energy release was observed with milling time approaching 4 hours, due to complete alloying in the ball-milling process. Fig. 21 also shows the reaction onset temperature as a function of milling time. While unmilled Ni-Al powder mixture showed only the single exotherm with melting of Al, the ball-milled mixtures showed only exotherms corresponding to the solid-state diffusion reaction. A significant shift in reaction onset temperature towards lower temperatures was also observed with increasing ball-milling time.

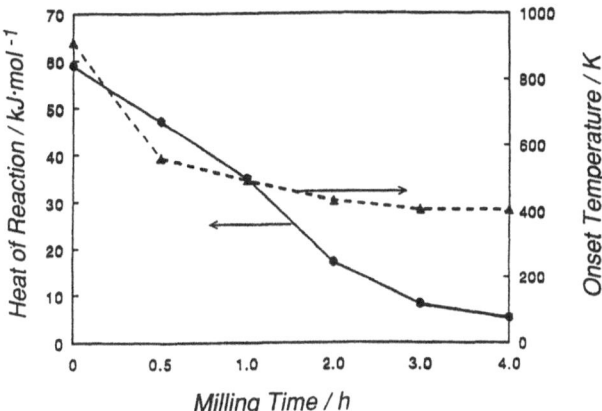

Fig. 21 Variation of heat of reaction and reaction onset temperature for Ni-Al powder mixtures ball-milled for various time periods.

Thus, the extensive cold-working and fracturing of powder particles and their intimate mixing and rewelding during ball milling, promotes solid-state diffusivity and causes reactions to occur at significantly lower temperatures until complete alloying occurs during milling itself.

4.2 Shock-Compression Processing

The mechanochemical activation effects, similar to those occurring in ball-milling of powder mixtures, are also prominently observed in shock-processed powder mixtures. During shock-compression, the plastic deformation of powders due to void collapse, relative mass motion of individual components forced by large acceleration forces, and mixing of components over large distances due to engulfing of components of different deformabilities, result in an highly activated configuration that has significant affects on the solid-state reactivity of powders.

The post-shock chemical reaction behavior of shock-treated, multi-component powders was first studied by Hammetter et al [40] on Ni-Al powder mixtures, using differential thermal analysis. They investigated two types of shock-treated Ni-Al powder mixtures, mechanically mixed and composite powders (Al core with Ni crust) of same bulk chemical composition (85 wt% Ni, balance Al). As shown in the DTA traces in Fig. 22, while the unshocked Ni-Al powder mixture exhibited a single exotherm at 650°C, the shocked mechanical powder mixture revealed two exotherms, the main exotherm at 650°C and another "pre-initiation" exotherm at ≈550°C. The shocked composite powder, unlike the shocked mechanical mixture, showed only the main exotherm at 650°C.

The "pre-initiation" exothermic event observed only in the shock-treated Ni-Al mechanical mixture is similar to the solid-state diffusion reaction observed by Bordeaux and Yavari [36] during DSC analysis of colaminated Pd-Sn composites as well as in DTA traces of ball-milled powders. Hammetter et al [40] attributed the occurrence of the "pre-initiation" event in shock-processed Ni-Al mechanical mixtures, to solid-state effects produced due to fine-scale mechanical mixing, generation of intimate contacts, and surface conditioning of elemental powders during shock compression.

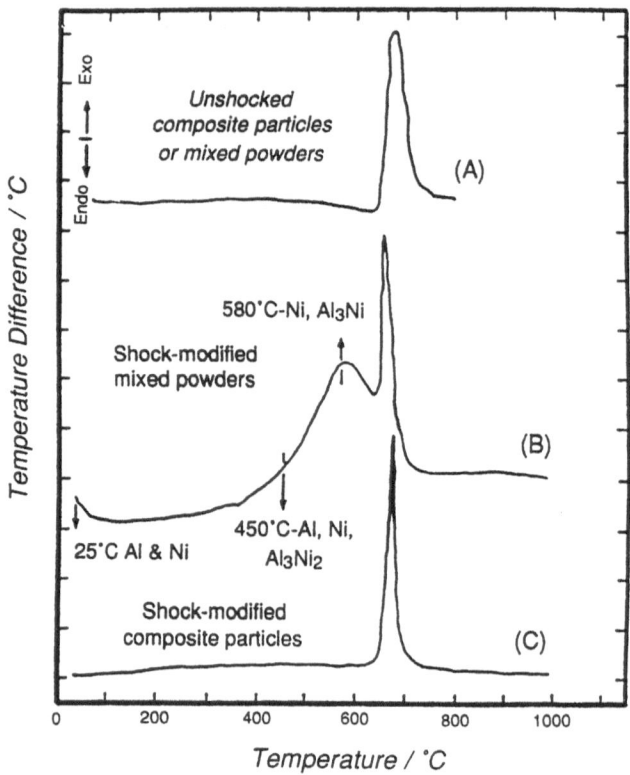

Fig. 22 DTA traces of (A) unshocked, (B) shock-treated mechanical mixtures, and (C) shock-treated composite powders (after Hammetter et al [40]).

The "pre-initiation" exothermic event is not observed in shocked composite powders due to lack of similar mixing and conditioning. Microstructural analysis also showed direct evidence for more intense mixing occurring in the mechanical powder mixtures. Electron microprobe analysis (EPMA) revealed that Ni in some cases was present many tens of microns deep inside the Al in the case when mechanical mixtures were used, while mixing at such a scale was never evident in coated composite powders.

In order to further characterize the solid-state "pre-initiation" event, Dunbar et al [41] performed a detailed study to elucidate the differences in reaction behavior of different morphology Ni-Al powder mixtures, mixed in different volumetric distributions, all shocked under identical conditions and to the same degree of densification. Fig. 23 shows DTA traces of shock-processed coarse, flaky and fine morphology Ni-Al powders, mixed in a volumetric distribution corresponding to Ni_3Al stoichiometry. The DTA results indicate that mixtures of all three morphologies exhibit an exothermic "pre-initiation" event (corresponding to solid-state diffusion reaction) prior to the main exotherm at about 650°C (Al melt temperature), similar to Hammetter et al's [40] observations.

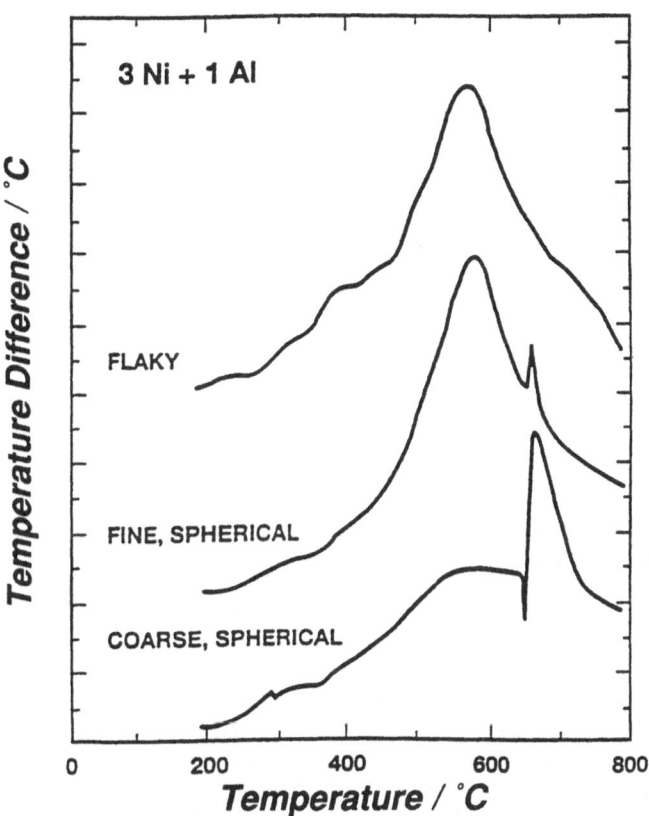

Fig. 23 DTA traces of shock-processed samples of 3Ni+1Al mixtures of coarse, flaky and fine morphology powders.

However, the magnitude of the solid-state "pre-initiation" event is strongly dependent on powder morphology. Coarse morphology powder mixtures exhibit the main exotherm at 650°C, and a small "pre-initiation" peak. The fine morphology powder mixtures exhibit a significantly larger "pre-initiation" peak, and a main exotherm of much smaller magnitude. The flaky powder mixture, on the other hand, shows only a broad solid-state "pre-initiation" exothermic event, and no trace of main exotherm corresponding to the liquid-state reaction with melting of Al.

The effect of morphology on the reaction behavior exhibited in the DTA traces has been explained by effects observed in optical micrographs shown in Fig. 24 [41,42]. The final shock-compressed configuration shows different levels of deformation and mixing of Ni and Al particles in the coarse, fine, and flaky morphology powder mixtures.

Flaky powder mixtures show more extensive deformation of both components, intimate mixing, and a greater surface area contact. Such a configuration results in solid state diffusion reaction at temperatures much below the melting of Al, during subsequent heating in the DTA. Coarse morphology powders show extensive deformation only at Ni- Ni particle contact areas, while fine morphology powders show much less deformation, but significantly greater surface area contacts.

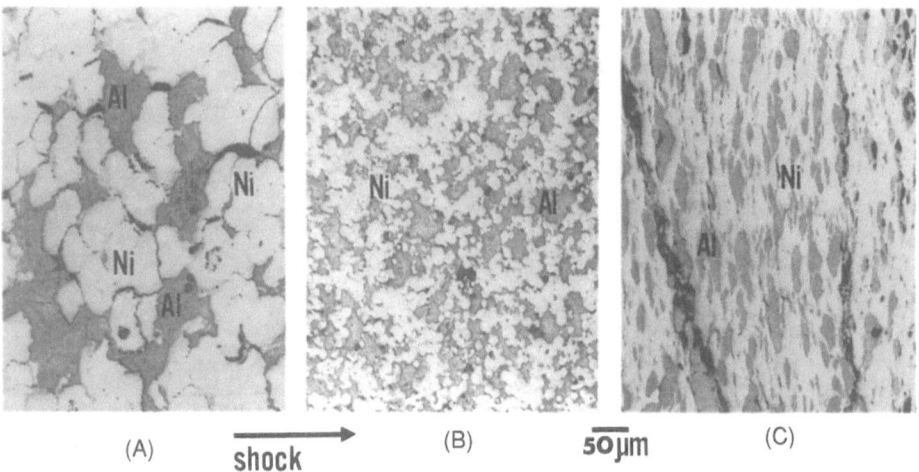

(A)   shock   (B)   50 μm   (C)

Fig. 24 Optical micrographs of shock-processed compacts of 3Ni+1Al mixtures of (A) coarse, (B) fine and (C) flaky morphology powders.

The DTA traces for the coarse/rounded morphology powders mixed in volumetric distributions corresponding to Ni_3Al, $NiAl$, and $NiAl_3$ atomic stoichiometries, also exhibit the occurrence of the solid-state pre-initiation event as shown in Fig. 25. The "pre-initiation" event is most dominant in the case of mixtures with Ni-rich volumetric distribution.

The corresponding optical micrographs (Fig. 26(A)) also show a much greater overall deformation in the Ni-rich mixtures, in contrast to those with Al-rich volumetric mixtures. The shock-processed configurations of the Al-rich powder mixture (Fig. 26 (C)) do not exhibit a similar degree of deformation, flow, and mixing, and thus, while no pre-initiation event is observed, the melting of Al (indicated by the large endotherm), overrides the thermal behavior. The equiatomic stoichiometry powder mixture shows results similar to that of the Al-rich mixture, only the Al melting endotherm is not as large because of lesser amount of Al, and neither does the shock-processed configuration (Fig. 26(C) show extensive deformation and mixing of the individual components.

The effect of shock activation on enhancing the solid state chemical reactivity is not just limited to Ni-Al powder mixtures. Similar results have also been obtained in shock-compressed Ni-Si [42] and Ti-Si [43] powder mixtures. Unlike the Ni-Al system, the silicide based systems form many different types of intermetallic compounds and solid solutions. Thus, the solid state diffusion reactions and those initiating with melt (eutectic) formation occur almost simultaneously, and are inseparable. In spite of that, the effect of shock-processing in causing reactions to occur at lower temperature can be clearly observed in the Ni-Si system [42]. As shown in Fig. 27, while unshocked flaky morphology Ni-Si powder mixtures undergo chemical reactions at a fixed temperature ($\approx$ 720°C) irrespective of volumetric distribution of components (traces (A) and (B)), the reaction onset temperature in shock-treated powder mixtures is shifted to $\approx$ 540°C in Ni+Si mixture (trace (C)) and to 620°C in 2Ni+Si powder mixture (trace (D)).

The endothermic peaks in the different traces in Fig. 27 correspond to melting of the product/products formed in the reactions. Traces (A) and (B) of unshocked and shocked Ni+Si powder mixtures show two endotherms at 830°C and 960°C, that correspond to melting of the ε-phase and equiatomic phase Ni-Si compounds; and traces (C) and (D) of shocked and unshocked 2Ni+Si powder mixtures show only the endotherm at $\approx$ 1,250°C corresponding to the melting of the γ-phase Ni-Si compound.

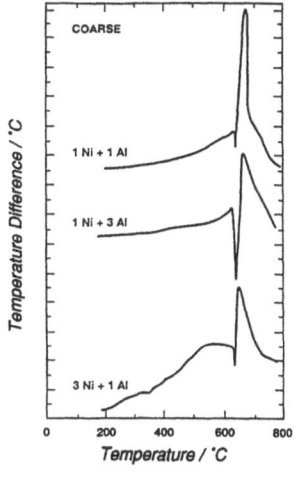

Fig. 25 DTA traces of shock-processed samples of coarse morphology powders mixed in three different volumetric ratios corresponding to 3Ni+1Al, 1Ni+1Al, and 1Ni+3Al atomic stoichiometry compounds.

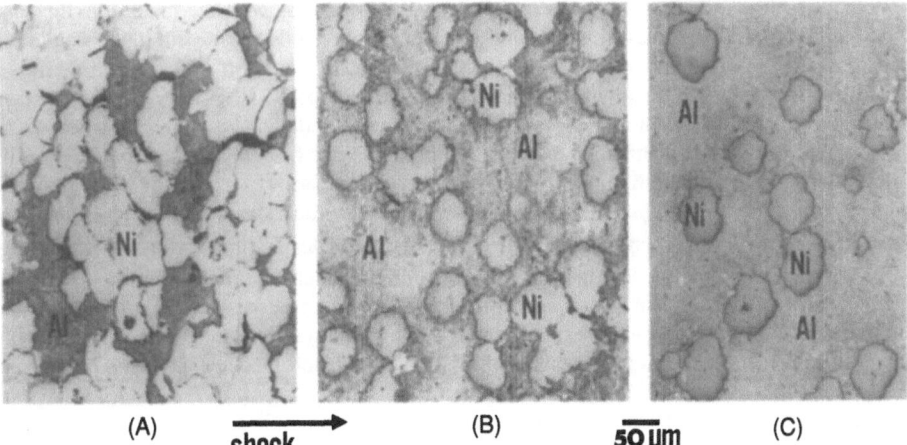

Fig. 26 Optical micrographs of shock-processed compacts of coarse morphology powders mixed in three different volumetric ratios corresponding to (A) 3Ni+1Al, (B) 1Ni+1Al, and (C) 1Ni+3Al atomic stoichiometry compounds.

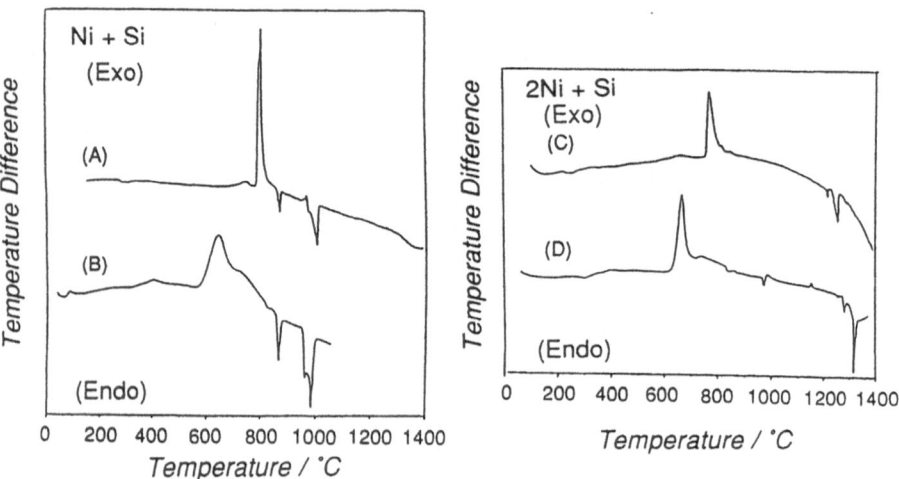

Fig. 27 DTA traces showing reaction exotherms and reaction-product melting endotherms of flaky powder mixtures of (A) unshocked Ni+Si; (B) shocked Ni+Si; (C) unshocked 2Ni+Si, and (D) shocked 2Ni+Si.

Differential thermal analysis provides a valuable tool to directly infer the chemical reaction behavior of powder mixtures, and compare the effects of different treatments on reactivity enhancement. It also resembles the "thermal explosion /reactive sintering" mode of self-sustaining reactions which is being popularly developed as a processing technique for synthesis of materials. Self-sustaining "combustion" reactions occurring by igniting a green compact at one end with a high-temperature source to produce a reaction front that propagates through the compact and synthesizes it, can also be used to establish the effect of shock compression on reactivity enhancement.

In Ti+C powder mixtures, the combustion (reaction) front propagates at a velocity which is dependent on Ti-C stoichiometry, inert diluent content, packing density, and initial temperature [34]. The combustion velocities generally decrease with inert addition, increase with higher initial temperatures, and initially increase and then decrease with increasing green densities. An example of combustion front velocity measurements for Ti-C powder mixtures, as a function of inert TiC diluent addition, is shown in Fig. 28 [44]. Typical combustion velocity for the Ti-C system at 65% green density, ambient temperature, and without additives is measured to be approximately 7.8 mm/s, which decreases to 2 mm/s with about 30% TiC diluent addition, at which point reaction propagation stops.   Theoretically predicted combustion velocities are generally 20% higher than measured values. The combustion front velocity for shock-processed ($\approx$ 1 GPa pressure) Ti+C powder mixture (point marked in Fig. 28) has been measured to be approximately 50% higher than that measured for unshocked powder mixtures. The velocity increase in shock-processed powders can be attributed to not only the generation of dense packing characteristics which permit efficient heat transfer, but also to the presence of defect states that promote the reaction mechanisms by favoring and increasing diffusion and/or precipitation kinetics.

In general, the enhancement in solid state reactivity of multiple components due to shock-compression effects occurs because of the formation of a unique dense packed and highly defective configuration. The intense degree of mechanical mixing and activation produced due to high-rate plastic deformation and flow of material into and around the voids, results in a configuration that can favor mass transport by solid-state mechanisms at conditions much below the threshold.

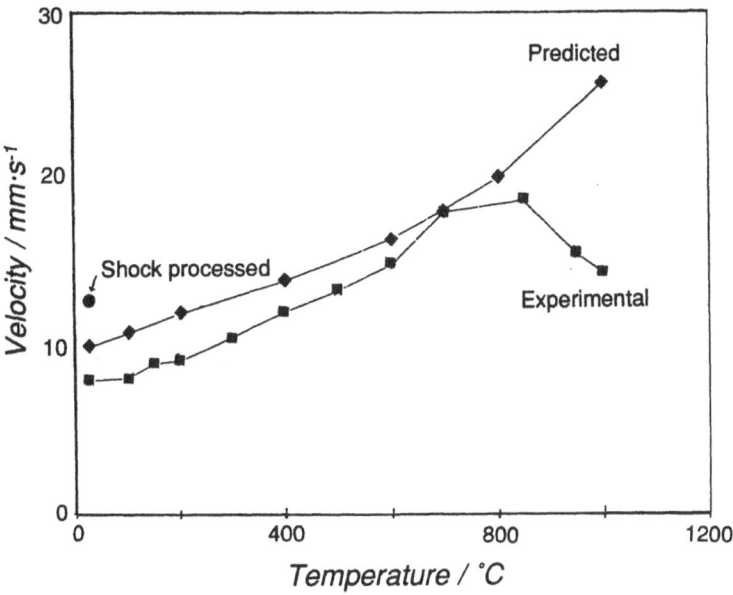

Fig. 28 Typical example showing predicted and experimentally measured combustion front velocity for Ti-C powder mixtures as a function of percent TiC inert addition. The single point corresponds to measured velocity of shock-activated Ti-C powder mixture (after Advani et al [44]).

5. Summary and Concluding Remarks

The present review brings together a selected group of individual studies which were undertaken to characterize enhanced solid state reactivity of shock-processed, single-and multiple-component solids. If we view an individual materials study as a fiber, the present collection of fibers may be viewed as a thread. We can use this thread, along with other such threads to weave a cloth, from which the goal of fabricating a fine suit representing the use of shock-processing techniques in materials fabrication technology, can be finally accomplished. What is the status of our quest for this new suit ? What is required to complete the quest?

Overall, it is clear that the enhancement in solid state reactivity is ubiquitous in shock-processed solids. Such behaviors are consistent with prior observations of highly defective material states in shock-processed solids. It is also clear that the effects of shock processing on solid state reactivity are large, typically influencing any diffusion-controlled process by many orders of magnitude. Processes controlled by lattice energy also show strong effects due to the residual strain resulting from the defects. Even though enhanced solid state reactivity can now certainly be expected as a first-order consequence of shock processing, both quantitative and qualitative details must be considered and developed for each particular process.

From a materials science perspective, these highly defective materials, with their interesting physical, structural, and chemical properties, represent a singular opportunity to characterize and understand a unique materials condition. Beyond these scientific interests what significant influences might be expected from further shock-processing work ? Perhaps one of the most important consequences of shock processing is the ability to introduce and control the defect states of inorganic solids systematically and quantitatively. Such an effort is well illustrated in the work of Casey and coworkers [31]. Based on prior work, large concentrations of dislocations, point defects, and other higher-order defects were introduced into rutile. With annealing processes, individual defect states were altered in a controlled manner such that specific defect influences could be separated. It appears that similar approaches could well be followed with other materials. A good example in which shock processing might be extended is in synthesis and fabrication of ceramic-oxide superconductors. Since superconducting properties largely depend on the defect states present in the materials, shock-processing can be utilized to form particular types of defects, the configuration of which can be subsequently tailored by post-shock thermal treatments. The work of Iqbal et al [46] and Syono [47] provide such examples.

Studies of solid state chemistry under high pressure shock compression have been of interest to study materials synthesis [48-50]. Studies of such processes show that the solid state reactivity of the reactants in the shocked state can play a critical role in the subsequent chemical reactions. Shock-processing may be used to alter and thus control the mechanisms and kinetics of a materials synthesis process, or to produce a tailored source of raw material for subsequent (post-shock) thermal/mechanical processing to fabricate a material with controlled microstructure and properties. The work summarized in the present paper, provides guidance on the expected behavior of a material in the shocked state.

The tailoring of the suit from the cloth of many materials studies is the technological goal of much of the shock-processing work. It is perhaps premature to come to firm conclusions about the role that shock processes can play in commercial materials processing. It appears that there is potential for direct chemical synthesis and the use of shock-activation processes to condition materials for more conventional materials processing. There appears to be potential in shock-activated sintering and shock-activated combustion (reaction) synthesis. Enhancement and control of solid state reactivity affords a degree of control not necessarily available from other sources.

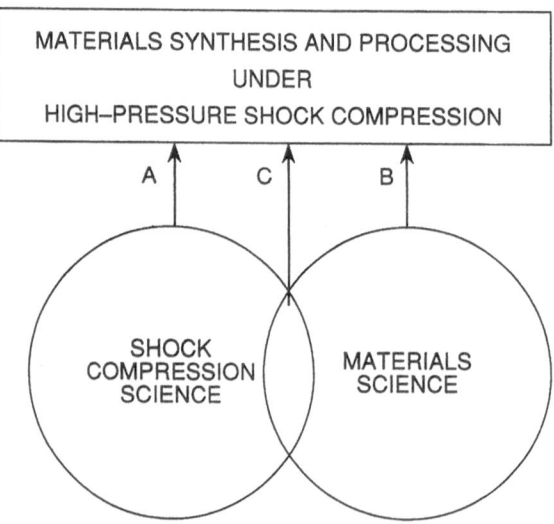

Fig. 29 Paths for achieving the goal of materials synthesis and processing technology by high-pressure shock compression.

Based on the work reported in the present review, it is clear that it is no longer particularly relevant to consider whether enhanced solid-state reactivity is a significant process. The issues are quantitative, and can only be addressed with consistently detailed studies, with a strong measure of control of shock-processing conditions. There is a tendency to regard the shock processes and their effects without consideration of the shock conditions used. Clearly a wide range of behaviors are to be expected, and the behaviors will be sensitive to first order to the quantitative details of the shock process.

If a beautiful coat is to be produced from the fibers, threads, and the cloth of work in shock processing, prior work reviewed here demonstrates that it must be recognized that the process is complex. Solution of the technological problems, requires focus on the relevant materials issues to a degree not achieved to date. As characterized by Fig. 29, successful effort requires a carefully coordinated activity between shock-compression science and materials science. Much of the prior work has been characterized by attempts by materials scientists or shock-compression scientists to go directly to achieve materials technology goals without sufficient involvement with each other. Such activities are indicated by paths 'A' and 'B' shown in Fig. 29. Prior work clearly demonstrates that the technological problems require a much greater degree of common involvement than has been characteristic in the past. Successful tailoring of the coat will thus require path 'C' stemming from the combined, strongly interacting efforts from both sciences.

References

1. Graham RA, Webb DM (1984) *Shock Waves in Condensed Matter-1983*: edited by Asay JR, Graham RA, Straub GK, North Holland, Amsterdam, 211.
2. Graham RA, Webb DM (1986) *Shock Waves in Condensed Matter-1985*: edited by Gupta YM, Plenum NY, 831.
3. Graham RA, Morosin B, Horie Y, Venturini EL, Boslough M, Carr MJ, Williamson DL (1986) *Shock Waves in Condensed Matter - 1985*: edited by Gupta YM, Plenum, 693.
4. Graham RA, Morosin B, Venturini EL, Carr MJ (1986) *Ann. Rev. Mater. Sci. 16*: 315.
5. Morosin B (1986) *High Pressure Explosive Processing of Ceramics*: edited by Graham RA, Sawaoka AB, Trans Tech, 285.
6. Carr MJ, Graham RA, Morosin B, Venturini EL (1984) *Defect Properties and Processing of High-Technology Nonmetallic Materials*: edited by Crawford JH Jr., Chen Y Sibley WA, North Holland, 343.
7. Carr MJ (1986) *High Pressure Explosive Processing of Ceramics*: edited by Graham RA, Sawaoka AB, Trans Tech, 341.
8. Schmalzried H (1981) *Solid State Reactions*: Verlag Chemie.
9. West AR (1984) *Solid State Chemistry and its Applications*: Wiley
10. Carr MJ, Graham RA (1986) *Metallurgical Applications of Shock Waves and High-Strain-Rate Phenomena*: edited by Murr LE, Staudhammer KP, Meyers MA, Marcel Dekker, 1037.
11. Morosin B, Graham RA (1985) *Materials Letters 3*: 119.
12. Zhang Y, Stewart JM, Morosin B, Graham RA, Hubbard CR (1989) *Applied Physics Communications 9*: 183.
13. Hammetter WF, Hellmann JR, Graham RA, Morosin B (1984) *Shock Waves in Condensed Matter-1983*: edited by Asay JR, Graham RA Straub GK, North Holland, 391.
14. Beauchamp EK, Carr MJ (1990) *J. American Ceramic Soc., 73*: 49.

15. Hankey DL, Graham RA, Hammetter WF, Morosin B (1982) *J. Materials Science Lett. 1*: 445.
16. Beauchamp EK (1987) "Shock-Activated Sintering, *High-Pressure Explosive Processing of Ceramics*: eds. Graham RA, Sawaoka AB, Trans Tech Publications, 139.
17. Bergmann OR, Barrington J (1966) *J. Amer. Ceram. Soc. 49*: 502.
18. Barrington J, Bergmann OR (1968) *Preparation of Brittle Inorganic Polycrystalline Powders by Shock-Wave Technique* : U.S. Patent 3,367,766.
19. Heckel RW, Youngblood JL (1968) *J. Amer. Ceram. Soc. 51*: 398.
20. Prümmer RA, Ziegler G (1977) *Powder Metallurgy International 9:* 11.
21. Hoeing CL Yust CS (1981) *Bull. Amer. Ceram. Soc. 60*: 1175 and 1221.
22. Beauchamp EK, Carr MJ (1987) *High-Pressure Explosive Processing of Ceramics:* edited by Graham RA, Sawaoka AB, Trans Tech Publ., 175.
23. Beauchamp EK, Graham RA, Carr MJ (1984) *Mater. Res. Soc. Symp. Proc. 24* : edited by Crawford JH Jr., Chen Y, Sibley WA, 281.
24. Morosin B, Graham RA (1984) *Mater. Res. Soc. Symp. Proc. 24*: edited by Crawford JH Jr., Chen Y, Sibley, WA, 335.
25. Carr MJ, Beauchamp EK (1983) *Shock Waves in Condensed Matter-1983*: edited by Asay JR et al, 403.
26. Morosin B, Graham RA (1984) *Shock Waves in Condensed Matter 1983*: edited by Asay JR et al, 355.
27. Beauchamp EK, Carr MJ, Graham RA (1985) *J. Amer. Cer. Soc. 68*: 696.
28. Beauchamp EK, Carr MJ, Graham RA (1987) *Adv. Ceram. Matls. 2*: 79.
29. Williamson DL, Venturini EL, Graham RA, Morosin B (1986) *Phys. Rev. B34*: 1899.
30. Beauchamp EK, Loehman RE, Graham RA, Morosin B, Venturini EL (1984) *Emergent Process Methods for High-Technology Ceramics*: edited by Davis RF, Palmour III H, Porter RL, Plenum, 735.
31. Casey WH, Carr MJ, Graham RA (1988) *Geochimica et Cosmochimica Acta*, 52: 1545.
32. Golden J, Williams F, Morosin B, Venturini EL, Graham RA (1982) *Shock Waves in Condensed Matter-1981*: edited by Nellis WJ, Seaman L, Graham RA, American Institute of Physics, 72.
33. Williams FL, Lee YK, Morosin B, Graham RA (1986) *Shock Waves in Condensed Matter-1985*: edited by Y. M. Gupta, Plenum, 791.
34. Merzhanov AG (1988) *Combustion and Plasma Synthesis of High Temperature Materials*: Holt JB, Munir ZA, American Ceramic Society, October, 1.
35. Munir ZA, Anselmi-Tamburini U (1989) *Mater. Sci. Rep. 3* : 277.
36. Bordeaux F, Yavari AR (1990) *J. Mater. Res. 5* : 1656.
37. Ma E, Thompson CV, Clevenger LA (1992) *MRS Symp. Proc. on Kinetics of Phase Transformations*: edited by Thompson MO, Aziz MJ, Stephenson Brian G, Vol. 205, Materials Research Society, Pittsburgh, 203.
38. Hida GT, Lin IJ (1988) *Combustion and Plasma Synthesis of High Temperature Materials*: edited by Holt JB, Munir ZA, American Ceramic Society, October, 246.
39. Thadhani NN, Srinivasan S, Schwarz RB, unpublished results.
40. Hammetter WF, Graham RA, Morosin B, Horie Y (1988) *Shock Waves in Condensed Matter-1987*: edited by Schmidt SC, Holmes NC, North Holland, 431.
41. Dunbar E, Thadhani NN, Graham RA (1991) *J. Mater. Sci.* in Press.
42. Dunbar E (1992) Master Thesis, New Institute of Mining and Technology.
43. Vreeland T, Mutz A "Shock Synthesis of Ti-Si Powder Mixtures, unpublished results.
44. Advani AH, Thadhani NN, Grebe HA, Heaps R, Coffin C, Kottke T (1992) *J. Mater. Sci.*, 27: 3309.
45. Grebe HA, Thadhani NN, unpublished results.

46. Iqbal Z, Thadhani NN, Chawla N, Ramakrishna BL, Sharma R, Skumeyev S, Reidinger F, Eckhardt H (1989) *Appl. Phys. Letts.* 55: 2339.
47. Syono Y, this volume.
48. Graham RA (1989) *Proc. of 3rd Int. Symp. on High Dynamic Pressure*: edited by Cheret R, Commissariat a l'Energie Atomic, Paris, 175.
49. Thadhani NN (1993) *Prog. Mat. Sci. 37*: 117.
50. Graham RA (1993) *Solids Under High Pressure Shock Compression: Mechanics, Physics, and Chemistry*, Springer-Verlag.

Chapter 4
Shock-Induced Chemical Reactions in Inorganic Powder Mixtures

Y. Horie

1. Introduction

The field of shock compression chemistry in material synthesis was described in 1985 as being in an early stage of development and an exercise in old-fashioned chemistry with a poorly characterized process [1]. The field was said to hold promise for future technological developments, but there are too many unsolved problems, including those that are not yet well posed. Even today, this assessment of the field is probably still accurate in spite of progress that has been made over the last decade. Since most shock chemistry studies involving nonenergetic materials have been carried out using recovery experiments with powder mixtures, the analysis and interpretation of observed chemical reactions have not always been unambiguous because of one or more of the following: inherent heterogeneity in composition, phase, morphology, material deformation, and chemical reactions, etc. As a result the delineation of basic mechanisms and process parameters has not been easy, nor always possible. However, the advantage of inorganic powders over energetic materials is that much of products can be recovered intact for post-shock characterization.

Studies of chemical reactions in inorganic powder mixtures at NCSU began about ten years ago and grew out of earlier studies on shock consolidations of ceramic powders such as AlN and TiC [2] and thermodynamics of defects in shock loaded solids [3]. It was a natural evolution of study progressing from shock consolidation and activation to chemistry. However, we find it interesting to note as a kind of serendipity, (1) that there also appeared at that time a report of mechanical alloying of intermetallic compounds [4] and a comprehensive review of the chemical effects of ultrasound [5], and (2) that the pioneering work of both mechano- and sono-chemistries are more than 50 years old and date back to the beginning of shock synthesis attempts [5,6]. Table 1 shows for comparison the range of process parameters for these three synthesis techniques.

Our selection of research themes has been strongly influenced by the above mentioned NMAB report. Out of many issues discussed therein we have chosen the following two as our main themes.

1) Uniqueness of shock process in materials synthesis.
2) Potential as a process for production of new materials.

On the basis of comparisons such as Table 1, it can be easily seen that shock waves in condensed matter obviously occupy a regime of macroscopic loading conditions that cannot be achieved by other techniques. But, as stated earlier there has been very little understanding of the question " how unique is unique?" in terms of basic mechanisms. On some of the questions, such as whether the shock effect is "benign" or "catastrophic", we do not know yet how to define the problem.

To begin, we have adopted two methods of approach. The first is an in-situ characterization of recovered samples to find clues to possible mechanisms. This approach is very similar to that of geochemists or petrologists. Through multi-probe analytical characterization, residual microstructures can shed critical light on mechanisms. The second is computational modeling of chemical reactions in inorganic powder mixtures under shock compression so that we can address some important issues in mathematical forms. Continuum calculations leave out a great deal of microstructual effects, but the least appeal of such an approach is that theories are refutable.

Also, from the beginning we have recognized the need to look at shock compression chemistry as an interdisciplinary problem and tried to leverage limited (or lack thereof) resources through collaborations with other institutions. Without the latter the progress we have made would not have been possible (a summary acknowledgment of the collaborations is given at the end of the chapter).

This article contains a retrospective review of our work to date as well as more recent work on numerical simulation and our current thinking on the rapidly evolving subject of shock compression synthesis. The second section contains a review of three synthesis experiments involving refractory aluminides, diamond, and diamond composites. The third section describes numerical work with emphasis on recent results.

Table 1 Physical methods to change chemical composition

	Shock Comp.	Mech. Alloy.	Ultrasonics*
Velocity	1 - 7 km/s	2 - 20 m/s	100 m/s
Impact Energy	10^4J	10^{-3}–10^{-2} J	–
Max ΔT	50,000 K	100K	5,000 K
Max P	500 GPa	1 GPa	0.1 GPa
Time	10^{-6} s	$10^4 - 10^5$ s	10^{-6} s

* in solution

2. Materials Synthesis

2.1 Aluminides

There was an element of luck in selecting nickel and titanium aluminides for our first synthesis attempt. We had known of their fascinating mechanical properties and their potential for use in superalloys, but at the time we had no idea that mixtures of elemental Ni and Al powders would produce such a good model system for study. There were many other possibilities. We could have repeated earlier examples such as carbides [7] and borides, or chosen a much simpler, completely miscible system such as a mixture of Cu and Ni, or niobium aluminides for its superconducting properties. There were three deciding factors. The first is the properties of aluminide that "intrigues the imagination and challenge the ingenuity of the metallurgist" [8]. The second is that no prior attempt of their synthesis using shock technique had been made, and the third is the existence of a large body of information regarding reaction kinetics of nickel and aluminum based on other processing techniques such as diffusion couples, SHS, pack cementation, and flame/plasma sprayings. It was felt that if the reaction between nickel and aluminum were observable under shock loading, then there exists a solid basis on which to address the question of uniqueness and potential applications.

Recovery experiments of Ni/Al powders using explosive loading were done at Sandia National Laboratories using Sandia recovery systems [9]. These recovery fixtures allow samples to be shocked in a controlled, reproducible manner. A planar shock wave is generated by detonating a high explosive lens next to an explosive pad. The shock, after passing through an iron wave shaper, impinges on a cylindrical copper fixture containing a powder sample.

As with most recovery systems, loading conditions of the Sandia systems are complex due to an interplay of two dimensional shock wave interactions, including radial waves. However, with the Sandia systems, pressure and temperature histories have been well characterized through an extensive program of numerical simulation using a rutile equation of state. The calculated ranges of peak bulk pressures and temperatures are listed as 7.5-2.7 GPa and 50-1100 °C respectively for the fixtures called Bear and Bertha. This complexity makes interpretation difficult, but the system has enabled us to observe clearly various stages of reaction in a single shot and has made us realize the importance of incomplete reactions to address the question of chemical evolution.

We were all aware of the desirability of conducting plane shock wave experiments, but most of the early studies had little external support. So, the initial focus on recovery experiments was a matter of opportunity. Only a limited study with Ti/Al powders was carried out to measure Hugoniots using a gas gun at NCSU [10]. The measurements did not reveal evidence for an unambiguous signature of shock-induced reactions. However, reactions were observed in the recovery experiments [11]. There were very interesting microstructual features including the phenomenon of liquation, but their analytical resolution was much harder and less complete than that of the Ni/Al mixture. To this date the task remains incomplete. Therefore, in this article our comments will be limited to the study of Ni/Al powder mixtures [12-15].

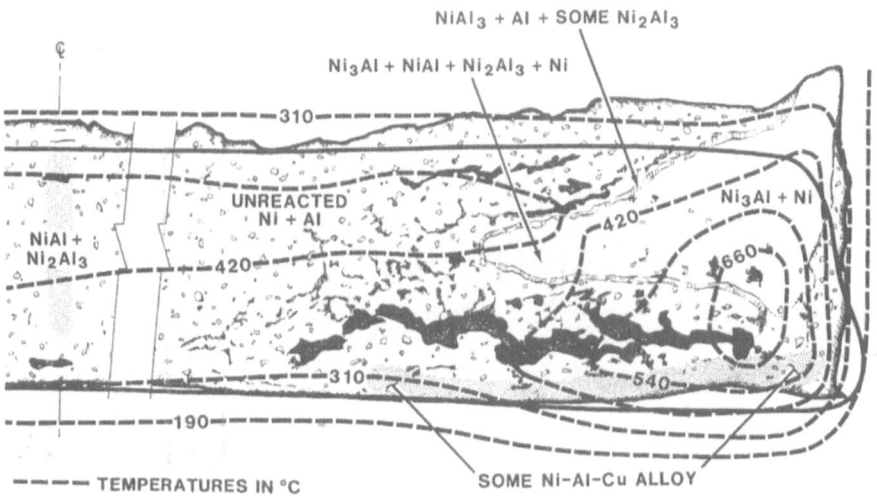

Fig. 1 Formation of nickel aluminides in a recovery experiment with the mixture consisting of 30 vol% Al and 70 vol% Ni and the initial porosity of 40% in comparison with calculated bulk temperature distribution. The calculations were performed with a porous rutile having the initial porosity of 45% (from {12} with permission).

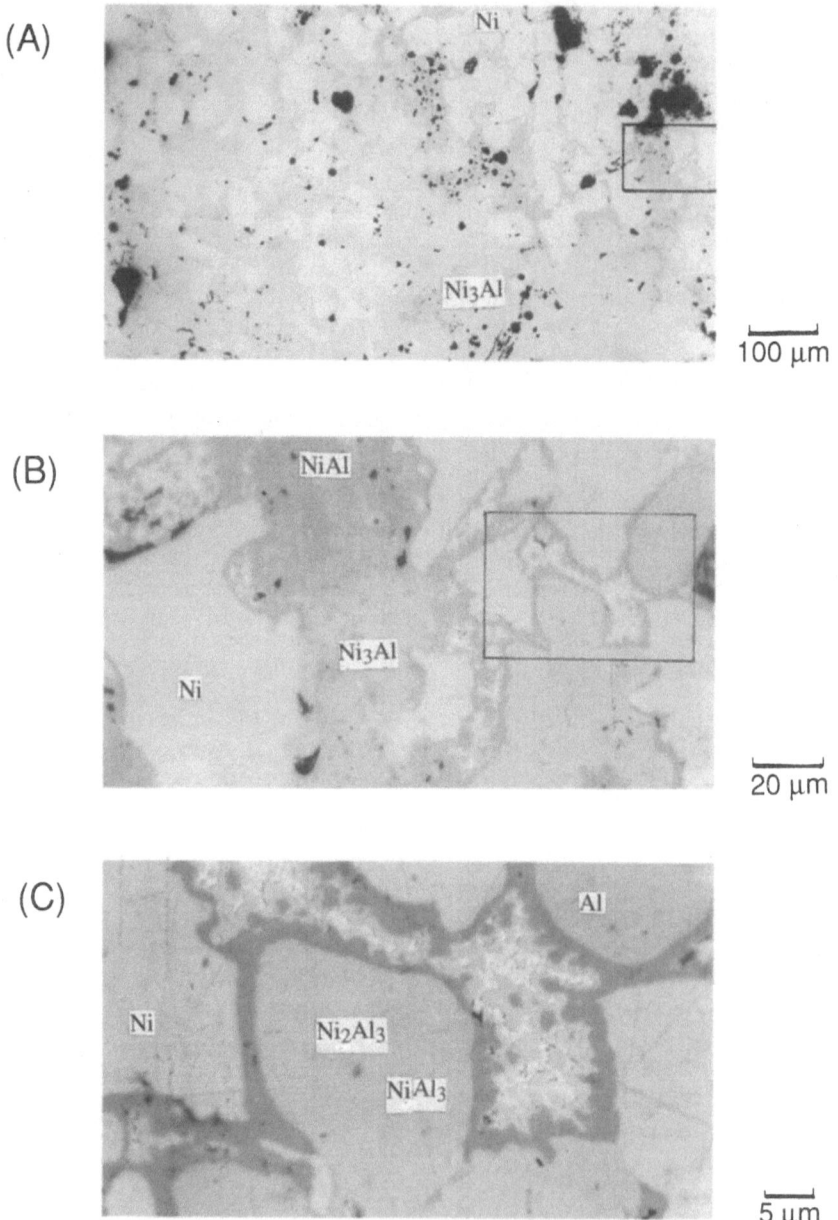

Fig. 2 Microstructures of the sample shown in Fig. 1 in a boundary region showing both strong and weak reactions. What one perceives from the microstructures depend on the level of magnifications.

Judging from what we know today, if we had tried to detect the initiation of chemical reactions with Ni/Al powders using the guns available at that time, we probably would not have succeeded. It is only very recently [16] that we have been able to obtain time resolved data that definitively indicated, in combination with post-shock metallographic analysis, the initiation of ultra-fast chemical reactions in Ni/Al powders. The results of this experiment will be described in a later section when we discuss their numerical simulation.

The excitement of studying Ni/Al powder mixture really began when we found that the metallography of chemical reactivity in a recovered sample showed a fair agreement with the two dimensional temperature calculations [12]. A schematic of the comparisons is duplicated in Fig. 1. The calculations were based on rutile, but this unexpectedly close agreement convinced us that at least a pin-hole window was opened allowing us to gain insight into the question of the distinctiveness of shock chemistry in material synthesis, or at least to help us formulate such questions. Although the agreement was seen in terms of temperature contours, we were not immediately predisposed to think that the temperature was the controlling factor in initiating the observed chemical reactions. At the time we were more eager to look for other "unique" mechanisms. We felt that temperature-controlled mechanisms such as conventional mass diffusion were too slow to be significant in powder particles of several tens of micrometers in the compression phase of a shock wave that may last only a few microseconds. The collapse of void may create hot spots which can be attributed to the initiation of localized limited reactions, but in bulk material, diffusional mass transport controlled by temperature was thought to be mostly responsible for late stage phenomena.

Recently, N. Thadhani of Georgia Institute of Technology has proposed the use of two terms, shock-induced reaction and shock-assisted reaction, to distinguish the two different mechanisms of chemical initiation under shock loading [17]. The former describe yet unclarified initiation mechanisms in the shock front. The latter describe much slower processes in which the effect of shock is purely that of heating, albeit with the possibility that they might take place in shock activated materials. There is increasing evidence that the former exist.

Subsequent to the finding reported in [12], a variety of analytical techniques has been used to study recovered samples of Ni/Al mixture. The results of these analytical observations revealed not only the complexity of shock effects on chemical reactions, but also recurring patterns of microstructure, giving hope that we might be able to delineate some mechanisms that are responsible for creating these microstructures. Fig. 2 shows typical microstructures of a transition region, from full reaction to no reaction, that are seen in all recovered samples of Ni/Al mixture. Calculated peak pressure and temperature for this sample are 19-22 GPa and 590°C respectively. The sample had a nominal composition of 30 vol% Al and 70% Ni. Average particle size of starting powders was 5-15 μm for Al and 44-74 μm for Ni.

Features that one can discern in microphotographs obviously depend not only on the scale of magnification, but also on what one wants to see. On the scale of hundreds of μm a demarcation may be seen running from the middle right-hand edge to the upper-left edge indicating a boundary between fully reacted and unreacted regions. A mottled feature in the left half (consisting of Ni and Ni_3Al) is thought to have been caused by heterogeneity in the initial mixing of starting powders. On the scale of a few μm (Fig. 2(B)), one sees features that can be clearly associated with diffusion controlled reactions: EMPA analysis has shown that products are arranged in the order of concentration between Ni particles and residual Al. On the intermediate scale of tens of μm, there was no obvious characteristic features, but features associated with the propagation of reactions may be discernible.

Occasionally, we have encountered very unusual features for which we have no explanation. For instance, Fig. 3 shows a completely reacted boundary layer between three nickel particles. The composition is that of Ni_3Al. Presently, it is a mystery as to how the subgrain structures in the product were formed and how chemical reactions propagated in the boundary layer, while retaining subgrain boundaries that are coherent to those found in the nickel particles. Fig. 4 is another example in which Al appear to have been forced (or "burst") into nickel particles under pressure. A better example of this feature is seen with a Ta/Si sample as shown in Fig. 5. Although scales are different, these pictures are reminiscent of those produced by molecular dynamical calculations of cold welding of two embedded-atom particles at a kinetic energy of a few percent of the melting energy [18]. However, in the case of the Ni-Al system, melting of Ni is a possible explanation. But, this explanation may not be applicable to Ta.

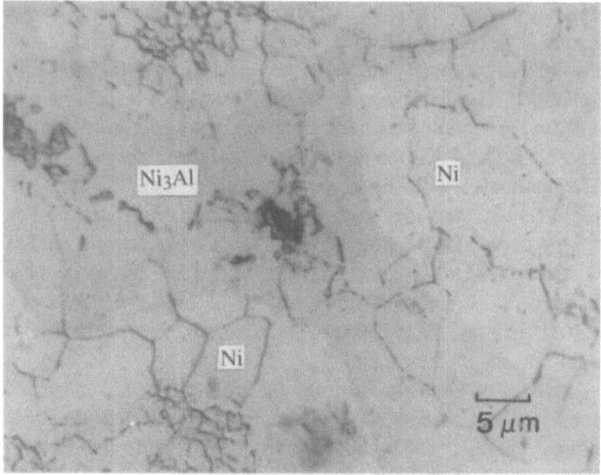

Fig. 3 Unusual subgrain structures observed in a Ni/Al recovery experiment.

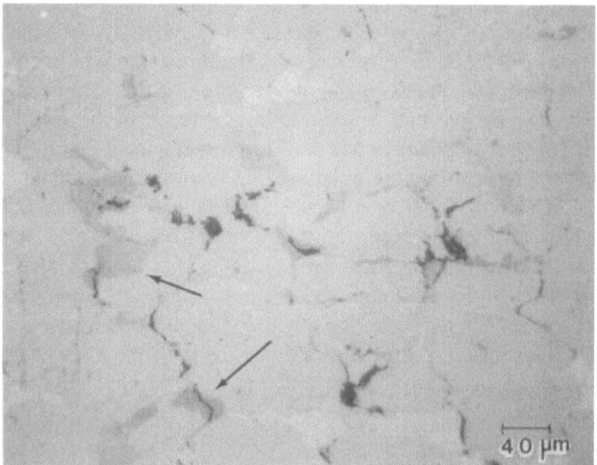

Fig. 4 An unusual pattern of "Al diffusion" into Ni particles under shock loading. A similar pattern is observed with a specimen consisted of Ta/Si as shown in Fig. 5. Arrows indicate affected regions.

(A)

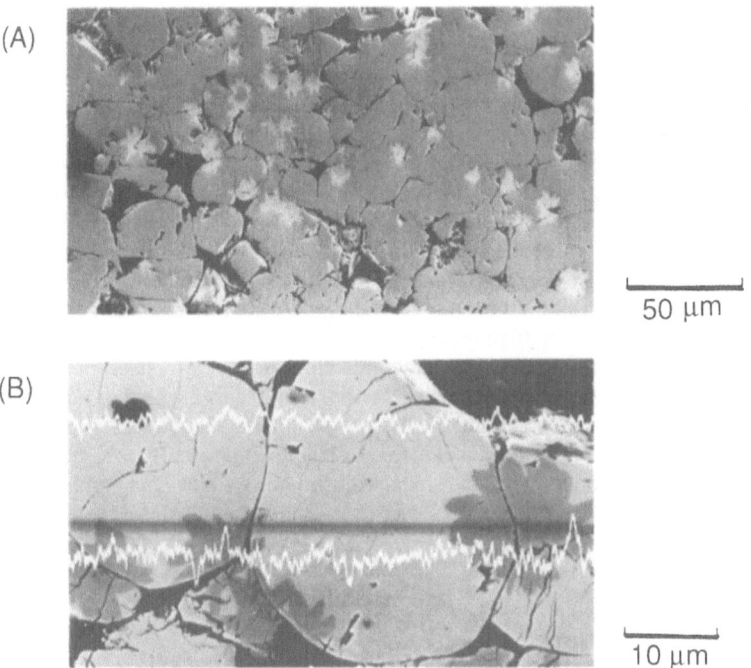

50 µm

(B)

10 µm

Fig. 5 A "bursting" pattern of "Si diffusion" into Ta similar to that shown in Fig. 4 for a Ni/Al mixture. (B) is an enlarged section of (A) and shows superposed EPMA lines for Ta. Bars are (A) 50 µm and (B) 10 µm.

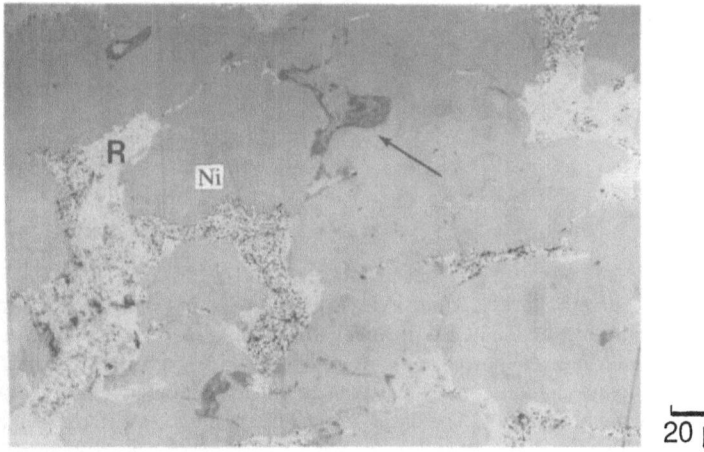

20 µm

Fig. 6 Microstructures of the sample shown in Fig. 1 in an "unreacted" region. The arrow indicates a localized spot of strong reaction showing a swirl-like microstructure. R indicates a region where Ni is "prereacted" with Al and electron microprobe analysis shows a Ni distribution having a eddy-like structure.

Indications of non-thermal mechanisms have come mostly from the study of regions where there was very little overall reaction. Fig. 6 shows a typical example of such a region found in the sample described in Fig. 1. Original boundaries of Ni particles are still identifiable. But there are also localized spots of strong reaction (the dark area in the upper middle section) having a vortex-like striation created by the colors of different aluminide products. These patterns are commonly found within spots of localized reactions. The most prevalent material found the inside was NiAl. Ni_3Al appears as a product of the interaction between Ni and NiAl. Although it is not apparent in Fig. 6, a turbulent like structure was also found in Al regions that show a smooth, light appearance by electron scanning microscopy [19]. A thermal analysis of the material taken out of this region shows a precursor reaction around 400°C. Transmission electron microscopy of a similar region suggests that the precursor reaction is a formation of amorphous nickel and aluminum mixture [20].

The vortex-like structure found in localized reactions does not by itself prove that it is a frozen flow-pattern created during shock compression. But at least the kind of randomness that exists in the reacted regions shown in Fig. 1 does not exist in these structures. More convincing microstructures were found in the results of a composite Ni/Al powder (Ni was deposited on Al particles by hydrogen reduction in aqueous metal salt solution). An example is shown in Fig. 7. Here, the reactions were clearly initiated by either interfacial instability or rotational large shear.

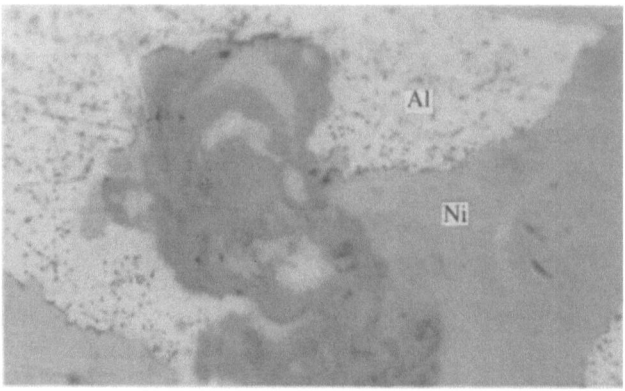

5 µm

Fig. 7 Chemical reactions observed in a composite Ni/Al powder. Interfacial instability appears to have caused the reactions.

The composite picture that emerges from the Ni/Al recovery experiments regarding chemical reactions under shock loading is as follows. Very fast reactions are initiated by localized flow instability associated with mechanisms such as shear, vortex motion, and interfacial instability. Since the powder mixture of Ni and Al has a higher Gibbs's free energy relative to its compound and amorphous phase, there exists an inherent tendency to lower the energy through chemical reactions. The onset of reactions is a matter of removing barriers for mass transport.

There is no doubt that there will be an enhanced mass transport under shock loading, but whatever the enhancement, a significant degree of reaction cannot be expected through mechanisms that only involve conventional atomic "hopping". A likely candidate is localized flow instability triggered by heterogeneity inherent in the powder mixture. Obviously, if there is a sufficient amount of the precursor reaction and heating (due to mechanical dissipation of energy as well as the heat of reaction), the reactions will go to completion depending on local initial compositions. Precursor reactions are likely to be an amorphous (or microcrystalline) mixture of nickel and aluminum, or NiAl. The formation of the latter may be attributed to its simpler configurational structure than other compounds. Subsequent to the very fast, initial reactions, the reactions that follow are likely to be diffusion controlled, but they are strongly dependent on the extent of the precursor reactions and do not appear to propagate far. This may be seen in Fig. 8 where the regions of partial reaction (upper left) to full reaction (lower right) are separated by a distance of no more than 100 μm. The abruptness of this transition suggests that the propagation of reactions due purely to thermal conduction is very localized and that strong chemical reactions occur mostly in regions where significant mass mixing had taken place. As a result, products under shock loading often have microstructures and properties that resemble those of rapidly solidified materials [20, 21]. Clearly, if a sample is overwhelmed by dissipative thermal energy, reactions will eventually go to completion by normal diffusion mechanisms.

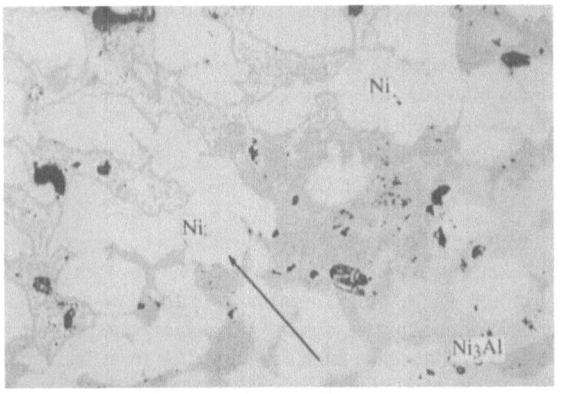

Fig. 8 A narrow boundary between regions of strong and weak reactions. The propagation of the reactions appear to strongly depend on the extent of localized shock-induced reactions. Diffusion controlled reactions do not appear to propagate over a long distance when shock-heating is not enough to melt the powder. The arrow indicates the direction of decreasing reaction.

Fig. 9 is a tentative picture of what might be happening in Ni/Al powder mixtures under shock loading. The selection of NiAl as first intermetallic product is a speculation, but is also based on the idea that a simpler structure emerges first under rapid loading.

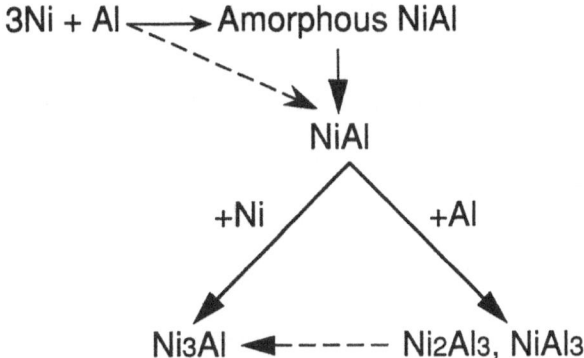

Fig. 9 Hypothetical reaction paths for the 3Ni/Al mixture based on recovery experiments.

2.2 Diamond

In the synthesis study of nickel and titanium aluminides, we have seen both equilibrium (e.g., Ni_3Al) and non-equilibrium products (e.g., amorphous phase). Properties of the equilibrium products resembled those of rapidly solidified materials. But, these products are also obtainable by other processing techniques such as mechanical alloying and SHS. The question of uniqueness is not really resolved, except for the possibility of very fast mass mixing in the compression phase of shock loading. Chemical reactions by themselves appear to be similar to those of other non-equilibrium techniques.

The diamond synthesis was intended to focus on a single mechanism and to study how it might be affected by shock excitation. Again, there was a unique circumstance that prompted us to undertake such a study using a Ni-coated graphite powder. They are

1) Commercial availability of Ni-coated graphite powder found during the course of the Ni/Al study.
2) Well characterized static synthesis of diamond using Ni as a catalytic/solvent agent.
3) A report of extraordinary graphite mass transport under shock loading [22].
4) Our own observation of mass mixing in the Ni/Al powders.
5) Possibility of diamond synthesis involving a sequence of phase transformations under a combined loading of shock and resistive heating [23].

Two kinds of powders were used in this study. The first is a commercially available Ni-coated graphite powder (for thermal spraying on jet engines) having 85 wt% Ni and a mean particle size of about 90 μm and the second is a mixture of the composite powder and Al powder of 20 μm in diameter. Aluminum was added to the composite powder to exploit the affinity of Al to carbon in nickel to control the latter's precipitation from the metal. The amount of Al in the latter is chosen to correspond to the stoichiometric ratio of Ni_3Al.

The shock loading was carried out using a Sawaoka recovery system in which 6-12 capsules measuring 12 mm in diameter and 5 mm in height were placed in a steel disk. A steel flyer plate accelerated by an explosive plane wave generator impinged on the capsules at the velocities of 1.1, 1.6, and 2.1 km/s. No matching calculations of the fixture are available, but the simulation based on 2.5 km/s impact velocity and the rutile equation of state shows peak temperatures and pressure of 2,000 °C to 4,600 °C and 60 to 100 GPa respectively. These calculations appear to be consistent with the post-shock observation that bulk melting has been observed only in the samples that were shock loaded at 1.6 and 2.1 km/s.

The results were reported in [24,25]. What follows is a summary of in-situ observations based on metallographic and transmission electron microscopy regarding mass diffusion and diamond synthesis.

1) Extensive formation of graphite spherulites were observed in the portion of nickel where it apparently melted under shock loading. Their structures were very similar to those of graphite spherulites in other metallic melts [26], which were crystallized upon cooling on gas bubbles. The presence of gas bubbles in the Ni-coated graphite is readily explained considering the probability of gases trapped in the graphite or even hydrogen containing gases from the nickel coating process.
2) Microcrystalline diamond is found in the center of spherulites that had a graphite ring and an incompletely filled interior.
3) Graphite nodules were found in a sample whose surface was melted by an acetylene torch at atmospheric pressures. But there was no diamond containing spherulites.

4) The formation of spherulites was significantly reduced by the addition of Al. There were a few spherulites that had a morphology suggestive of diamond, but they were too small to identify it.

5) In the Al containing sample, we observed in-situ the transition of graphite fibers to diamond phase.

6) In many places of the aluminum containing sample the normally parallel fibers of graphite flakes were disturbed as if they had experienced violent turbulence. A diffraction pattern of such an area shows mostly the full rings of random orientation. The diamond transition described above is found in the neighborhood of these perturbed areas.

Presently, there is no understanding of how polycrystalline diamond might be formed in the graphite spherulites. But, we believe that it happened under pressure. The turbulent perturbation of graphite flakes lends support to the idea that the initiation of strong chemical reactions in Ni/Al powder mixture is caused by localized flow instability in the shock front.

2.3 Diamond/Ceramics Composites

The goals of this project in which diamond was shock-consolidated in three reactive mixtures: Si/graphite, Ti/graphite, and Ti/Si, were manifold. The first was to conduct a follow-up experiment on the result reported in two U.S. patents that the microhardness of diamond compacts with SiC increased from 22 to 47 GPa when the binder SiC was synthesized from elemental silicon and graphite [27]. The second was that this combination of materials appeared to be a good system to study the question of how chemical reactions under shock compression can be controlled and quenched for post-shock analysis. Diamond was considered for three reasons, first to control local temperature similar to the technique used by W. Nellis et al [28], secondly to change the overall stiffness of the mixture to vary pressure. We envisioned the experiment to be a kind of "diamond anvil test". The third goal was to develop a computational model that can deal with such control mechanisms in reactive inorganic powder mixture.

In this section we will present a brief description of the experiments and two noteworthy results. The computational model is discussed in the next section. Details may be found in [29].

The recovery experiments were carried out using a powder gun at the Tokyo Institute of Technology. Sample powders were pressed into a single cell Sawaoka-capsule to the density of 60% of solid density and impacted by a flyer plate at the velocity of 1.8 km/s. The capsule is confined by a steel tube and backed by a steel momentum trap. The composition of reactive components was stoichiometric to form SiC, or TiC. The ratio of Ti to Si was also one to one. The volume percents of diamond powder tested were 40 and 60%.

The first noteworthy result is the microstructure of shock-synthesized SiC and TiC shown in Figs. 10 and 11. It shows two TEM microphotographs of the product, SiC or TiC, in a central region of the recovered samples that initially contained 60 vol% diamond. It is believed that these ultrafine, nanosized particles of SiC and TiC were a result of rapid cooling by diamond particles. The numerical simulation indicates a cooling rate of about 1,500 K/μs.

The second noteworthy result is the interesting correlation shown in Fig. 12 between the distribution of microhardness in the Si/C sample that contained 40 vol% diamond and the calculated profile of chemical reaction. As described in the next section, the modeling involve a number of assumptions and educated guesses about material properties, but there is a qualitative agreement between the two. The reactions are described by an Arrhenius equation, and the threshold condition was set arbitrarily at 1,000°C. However, no synergetic effect reported in the patents is observed. The maximum hardness measured (27 GPa) is close to that of conventionally fabricated SiC. Nevertheless, the sample was very difficult to machine.

The heterogeneity of material properties is inherent with any powder mixture and makes it difficult to conduct a controlled experiment and to interpreat results unambiguously. But, the above results show that the situation may not be hopeless.

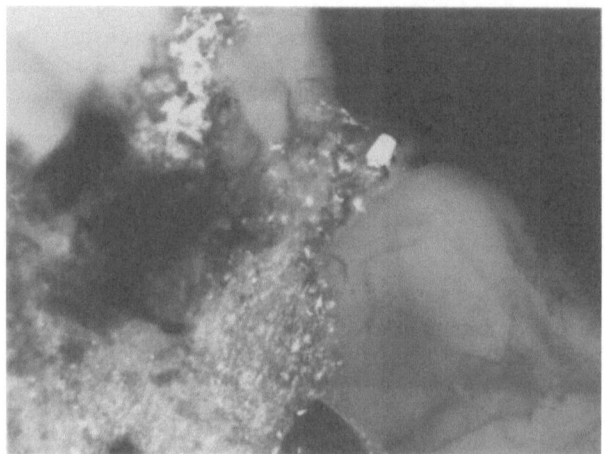

200 nm

Fig. 10 A dark field TEM microphotograph of the diamond/Si/C mixture indicating the formation of ultrafine microcrystalline SiC. It is thought that rapid cooling by diamond may be responsible for the nano-structure. Diamond particles appear to contain many dislocations. The diffraction pattern of the same region indicate the existence of unreacted graphite.

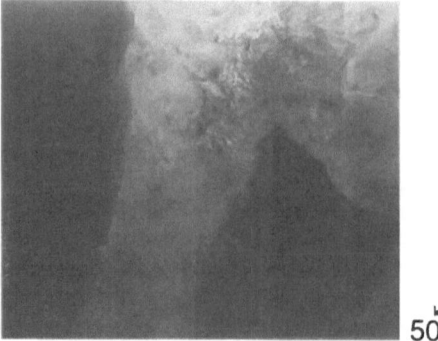

50 nm

Fig. 11 A bright field TEM microphotograph of the diamond/Ti/C mixture showing the formation of ultrafine microcrystalline TiC. SEM analysis shows a great deal of reaction-related voids likely to have been created after the passage of shock wave, because of strong exothermicity of the Ti/C reaction.

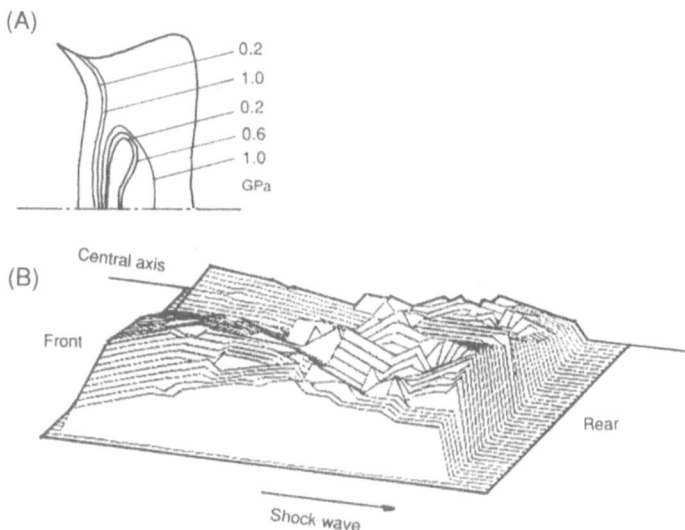

Fig. 12 (A) a calculated distribution of chemical reaction using the constitutive model called VIR and a two dimensional hydro-code, and (B) Vickers microhardness in the diamond/Si/C mixture where the maximum and minimum hardness were 56 and 24 GPa, respectively.

3. Computational Modeling

Sometimes, the formal development of constitutive models can be (and has been) carried out without input from experimental observations, but in our study actual involvement in analytical characterization of recovered samples has been immensely valuable, if not indispensable. Seeing literally hundreds of microphotographs of recovered samples at all levels of magnification has shown us that although observed features are very complex and the list of parameters that affect the basic mechanisms is extensive, critical features of the chemical reactions can be described by a small number of certain variables.

Our first attempt [30] has considered only an effective single distended solid for a powder mixture which undergoes an irreversible chemical reaction at appropriate conditions of temperature and pressure. It was a simple model intended to explore significant parameters and physical trends that are thought to control the outcome of early stages of chemical reactions. Although comparable experimental data were not available at that time, calculations were useful for development of the more comprehensive model called VIR. They have shown, or brought into focus:

1) The need to isolate early stage reactions from late stage reactions. Probably there is no single calculation that can cover the entire range from submicroseconds to minutes at the moment.

2).The importance of localized quenching of chemical reactions. A corollary is that a significant mass mixing must be assumed to explain some of the features seen in post-shock samples.

3) The assumption of impenetrability in continuum theories. This means that mass mixing must be handled as a constitutive relationship on hydro-codes.

4) The balance between complexity and model verification.

In this section we will review our recent model development and illustrate features of the model called VIR using simulation of an instrumented recovery experiment in which plane shock wave profiles indicated clear evidence of very fast chemical reactions in Ni/Al powder mixtures.

Lagrangian VIR

This model was first developed to interpret recovery experiments of diamond powder in reactive mixtures described earlier using a Lagrangian hydro-code. But, we now know that it can describe a wider class of reactive powder mixtures. However, the model is still focused on features that are thought to be important in the first few, if not tens of microseconds, of shock loading. These features are the adiabatic shock compression of multimaterial powder mixtures, the existence of both chemically active and inactive species which may exist in the state of different pressures and temperatures, mass mixing, localized heat conduction among constituents, and void collapse. Interactions of these basic mechanisms are simplified into basic four blocks as shown in Fig. 13.

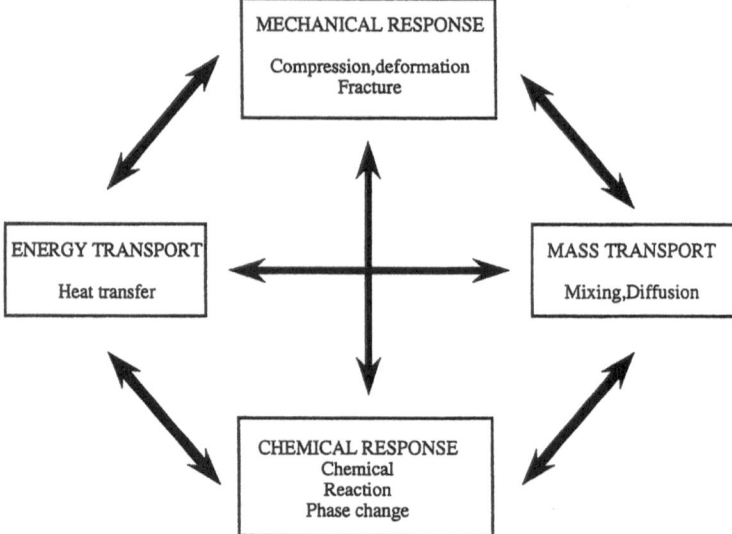

Fig. 13 Four primary interacting mechanisms that are considered in the VIR model.

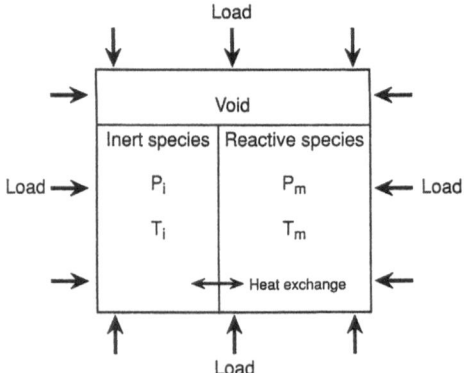

Fig. 14 A Schematic representation of the VIR model.

To develop a mathematical model we used a diagramatic representation of the reactive powder mixture shown in Fig. 14. The mixture is subdivided into two material subsystems and void. The materials subsystems are further divided into inactive and reactive subsystems depending on the possibility of chemical reactions. In addition, these material subsystems are assumed to be in separate, locally equilibrium states. The acronym VIR stands for Void, Inactive, and Reactive materials.

The development of governing equations depend on additional simplifying assumptions as well as specific expressions for the basic interacting mechanisms shown in Fig. 13. Currently, there are three major simplifying assumptions:

1) The mixture as a whole is assumed to be a thermodynamically closed system because of the time scale of interest. No external heat and mass transport is considered in the VIR model. The relaxation of the latter restriction is straight forward.
2) The material subsystems are in mechanical equilibrium relative to each other and hence there is only one motion for the entire mixture, dictated by the choice of material coordinates. The relaxation of this assumption to allow heterogeneous flow in the mixture is considered in the next section.
3) Thermal exchange between the material subsystems is included, but not mass transport.

Constitutive equations for the interacting mechanisms are based on the following phenomenological equations:

1) Void collapse is assumed to take place only in the reactive systems and obeys the P-α model. Again, the relaxation of this restriction to allow rate effects and/or a proportional void collapse in each of the material system can be easily accomplished.
2) Chemical reactions and mass transport in the reactive system are described by a single Arrhenius equation. A more elaborate system of equations such as those developed in the first model [30] can be introduced very easily. But it was felt that at this stage in the model development the simpler, the better.
3) Heat conduction between the material systems is described by the Newtonian equation of heat conduction that includes the effect of particle size.
4) The equation of state for each species is described by the Birch-Murnaghan equation.

The following is a summary of the governing equations based on the above described postulates and interaction mechanisms. Details of the derivation is given in [29].

First law of thermodynamics

$$de = - (p + \eta)dv \tag{1}$$

$$de_i = \delta q_i - (p + \eta)dv_i \tag{2}$$

$$de_m = \delta q_m - (p + \eta)(dv - dv_i) \tag{3}$$

where e = specific internal energy, p = hydrostatic pressure acting on the mixture, η = artificial viscosity, δq = specific heat exchanged between the material subsystems. Those without subscript are for the whole mixture. The subscripts "i" and "m" represent inactive and reactive subsystems respectively. There are summation rules for the energy equations

$$e = \lambda_i e_i + \lambda_m e_m \tag{4}$$

$$\lambda_i \delta q_i + \lambda_m \delta q_m = 0 \tag{5}$$

where λ_k are mass fractions defined by

$$\lambda_i = M_i / M \tag{6}$$

$$\lambda_m = M_m / M \tag{7}$$

$$M = M_i + M_m \tag{8}$$

where M_m and M_i are masses of the active and inactive systems respectively. Eq. (5) expresses the adiabatic conditions for shock compression process. Similarly, the mass fraction of species j in the reactive subsystem is defined by

$$\zeta_j = M_j / M_m \tag{9}$$

Then, it may be seen that the specific volumes must satisfy the following summation rules

$$v_s = \lambda_i v_i + \lambda_m v_m \tag{10}$$

$$v_m = \sum_j \zeta_j v_j \tag{11}$$

$$v = v_s + v_{void} \tag{12}$$

where v_s is the specific volume of solid components.

Second law of thermodynamics
Since local equilibrium is assumed for each of the two subsystems, they satisfy

$$de_i = T_i ds_i - p_i dv_i \tag{13}$$

$$de_m = T_m ds_m - p_m dv_m + \sum_j G_j d\zeta_j \tag{14}$$

where T = temperature, s = specific entropy, and G_j = specific Gibbs free energy of species j.

Equation of state
The following Birch-Murnaghan equation is used for species equations of state.

$$p(v,T) = \beta_{T0}[(v / v_0)^{-n} - 1] + (\Gamma / v)C_v(T - T_0) \tag{15}$$

$$C_v = \text{constant} \tag{16}$$

$$\Gamma/v = \text{constant} \tag{17}$$

$$\beta_T = \beta_{T0}(v_0 / v)^n / n \tag{18}$$

where β_{T_o} = isothermal bulk modulus, C = specific heat at constant volume, Γ = Grünieisen constant, n = material constant.

Since mechanical equilibrium is assumed between the subsystems, we have

$$p_i = p_m \tag{19}$$

Pore collapse

Currently, a parabolic function is used to connect the initial distention ratio to the state of complete collapse. Mathematically, the model is described by

$$\alpha = \begin{cases} \alpha_0 & p < p_e \\ 1+(\alpha_0 - 1)\left(\dfrac{p_s - p}{p_s - p_e}\right)^2 & p_e \le p \le p_s \\ 1 & p_s < p \end{cases} \tag{20}$$

where α = the distention of the mixture defined by $\alpha = v / v_s$, a_o = initial distention, P_e = elastic limit, P_s = crush-up pressure where the powder becomes dense solid.

Chemical kinetics

Effective mass transport and a single irreversible chemical reaction is described by an Arrhenius type equation.

$$\sum_j G_j \dot{\zeta}_j = H_0 \dot{\xi} = H_0 \dot{\xi}(1 - \xi)\exp(-E_m / RT_m) \tag{21}$$

where "." signifies total time derivative, H_o = heat of formation, ξ = reaction coordinate, $\dot{\xi}$ =reaction constant, E_m = "activation" energy, R = gas constant. The modification of the above equation to include multi-reactions and mass transport is straightforward [30].

Lacking specific information, the initiation of chemical reaction is prescribed by a threshold conditions on temperature (or conditions, if pressure is also thought to play a role). The question of initiation will be considered again when we discuss the calculations of an instrumented recovery experiment.

When a chemical reaction occurs, the mass fractions of the reactive system must be readjusted based on its stoichiometric relationship. For instance, in the case of the Si/graphite system, they are given by

$$\zeta_{Ri} = (1 - \xi)M_{Ri} / M_m \tag{22}$$

$$\zeta_{Pj} = (\xi)M_{Pj} / M_m \tag{23}$$

where ζ_{Ri} = mass fraction of reactant i, ζ_{Pj} = mass fraction of product j, M_{Ri} = atomic weight of reactant i, and M_{Pj} = atomic weight of product j.

Heat transfer between the subsystems

Heat transfer is described by the following Newtonian equation.

$$\dot{q}_i = Ah \ (T_m - T_i) \tag{24}$$

where q_i = specific rate of heat transfered to the inert subsystem, h = heat transfer coefficient, A = specific area = $6 \ v_i/Di$, Di = average diameter of the inert particles. Also, the above heat transfer must satisfy the summation rule (5). In the continuum theory discussed in section III this equation is derived based on the entropy inequality equation.

Computational constitutive equations

For purpose of numerical simulation using a Lagrangian hydro-code, it is convenient to reorganize the above set of equations so that the changes in pressure and temperature are given in terms of specific volumes. This transformation yields the following equations

$$\dot{P}_i = \{ - \beta_{si} + \Gamma_m[\ (\alpha - 1)P - \eta \ ] \} \frac{\dot{v}_i}{v_i} + \frac{\Gamma_i}{v_i}\dot{q}_i \tag{25}$$

$$P_m = - (\beta_{sm} + \Gamma_m \alpha \eta \ )\frac{\dot{v}_m}{v_m} + \frac{\lambda_i}{\lambda_m}\Gamma_m (1 - \alpha \ )(P \ + \eta \ ) \frac{\dot{v}_i}{v_m}$$
$$- \frac{\Gamma_m(P + \eta \ )}{\lambda_m v_m} v_s \dot{\alpha} + \frac{\Gamma_m}{v_m} (\dot{q}_m + H_0 \dot{\xi} \ ) \tag{26}$$

$$\dot{T}_i = \left[\frac{(\alpha - 1)P - \eta}{C_{vi}} - \frac{\Gamma_i T_i}{v_i} \right]\dot{v}_i + \frac{\dot{q}_i}{C_{vi}} \tag{27}$$

$$\dot{T}_m = - \left(\frac{\alpha \eta}{C_{vm}} + \frac{\Gamma_m T_m}{v_m} \right)\dot{v}_m + \frac{\lambda_i}{\lambda_m} (1 - \alpha \ )(P + \eta \ ) \frac{\dot{v}_i}{C_{vm}}$$
$$- \frac{P + \eta}{\lambda_m C_{vm}} v_s \dot{\alpha} + \frac{\dot{q}_m + H_0 \dot{\xi}}{C_{vm}} \tag{28}$$

For simplicity only one component is included in the inactive system in the above derivation. Mixture properties such as C_{vm} and Γ_m are listed below.

$$C_{vm} = \sum_j \zeta_j C_{vj} + \sum_j \zeta_j \left[p_m - T_m \Gamma_j C_{vj} / v_i \right. $$
$$\left. \frac{\sum_k (C_{vk}\Gamma_k / v_k - C_{vj}\Gamma_j / v_j)(\zeta_k v_k / \beta_{Tk})}{\beta_{Tj} / v_j \sum_k \zeta_k v_k / \beta_k} \right] $$

$$\Gamma_{vm} = \frac{v_m}{C_{vm}} \left[\frac{C_{vj}\Gamma_j}{v_j} + \frac{\sum_k (C_{vk}\Gamma_k / v_k - C_{vj}\Gamma_j / v_j)(\zeta_k v_k / \beta_{Tk})}{\sum_k \zeta_k v_k / \beta_{Tk}} \right] \tag{29}$$

$$\beta_{Tm} = \frac{v_m}{\sum_j \zeta_j \, v_j / \beta_{Tj}}$$

$$\beta_{sm} = \beta_{Tm} + C_{vm} \Gamma_m^2 T_m / v_m$$

The extension to handle multicomponents in the inactive system can be easily done using the above listed mixture properties. We also note that in the above derivation, the following identity equations were used based on the assumption of local equilibrium.

$$dp_i = \left(\frac{\partial p_i}{\partial v_i}\right) d\,v_i + \left(\frac{\partial p_i}{\partial T_i}\right) dT_i \qquad ds_i = \left(\frac{\partial s_i}{\partial v_i}\right) d\,v_i + \left(\frac{\partial s_i}{\partial T_i}\right) dT_i$$

$$dp_m = \left(\frac{\partial p_m}{\partial v_m}\right) d\,v_i + \left(\frac{\partial p_m}{\partial T_m}\right) dT_m \qquad ds_m = \left(\frac{\partial s_m}{\partial v_m}\right) d\,v_m + \left(\frac{\partial s_m}{\partial T_m}\right) dT_m$$

Eqs. (25)-(28) are in form where the physical meaning of individual terms can be easily discerned. But, numerically each specific volume must be calculated so that they are consistent with their own equation of state and over all kinematic constraint, eqs. (10)-(12). Therefore, in our actual calculations discussed later, eqs. (25)-(28) are retransformed so that they satisfy the kinematic conditions automatically. The results are

$$
\begin{bmatrix}
\dfrac{P_m-(P+\eta)}{C_{vi}} - T_i\dfrac{\Gamma_i}{v_i} & 0 & \dfrac{\beta_{si}}{v_i} - \dfrac{\Gamma_i}{v_i}[P_m-(P+\eta)] \\[2ex]
\dfrac{P_m}{C_{vm}} - T_m\dfrac{\Gamma_m}{v_m} + \dfrac{\lambda_i}{\lambda_m}\dfrac{P+\eta}{C_{vm}}\dfrac{\beta_{tm}}{\beta_{ti}}\dfrac{v_i}{v_m} & \dfrac{\beta_{sm}}{v_m} - P_m\dfrac{\Gamma_m}{v_m} & -\dfrac{\lambda_i}{\lambda_m}\dfrac{P+\eta}{C_{vm}}\dfrac{\beta_{tm}}{\beta_{ti}}\dfrac{v_i}{v_m}C_{vi}\left(\dfrac{\Gamma_i}{v_i}\right) \\[2ex]
\dfrac{\lambda_i}{\lambda_m}\dfrac{v_i}{\beta_{ti}} + \sum_j \dfrac{\zeta_j v_j}{\beta_{tj}} & -\sum_j \dfrac{\zeta_j v_j}{\beta_{tj}}C_{vj}\left(\dfrac{\Gamma_j}{v_j}\right) & -\dfrac{\lambda_i}{\lambda_m}\dfrac{v_i}{\beta_{ti}}C_{vi}\left(\dfrac{\Gamma_i}{v_i}\right)
\end{bmatrix}
\begin{Bmatrix}
\dot{P}_m \\[2ex] \dot{T}_m \\[2ex] \dot{T}_i
\end{Bmatrix} =
$$

$$
\begin{Bmatrix}
\dfrac{\beta_{ti}}{C_{vi}v_i}\dot{q}_i \\[3ex]
\dfrac{\beta_{tm}}{C_{vm}v_m}\left(-\dfrac{P+\eta}{\lambda_m}\dot{v}+\dot{q}_m+H_0\dot{\xi}\right) \\[3ex]
\sum_j \zeta_j v_j - \dfrac{1}{\alpha\lambda_m}(\dot{v}-\dot{\alpha}v_s)
\end{Bmatrix}
\tag{30}
$$

Although the right hand side of eq. (30) still contains time dependent expressions, they can be replaced by functions of state and other variables that do not include time derivatives.

Eulerian VIR Model
The Lagrangian VIR cannot in principle deal with mass flux and heterogeneous flow of powder constituents because of the use of the coordinates that are attached to materials. Also, the construction of interaction mechanisms was done using ad hoc diagrams such as ones shown in Figs. 13 and 14. Therefore, there were two-fold goals in development of an Eulerian VIR. The first is to "reframe" the Lagrangian VIR model to construct an Eulerian VIR model to deal with heterogeneous flow in powder mixture through use of continuum mixture theory (CMT). The second is to reexamine the assumptions used in the Lagrangian VIR, because CMT is said to bring better mathematical consistency and greater abstraction to the formulation of constitutive relations. The particular CMT we followed is that used extensively at Sandia National Laboratories for the description of granular explosives [31].

Continuum Mixture Theory
There are many excellent monographs and review articles on CMT, sō we shall list selectively only those basic equations and concepts that are necessary for ensuing discussions. For details, see [31,32]. Also, for simplicity, the equations will be confined to one spacial dimension.

The conservation equations of the continuum mixture theory are written in form similar to those of a single material except those terms that express mutual interactions within the mixture. They are

$$\dot{\rho}_a = -\rho_a \frac{\partial v_a}{\partial x} + c_a^+ \tag{31}$$

$$\rho_a \dot{v}_a = -\frac{\partial \pi_a}{\partial x} + m_a^+ + c_a^+ v_a \tag{32}$$

$$\rho_a \dot{e}_a = -\pi_a \frac{\partial v_a}{\partial x} - \frac{\partial q_a}{\partial x} + \rho_a r_a + e_a^+ - m_a^+ v_a - c_a^+ \left(e_a - \frac{1}{2} v_a^2 \right) \tag{33}$$

where v_a is the velocity of constituent a, c_a is the mass production of constituent a due to interactions with other constituents, the partial pressure associated with a (called peculiar stress by Treusdell [32]),m_a momentum production due to interactions with other constituents, e_a the specific internal energy of a, r_a radiative external heat supply, q_a conductive heat flux, and e_a energy production due to interactions with other constituents.

Additionally, it is assumed that when properly weighted and summed the species conservation equations satisfy the balance equations for the mixture as a whole. Then, the following summation rules must accompany the above described species balance equations.

$$p + \rho v^2 = \sum_a (\pi_a + \rho_a v_a)$$

$$\rho b = \sum_a \rho_a b_a$$

$$\rho \left(e + \frac{1}{2} v^2 \right) = \sum_a \rho_a \left(e_a + \frac{1}{2} v_a^2 \right) \tag{34}$$

$$pv - q - \rho \left(e + \frac{1}{2} v^2 \right) = \sum_a \left\{ \pi_a v_a - q_a - \rho_a \left(e_a + \frac{1}{2} v_a^2 \right) v_a \right\}$$

$$\rho r = \sum_a \rho_a(r_a)$$

$$\sum_a c_d^+ = \sum_a m_d^+ = \sum_a e_d^+ = 0$$

The density of the mixture is defined by the sum of partial densitites, ρ_a,

$$\rho = \sum_a \rho_a \tag{35}$$

where the partial density is related to the intrinsic density and the volume fraction as follows.

$$\rho_a = \alpha_a \gamma_a \tag{36}$$

Atkin and Crain [33] points out that the total mass density above is a purely mathematical entity and has no physical meaning except in a special case. If the mixture is saturated, then

$$\sum_a \alpha_a = 1$$

If the mixture contains voids, then

$$\alpha = \frac{V}{V_s} = \frac{1}{1 - \alpha_{void}} = \frac{1}{\alpha_i + \alpha_m} \tag{37}$$

$$\dot{\alpha} = \frac{\dot{\alpha}_{void}}{(1 - \alpha_{void})^2} = -\frac{(\dot{\alpha}_i + \dot{\alpha}_m)}{(\alpha_i + \alpha_m)^2}$$

Using the definitions, one can derive the following useful relationship between the mass fraction and the density of the mixture.

$$\alpha_a = \frac{V_a}{V} = \frac{m_a v_a}{m v} = \lambda_a \frac{\rho}{\gamma_a} \tag{38}$$

> where V_a : volume of constituent a
> V : total volume
> m_a : mass of constituent a
> m : mass of the mixture
> v_a : specific volume of constituent a
> v : ($=1/\rho$) specific volume
> λ_a : mass fraction of constituent a

The development of constitutive equations in the CMT is guided by the overall entropy ineqality equation for the mixture,

$$\sum_a \left\{ \rho_a \dot{s}_a + c_d^+ s_a + \nabla \cdot \left(\frac{q_a}{T_a} \right) - \frac{\rho_a r_a}{T_a} \right\} \geq 0 \tag{39}$$

where T_a is the temperature of constituent a, s_a is the entropy of a, and q_a/T_a the entropy flux due to external heat flux. Strictly speaking, the construction of constitutive equations is out side of the frame work of classical thermodynamics. But , the inequality equation, when couple with educated guess about physical mechanisms, has been found to specify the admissible functional form for such mechanisms. Typically, this involves the introduction of Helmholtz free energy defined by

$$\psi_a = e_a - T_a s_a \tag{40}$$

Upon substitution of eq. (40) into eq. (39), one finds

$$\sum_a \frac{1}{T_a} \left\{ -\rho_a(s_a \dot{T}_a + \dot{\psi}_a) - \pi_a \frac{\partial v_a}{\partial x} - \frac{q_a}{T_a} \frac{\partial T_a}{\partial x} - e_a^{\ddagger} - c_a^{\ddagger} \left(\psi_a - \frac{1}{2} v_a^2 \right) \right\} \geq 0 \tag{41}$$

The positivity of this equation can be used to assert certain relationships among conjugate thermodynamic variables expressing dissipative processes such as heat conduction and momentum and energy exchanges. However, as we will see shortly, the "art" of generating constitutive equations depends, to a large extent, on how one perceives the dependence of the Helmholtz function in terms of conjugate variables.

Derivation of the VIR Model Using the CMT

One approach, out of several, to examine the VIR model using the CMT is to consider the material subsystems as species. This representation will greatly simplify the conservation equations. Since, there is no mass and momentum exchange between the two, the resulting equations are basically those of standard hydrodynamic equations except the terms expressing energy interaction. The resulting equations

$$\dot{\rho}_m = -\rho_m \frac{\partial v_m}{\partial x} \tag{42}$$

$$\dot{\rho}_i = -\rho_i \frac{\partial v_i}{\partial x} \tag{43}$$

$$\rho \dot{v} = -\frac{\partial p}{\partial x} \tag{44}$$

$$\rho_i \dot{e}_i = -\pi_i \frac{\partial v_i}{\partial x} + \frac{\partial q_i}{\partial x} + e_i^{+} \tag{45}$$

$$\rho_m \dot{e}_m = -\pi_m \frac{\partial v_m}{\partial x} + \frac{\partial q_m}{\partial x} + e_m^{+} \tag{46}$$

where p and v satisfy the summation rules given in eq. (34),

$$\pi_m = \sum_a \pi_a \quad q_m = \sum_a q_a \quad e_m^{+} = \sum_a e_a^{+}$$

and the summation is over the components in the reactive system.

The deduction of constitutive equations involve several steps. First, we begin with the assumption that the Helmholtz energy is a function of four variables: T_a, v_a, γ_a and $\partial T_a/\partial x$. This is a startling choice to those who have learned thermodynamics in a traditional way, because there is no physical basis to include the velocity and the temperature gradient in the Helmholtz function. But this is a common practice in CMT literature. A conceivable purpose is to show that the procedure has a self-correcting logic to exclude the dependency of wrongly guessed variables.

$$\dot{\psi}_a = \frac{\partial \psi_a}{\partial T_a}\dot{T}_a + \frac{\partial \psi_a}{\partial \gamma_a}\dot{\gamma}_a + \frac{\partial \psi_a}{\partial v_a}\dot{v}_a + \frac{\partial \psi_a}{\partial(\partial T_a/\partial x)}\overline{(\partial T_a/\partial x)} \tag{47}$$

The expansion of ψ in terms of the assumed variables yields

$$\sum_a \frac{1}{T_a}\left\{-\rho_a\left(s_a + \frac{\partial \psi_a}{\partial T_a}\right)\dot{T}_a - (\pi_a - \alpha_a p_a)\frac{\partial v_a}{\partial x} + \dot{\alpha}_a p_a - \rho_a\frac{\partial \psi_a}{\partial v_a}\dot{v}_a \right.$$
$$\left. - \rho_a\frac{\partial \psi_a}{\partial(\partial T_a/\partial x)}\overline{(\partial T_a/\partial x)} - \frac{q_a}{T_a}\left(\frac{\partial T_a}{\partial x}\right) + e_a^{\pm} - c_a^{\pm}\left(\psi_a + \frac{p_a}{\gamma_a} - \frac{1}{2}v_a^2\right)\right\} \geq 0 \tag{48}$$

where p_a represents the phase pressure defined by

$$p_a = \gamma_a^2 \frac{\partial \psi_a}{\partial \gamma_a} \tag{49}$$

Now, we assert that the inequality conditions (48) must be satisfied for all possible thermodynamic processes. Then, it may be seen that the coefficients of $\dot{T}_a$, $\partial v_a/\partial x$, $\dot{v}_a$, $\overline{(\partial T_a/\partial x)}$ must be vanish so that

$$\psi_a = \Psi_a(\gamma_a, T_a) \tag{50}$$

$$s_a = -\frac{\psi_a}{\partial T_a} \tag{51}$$

$$\pi_a = \alpha_a p_a \tag{52}$$

The first two equations are standard thermodynamic relationships and the third provides the expression for the partial pressure in terms of the intrinsic pressure and the volume fraction of that phase. What remains is called the net inequality equation and represents the net dissipation of energy for the mixture.

$$\sum_a \frac{1}{T_a}\left\{\dot{\alpha}_a p_a - \frac{q_a}{T_a}\left(\frac{\partial T_a}{\partial x}\right) + e_a^{\pm} - c_a^{\pm}\left(\psi_a + \frac{p_a}{\gamma_a} - \frac{1}{2}v_a^2\right)\right\} \geq 0 \tag{53}$$

Secondly, we expand the net inequality equation in terms of each constituents and recombine them using the idea [31] that "in a mixture of discrete phases, the free energy, the pressure, the entropy, and heat flux of a given constituent depend on the properties and the thermodynamic states of that constituent, and that the interactions between the two constituents depend on the properties and thermodynamic states of both components". But, finding right conjugate variables is still an art, a great deal depends on expected outcomes and educated guess. The result is

$$-\frac{q_i}{T_i^2}\left(\frac{\partial T_i}{\partial x}\right) - \frac{q_m}{T_m^2}\left(\frac{\partial T_m}{\partial x}\right) + \left(\frac{1}{T_i} - \frac{1}{T_m}\right)(e_i^+ + \dot{\alpha}_i p_i)$$
$$-\frac{c_R^+(G_R - G_p)}{T_m} - p_m\frac{\dot{\alpha}}{T_m}(\alpha_i + \alpha_m)^2 \geq 0 \tag{54}$$

where we used

$$G_a = \psi_a + \frac{p_a}{\gamma_a} \tag{55}$$

$$e_i^+ = -e_m^+$$

$$c_R^+ = -c_p^+$$

$$e_m^+ = e_R^+ + e_p^+$$

Then, imposing again the requirement that the net inquality to hold for all permissible process, one finds

$$q_a = -k_a\frac{\partial T_a}{\partial x} \tag{56}$$

$$e_i^+ = h(T_m - T_i) - \dot{\alpha}_i p_i \tag{57}$$

$$c_R^+ = -\varepsilon(G_R - G_p) \tag{58}$$

$$\dot{\alpha} = -f p_m \tag{59}$$

where k_a, h, e, and f are non-negative functions of thermodynamic variables of each species. They are all familiar functions and represent heat conduction, Newtonian heat exchange, work flux due to change in the volume fraction, chemical reactions, and void collapse. a becomes a function of p_m because of the assumption that void is only associated with the reactive subsystem in the present VIR model.

Next, we examine the thermodynamics of the mixture in more familiar forms. To this end, we first rewrite eq. (33) using eqs. (56) - (59) to obtain the first law of thermodynamics similar to those in the derivation of the Lagrangian VIR.

$$de_i = -p_i dv_i + \frac{v_i}{\alpha_i}\left\{\frac{\partial q_i}{\partial x} + h\,(T_m - T_i)\right\}$$ (60)

$$de_m = -p_m dv_m + \frac{v_m}{\alpha_m}\left\{\frac{\partial q_m}{\partial x} - h\,(T_m - T_i)\right\} + \frac{p_m}{\alpha_m \gamma_m}(\dot\alpha_i + \dot\alpha_m)$$

$$= -\frac{p}{\lambda_m}dv + p_i\frac{\lambda_i}{\lambda_m}d\,v_i + \frac{v_m}{\alpha_m}\left\{\frac{\partial q_m}{\partial x} - h\,(T_m - T_i)\right\}$$ (61)

where it is used that

$$\frac{p_m v_m}{\alpha_m}\dot\alpha\,(\alpha_i + \alpha_m)^2 = \frac{p}{\lambda_m}d\,v - p_m\frac{\lambda_i}{\lambda_m}d\,v_i - p_m d\,v_m$$ (62)

Similarly, the entropy inequality equation can be transformed into the second law of thermodynamics by eliminating all irreversible processes (i.e. by imposing the assumption of local equilibrium). That is, the equality expression of eq. (41) yields

$$de_i = -p_i d\,v_i + T_i ds_i$$ (63)

$$\sum_a\left\{-\rho_a(\dot\psi_a + \dot T_a s_a) - \pi_a\frac{\partial v_a}{\partial x} + e_a^+\right\} = 0$$ (64)

The latter can be further simplified by the following identity equation.

$$\rho_m\dot\omega_m = \sum_a \rho_a\dot\omega_a + \sum_a c_a^+\omega_a$$ (65)

where ω_a is any property of constituent a and ω_m is the mass weighted average value of ω. Substituting eq. (65) into (64), we obtain

$$\dot e_m - T_m\dot s_m + p_m\dot v_m - \sum_a \frac{c_a^+}{\rho_m}(e_a - T_a s_a + p_a v_a) = 0$$ (66)

Now, by equating

$$\dot\varsigma_a = c_a^+ / \rho_m$$ (67)

we find the second law shown in eq. (14).

$$de_m - T_m ds_m + p_m d\,v_m - \sum_a \dot\varsigma_a G_a = 0$$ (68)

Finally, rearranging eqs. (60), (61), (63), and (68) as before, one can derive the computationally convenient equations. The results are

$$\dot{p}_i = \{-\beta_{si} - \Gamma_i \eta\} \frac{\dot{v}_i}{v_i} + \frac{\Gamma_i v}{\lambda_i v_i} \left\{\frac{\partial q_i}{\partial x} + h\ (T_m - T_i)\right\}$$

$$\dot{p}_m = -\{\beta_{sm} + \alpha \Gamma_m \eta\} \frac{\dot{v}_m}{v_m} + \frac{\lambda_i}{\lambda_m} \Gamma_m (1-\alpha)\eta \frac{\dot{v}_i}{v_m} - \frac{\Gamma_m\ (p + \eta)v_s}{\lambda_m v_m}\dot{\alpha}$$
$$+ \frac{\Gamma_m v}{\lambda_m v_m}\left\{\frac{\partial q_m}{\partial x} - h\ (T_m - T_i)\right\} - \frac{\Gamma_m}{v_m}\sum_a \zeta_a G_a$$

$$\dot{T}_i = \left\{-\frac{\eta}{c_{vi}} - \frac{\Gamma_i T_i}{v_i}\dot{v}_i\right\} + \frac{v}{\lambda_i c_{vi}}\left\{\frac{\partial q_i}{\partial x} + h\ (T_m - T_i)\right\}$$

$$\dot{T}_m = -\left[\frac{\alpha\eta}{c_{vm}} + \frac{\Gamma_m T_m}{v_m}\right]\dot{v}_m + \frac{\lambda_i}{\lambda_m}(1-\alpha)\eta \frac{\dot{v}_i}{c_{vm}} - \frac{(p + \eta)v_s}{\lambda_m c_{vm}}\dot{\alpha}$$
$$+ \frac{v}{\lambda_m c_{vm}}\left\{\frac{\partial q_m}{\partial x} - h\ (T_m - T_i)\right\} - \frac{1}{c_{vm}}\sum_a \zeta_a G_a$$

These equations differ from the Lagrangian VIR equations in two respects. The first is that the latter does not contain external heat conduction terms by assumption. This is not a difference, because this term can be easily restored in the VIR. The second is the expressions of work terms. This can be easily seen if the energy equations are written just for that part of the work that results from pressures. In the VIR model, the work terms are

$$de_i = -p\ d\ v_i$$

$$de_m = -p\left(\frac{1}{\lambda_m}d\ v - \frac{\lambda_i}{\lambda_m}d\ v_i\right)$$

On the other hand in the CMT model, the corresponding terms are

$$de_i = -p_i d\ v_i$$

$$de_m = -\frac{p}{\lambda_m}d\ v + p_i \frac{\lambda_i}{\lambda_m}d\ v_i$$

That is, in the CMT model, work results only from the phase pressure. In contrast, in the VIR model it includes the effect of void collapse. Presently we have no criteria to chose either one of them over the other. The criteria for selecting the models may depend on the response behavior of the system under consideration.

4. Model Simulation of Ni/Al Experiments

To illustrate the capabilities of the VIR model, we would like to show the simulation of a recent instrumented recovery experiment conducted at NCSU that indicated strong evidence for chemical reactions in the shock front [32].

The experimental configuration is shown in Fig. 15. The gauge records the profile of a reflected plane shock --- reflection of an incident shock in the sample from the left cover disk. The sample size is approximately 15 mm in diameter by 4 mm thick and permits, at most, one microsecond of measurement time. But this duration is sufficient to determine the strength of reflected shocks. The initiation of chemical reaction is monitored through reflected shocks by varying impact velocity. Fig. 16 shows an example of such measurements for the mixture consisting of $20\mu m$ Al particles and 3-7 μm Ni particles. The initial density was 55 % of solid density, a molar ratio of Ni and Al was 2.604, and the calculated Hugoniot was based on simple mass-averaged mixture properties. The strength of measured shocks clearly indicates a jump at the impact velocity of about 1.07 km/s, which is very similar to that observed by Batsanov et al on a Sn/S. The jump in pressure is interpreted as the result of strong chemical reactions in the reflected shock front. In fact, metallography of the recovered samples showed that all aluminum particles reacted fully when the impact velocity exceeded the threshold value. To strengthen the interpretation, we conducted another experiment where the back steel disk is replaced by a lexan disk to eliminate a reflected shock wave. There was no sign of chemical reactions in the recovered sample.

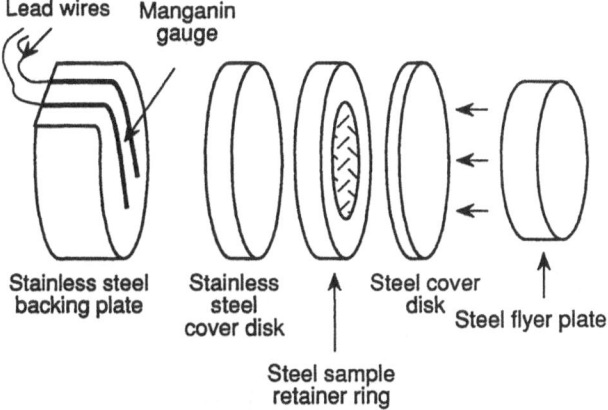

Fig. 15 A Schematic arrangement of the instrumented recovery experiment in which wave profiles indicated the initiation of strong chemical reaction in the reflected shock front (from [16] with permission).

The simulation of the above described experiments consist of two parts. The first is the one dimensional calculation of measured wave profiles. The second is an Eulerian calculation, again in one dimension, aimed at studying initiation mechanisms.

Fig. 17 shows the result of the first calculation for the sample loaded at the impact velocity of 1.4 km/s. Material and process parameters used in the calculation are shown in Table 2. The only unusual quantities in the table are the numbers associated with the rate of chemical reaction. They are arbitrarily chosen to yield a characteristic reaction constant of about 0.5 μs. Also, the threshold conditions for chemical reaction were set at 1,200 K and 12 GPa. Obviously, these numbers are not at all unique and represent only the orders of magnitude that are necessary to observe in calculation the effect of reactions in the reflected shock front. Nevertheless, it is our belief that the observed agreement is not totally arbitrary.

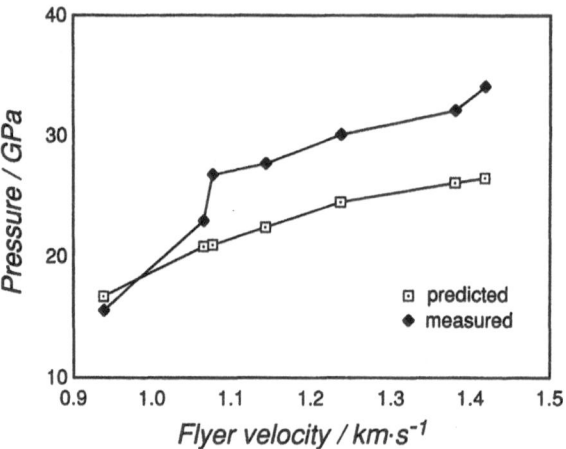

Fig. 16 Measured strength of reflected shock waves vs calculated values based on a simple mixture model.

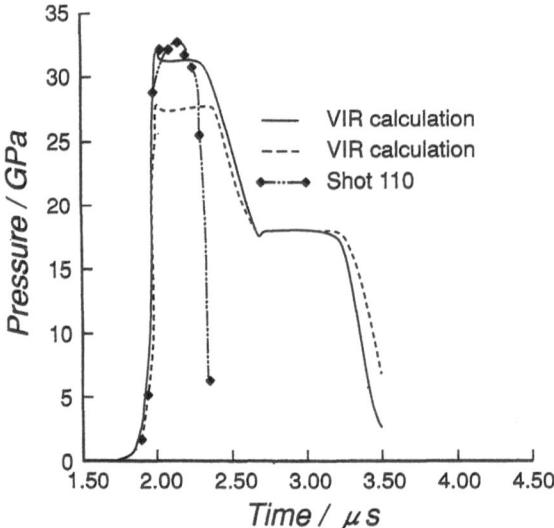

Fig. 17 Model simulation of the experiment described in Figs. 15 and 16. Experimental records show a premature failure of the gauge. The upper and lower calculated curves represent the profiles with and without chemical reaction respectively.

Wave interactions in the impact system are complex, even in one dimension. As an illustration, Fig. 18 shows propagating waves in the system without chemical reaction at the impact velocity of 1.07 km/s at successive times of 2.0, 2.1, 2.2, and 3.0 μs. This shows that the unloading in the calculated profiles shown in Fig. 17 are caused by rarefaction waves from the free back-surface of the flyer disk. The measured pressure drop after about 2.25 μs is attributed to a gauge failure.

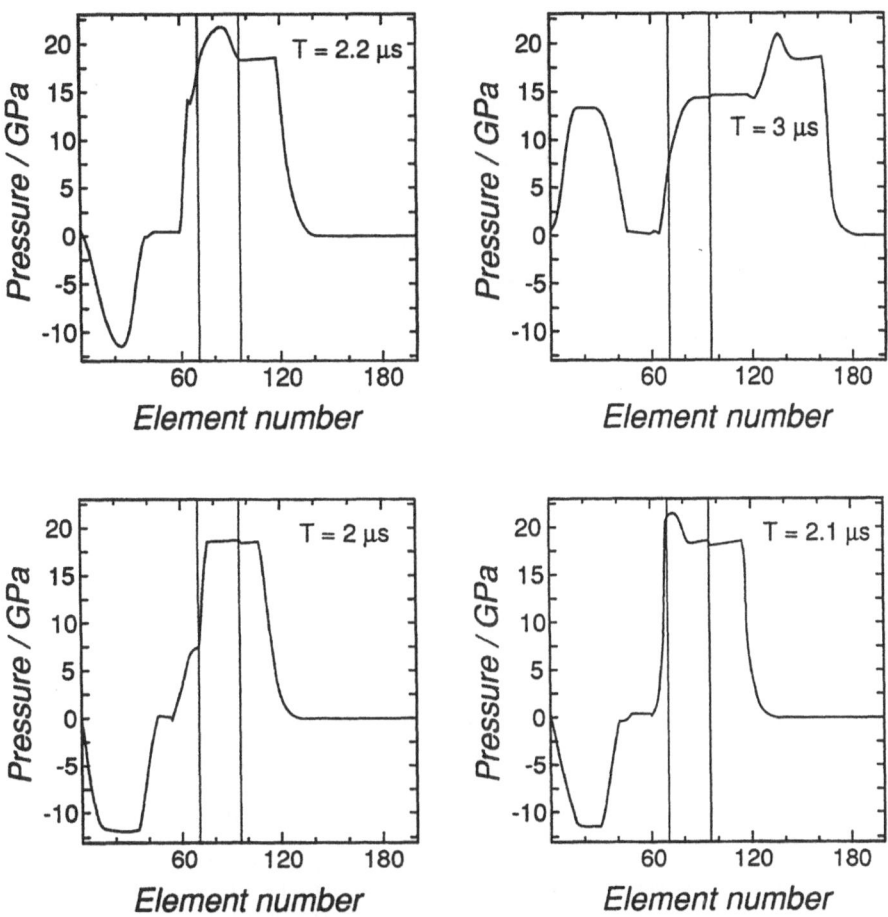

Fig. 18 Simulation of plane-shock propagation in the recovery system described in Fig. 15 without reaction at four successive times: 2.0, 2.1, 2.2 and 3.0 μs from impact. The two vertical lines in the middle represent the sample boundaries. The shock propagates from left to right.

Table 2 Material constants used for Ni/Al calculations

	ρ_0 (g/cm^3)	β_{t0} (Mb)	Γ_0	C_v (J/g–K)	n
Ni	8.875	1.925	1.91	0.4279	3.940
Al	2.790	0.792	2.00	0.9309	3.555
Ni$_3$Al	6.820	1.500	1.94	0.4850	1.000

$h = 1.0 \times 10^{-6}$ (J/cm^2–K-μs) = coefficient of heat transfer

As indicated above, the threshold conditions are only bounded by the experimental observation of strong reaction in the reflected shock. Presently, we have no theory as to how this happens in the powder mixture. Furthermore, it happens when voids are mostly collapsed. Our speculation is that the phenomena are related to the instability of hydrodynamic flow due to local heterogeneities, inherent to all powder mixtures, of not only material properties, but also flow itself.

The second calculation is undertaken to test the above hypothesis using an Eulerian model so that we do not have to assume the assumption of local mechanical equilibrium. We list a summary of conservation equations used in the calculation in the Appendix. They can be derived by repeating the procedure discussed for the Eulerian VIR, without imposing the assumption of local mechanical equilibrium. In this calculation, however, thermal interactions were excluded to focus on the effects of mechanical equilibration. Also, instead of simulating the entire impact event as we did in the first calculation, the problem of chemical initiation in the reflected shock is replaced by that of a rigid wall impact of solid mixture at a homogeneous velocity. This is a model calculation designed to search for a possible connection between chemical reaction and flow instability.

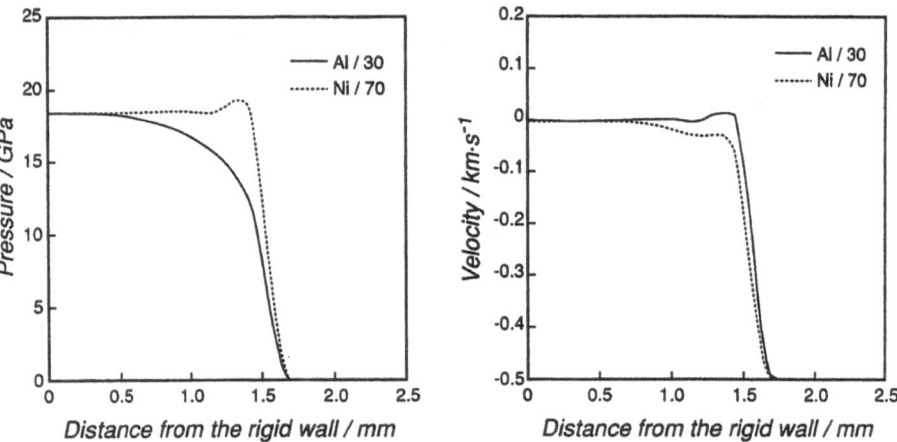

Fig. 19 Continuum-mixture-theory calculation of a rigid wall impact of Ni/Al solid mixture at 500 m/s. The wall is located at the left end and the mixture initially moves from right to left. Consequently, the shock of about 18 GPa propagates from left to right. The arrow indicates the reversal of the particle velocity in Al. Also, the result is schematically illustrated in Fig. 21.

Figure 19 shows the pressure and particle velocity profiles of a plane shock propagating into a solid powder mixture consisting of 70 vol% Ni and 30 vol% Al when the impact velocity is 0.5 km/s. The left boundary is the rigid wall. Clearly, neither the velocity, nor the pressure are in equilibrium. In the Eulerian calculation the equilibration of these variables is controlled by two new degrees of freedom that appear in the conservation equations. They are volume fraction and slip velocity, representing the momentum interaction between Ni and Al particles (see Appendix). Parametric studies show that wave profiles are very sensitive to these two variables. A large coefficient of slip velocity is found necessary to keep the two velocities close together. The above example is calculated with 10 TPa·s/km.

What is noteworthy in Fig. 19 is the appearance of reversing flow in Al. That is, Al particles appear to have been over-driven by an unequilibrated pressure which is larger than that generated by impact by itself. This results because these particles are "tied together" to act as a mixture through the summation rules. A far larger drag coefficient is necessary to eliminate the reversing flow. This means that if there exists a maximum for momentum transfer due to slip velocity, then there will be a threshold impact velocity above which flow of certain components will always be reversed. Fig. 20 shows such a demonstration for the drag coefficient of 10 TPa·s/km and the volume fraction of 80 Ni/20 Al. Such a reversal of motion can certainly lead to shear related instability, turbulence, vortex generation, etc. As shown in Fig. 21 pictorialy, the situation is analogous to driving at a very high speed in a crowed high way in the wrong direction. The term "catastrophic" is certainly not inappropriate to describe the consequence.

The idea that flow instability results in the onset of strong chemical reactions in powder mixture has certain appealing features that can be tested numerically as well as experimentaly. For instance, if we look at comparative properties of Al, Ni, and Ti shown in Table 3, one can surmise that the density of Ni made a critical difference in the initiation of chemical reactions under shock loading.

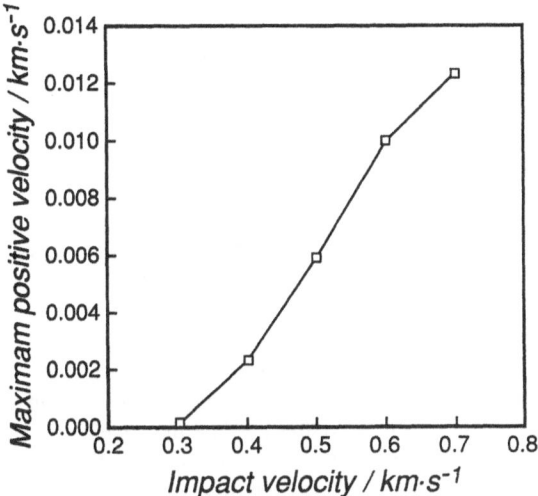

Fig. 20 The effect of impact velocity on the appearance of the flow reversal.

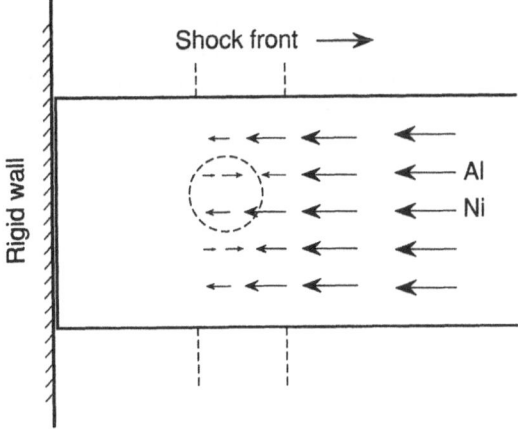

Fig. 21 A schematic description of the results shown in Fig. 19.

Table 3 Comparative properties of Al, Ni, and Ti

Element	Structure	Melting point(°C)	Hardness (GPa)	Thermal conductivity (W / mK)	Density (g/cm³)	Heat capacity at 800K (cal / mol)
Al	fcc	660	0.27	233	2.7	7.37
Ni	fcc	1,453	2.1	67	8.8	7.44
Ti	hexagonal	1,670	0.65	19	4.5	7.25

5. Conclusions

We have studied shock compression chemistry in inorganic powder mixtures using a variety of techniques to understand the question of uniqueness of such processes, if not to learn to formulate the right questions. It began with the recovery experiments of aluminides synthesis with nickel and titanium. Post shock-characterization of the nickel aluminides with modern analytical techniques such as TEM and DTA have provided grounds on which to speculate mechanistic understanding of shock-induced chemical reactions in powder mixtures. Key observations as they relate to chemical effects were the detection of both precursor and localized reactions and the finding of their turbulent-like microstructures.

Subsequent work on diamond and diamond/ceramics composites are a variation of the same theme with focus on industrial applications, but still emphasis is placed on the understanding of shock processes. It was shown in the diamond work that there is a new mechanism to transform graphite to diamond under shock loading and that it likely involves a solvent mechanism via molten nickel. Also, the transition of graphite to diamond was seen in regions where there was a large rotational deformation of graphite flakes.

The idea of controlling shock-induced chemical reactions and certain technological potential led to the study of diamond/ceramics composites and the development of the computational constitutive model called VIR. The model can deal with features that are thought to be important in the shock compression chemistry of reactive inorganic mixtures. They are multi-mixture properties, void collapse, and non-equilibrium temperature. The model is now extended to deal with non-equilibrium mechanical states in powder mixtures.

More recent experiments with Ni/Al showed strong evidence that chemical reaction does occur in a shock front. An Eulerian calculation of a rigid wall impact of this mixture suggests that the initiation of the reaction may be related to localized flow instability and that slip velocity may be a critical factor in understanding the relation of the two.

Appendix

Governing equations used for the calculations shown in Figs. 19 and 20.
Mass

$$\frac{\partial \rho_i}{\partial t} + \frac{\partial}{\partial x}(\rho_i v_i) = 0$$

$$\frac{\partial \rho_m}{\partial t} + \frac{\partial}{\partial x}(\rho_m v_m) = 0$$

Momentum

$$\rho_i\left(\frac{\partial v_i}{\partial t} + v_i\frac{\partial v_i}{\partial x}\right) = -\alpha_i\frac{\partial p_i}{\partial x} - \delta(v_i - v_m) + (p_m - p_i)\frac{\partial \alpha_i}{\partial x}$$

$$\rho_m\left(\frac{\partial v_m}{\partial t} + v_m\frac{\partial v_m}{\partial x}\right) = -\alpha_m\frac{\partial p_m}{\partial x} + \delta(v_i - v_m)$$

Energy

$$\rho_i\left(\frac{\partial e_i}{\partial t} + v_i\frac{\partial e_i}{\partial x}\right) = -\alpha_i p_i\frac{\partial v_i}{\partial x} + \frac{\partial}{\partial x}\left(k_i\frac{\partial T_i}{\partial x}\right) + h(T_m - T_i) - \dot{\alpha}_i p_i$$

$$\rho_m\left(\frac{\partial e_m}{\partial t} + v_m\frac{\partial e_m}{\partial x}\right) = -\alpha_m p_m\frac{\partial v_m}{\partial x} + \frac{\partial}{\partial x}(k_m\frac{\partial T_m}{\partial x}) - h(T_m - T_i)$$

$$- (v_i - v_m)\left\{p_m\frac{\partial \alpha_i}{\partial x} - \delta(v_i - v_m)\right\} + \dot{\alpha}_i p_i$$

Evolution of volume fraction

$$\left(\frac{\partial \alpha_i}{\partial t} + v_m\frac{\partial \alpha_i}{\partial x}\right) = \frac{\phi}{\mu}(p_i - p_m)$$

$$\alpha_m = 1 - \alpha_i - \alpha_{void}$$

$$\frac{\partial \alpha}{\partial t} + v_m\frac{\partial \alpha}{\partial x} = -f p_m$$

References

1. Duvall GE (1984) *Shock Compression Chemistry in Material Synthesis and Processing*: *NMAB-414*: Nat. Acad. Press, Washington, D. C.
2. Hoy DEP,Akashi M, Park JK, Horie Y, Whitfield JK (1984) *Shock Waves in Condensed Matter-1983* : edited by. Asay JR et al, Elsevier, New York, 451.
3. Horie Y (1984) ibid., 369.
4. Kock CC, Kim MS (1986) *J. Phsique 46*: C8-573.
5. Suslick KS (1986) *Modern Synthetic Methods 4*: 1
6. Kubo K (1978) *Introduction to Mechanochemistry* , Tokyo Kagaku Dojin, Tokyo, (Japanese).
7. Horiguchi Y, Nomura Y (1963) *Bull. Chem. Soc. Japan 36*: 486.
8. Westbrook JH (1959) *Mechanical Properties of Intermetallic Compounds*: Wiley, New York, 1959.
9. Graham RA, Webb DM (1984) *Shock Waves in Condensed Matter-1983* : edited by Asay JR et al, Elsevier, New York, 423.
10. Hoy DPE (1985) unpublished Ph.D thesis, North Carolina State University, Raleigh, NC.
11. Horie Y, Hoy DEP, Simonsen I, Graham RA, Morosin B (1986) *Shock Waves in Condensed Matter*: edited by Gupta YM, Plenum, New York, 749.

12. Horie Y, Graham RA, Simonsen IK (1985) *Mat. Letters 3*: 354.
13. Horie Y (1986) *Mat. Appl. of Shock Waves and High-Stain-Rate Phenomena*: edited by Murr LE, Staudhammer KP, Meyers MA, Marcel Dekker, New York, 1023.
14. Simonsen IK, Horie Y, Graham RA, Carr M (1987) *Mat. Letters 5*: 75.
15. Pak H, Horie, Y, Graham RA (1986), *Shock Waves in Condensed Matter*: edited by Gupta YM, Plenum, New York, 1986, 761.
16. Bennett LS, Sorrell FY, Iyer KR, Simmonsen IK, Horie Y(1992) *Appl. Phys. Letters 61*:
17. Thadhani N (1992) presentation at a Workshop, May 5-7, Los Alamos, NM.
18. Hoover W et al (1990) *Microscopic Simulations of Complex Flows*: Plenum, New York.,321.
19. Hammeter WF, Graham RA, Morosin B, Horie, Y (1988) *Shock Waves in Condensed Matter 1987*: edited by Schmidt SC, Holmes NC, Elsevier, New York, 431.
20. Simonsen IK, Horie Y, Akashi T, Sawaoka AB (1992) *Shock-Wave and High-Strain-Rate Phenomena in Materials*: edited by Meyers MA Murri LE, Staudhammer KP, Marcel Dekker, New York, 233.
21. Myers et al (1986) *Shock Waves in Condensed Matter*: edited by Gupta YM, Plenum, New York, 755.
22. Zemsky SV, Ryabchikov YA, Epshteyn GN (1979) *Fiz. Metal. Matalloved. 46*: 171.
23. Kleiman J Heimann RB, Hawken D, Salanky NM (1984) *J. Appl. Phys. 56*: 1440.
24. Simonsen IK, Chevacharoenkul S, Horie Y, Akashi T, Sawaoka A (1989) *J. Mat. Sci. 24*: 1486.
25. Simonsen I. K, Horie Y, Akashi T, Sawaoka AB (1992) *J. Mat. Sci. 27*: 1735.
26. Stadelmaier HH (1960) *Z. Metallkde. 51*: 601.
27. Sawaoka AB, Akashi T (1987) U. S. Patents 4,655,830 and 4,695,321.
28. Nellis W, Gourdin WH, Maple MB (1988) *Shock Waves in Condensed Matter -1987*: dited by Schmidt SC, Holmes NC, Elsevier, New York, 407.
29. Kunishige H, Oya Y, Fukuyama Y, Watanabe S, Tamura H, Sawaoka A, Taniguchi T, Horie Y (1990) *Rept. of the Res. Lab. of Engr. Mat.(No. 15)*: Tokyo Institute of Tech., 235.
30. Horie Y, Kipp M (1988) *J. Appl. Phys. 63*: 5718 .
31. Baer MR, Nunziato JW (1986) *Int. J. of Multiphase Flow 12*: 861.
32. C. Trusdell (1969) *Rational Thermodynamics*: McGraw-Hill, New York.
33. Atkin RJ, Craine RE (1976) *Q. Jl. Mech. Appl. Math. 29*: 209.

Chapter 5
Shock Effects on Structural and Superconducting Properties of High T$_C$ Oxides

Y. Syono and M. Kikuchi

1. Introduction

Shock compression technique has been applied for the research of high T$_C$ oxides in several different ways. Synthesis of high T$_C$ oxides from component oxides by means of shock compression [1-5], as revealed in successful synthesis of LSCO*, is promising to get new high T$_C$ materials. Enhanced reactivity was also noted in the shock-processed ceramic powders, which could be utilized for easier sinterability.

Fabrication of high-density bulk materials by shock compaction which has been successful for amorphous alloys [6] is also applicable for high T$_C$ oxides [7-15]. Because shock compaction is very fast process in both heating and quenching, localized heat generated only near grain boundaries upon shock loading easily leads to bonding of grains without affecting the inside of each grain. Consolidation of crystallographically oriented bulk materials, or formation of a monolithic piece of coil and disc of high T$_C$ oxides by shock loading is particularly interesting, if an industrial application of shock compression technique is premised.

Shock processing has been expected to control superconducting properties of high T$_C$ oxides. Particular attention has been paid on possible increase in critical currents via structure defects induced by shock loading [16-25]. However considerable degradation of superconductivity due to induced strain or oxygen loss by shock loading might be an drawback in the practical application. Therefore optimization is crucial for applying shock wave technique for processing high T$_C$ materials. Importance of annealing of the shocked materials has been pointed out. Under such circumstances, knowledge on the mechanism how the shock-induced defects work for flux pinning would be important.

All these features are closely related to the basic structure and composition of high T$_C$ oxides, which show very peculiar superconducting properties. The present article is intended to summarize the experimental studies which have been carried out so far and to elucidate the effect of shock waves on the structural and superconducting properties of high T$_C$ oxides from the basic viewpoint.

---

* Hereafter following abbreviations will be applied; LSCO for La$_{1.85}$Sr$_{0.15}$CuO$_4$, YBCO for YBa$_2$Cu$_3$O$_7$, BSCCO for Bi-Sr-Ca-Cu-O system and TBCCO for Tl-Ba-Ca-Cu-O system. Sometimes Bi2212 or Tl2201 will also be used for representing the Bi$_2$Sr$_2$CaCu$_2$O$_{8.2}$ or Tl$_2$Ba$_2$CuO$_6$ phases, respectively.

2. Specific Features of High T_C Oxides as Type II Superconductor

Discovery of high T_C oxides was received as a revolutionary event in both fundamental and application fields. Mechanism leading to such extraordinarily high critical temperature is expected to imply new concept in solid state physics. Extensive applications are naturally promised if the new materials could be used at liquid nitrogen temperature. However, rapid progress due to enthusiastic research efforts has uncovered various interesting, but not always favorable, features inherent to these new materials. These features are especially important if these materials are considered for practical application.

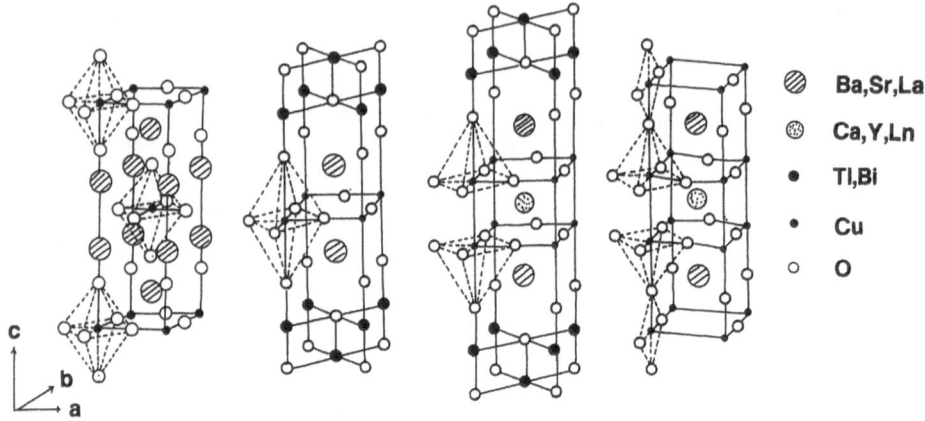

Fig. 1 Schematic illustration of crystal structures of typical high Tc oxides; LSCO, Bi (or Tl) 2212, Bi (or Tl) 2223 and YBCO (from left to right). Small closed and open circles are Cu and O respectively.

Firstly high T_C oxides show very strong anisotropy in conductivity and upper critical field, H_{c2}, which are originated from the very peculiar layered structure (Fig. 1) and concomitant with the two dimensional CuO_2 sheets serving as conduction path. Secondly high T_C oxides are characterized by extraordinarily short coherence length compared with conventional metallic superconductors. It is also very anisotropic; extremely short along the direction perpendicular to the CuO_2 sheets, i. e. only a few Å for BSCCO. Short coherence length causes weak link across the grain boundary, which limits intergrain currents. Small carrier concentration in oxide superconductor might also result in weak shielding against potential disturbance on the grain boundary.

Therefore assessment is needed among these characteristic features for practical applications, particularly to achieve high critical current density. For that purpose single crystal or well oriented textures is favored to ensure connectivity to get rid of weak links across the grain boundary. Introduction of strong pinning centers for preventing flux creep is also indispensable. Precipitation of finely dispersed normal or insulating impurities is proved to be effective to raise critical current density. Many kinds of defects, dislocations, or stacking faults are considered to be promising pinning centers, while grain or twin boundaries are also raised as possible candidates. All these features imply possibilities that shock-induced defects may work as promising pinning centers.

3. Mechanical and Chemical Effects of Shock Waves on High T_C Oxides

3.1 Shock Synthesis and Decomposition

Shock synthesis of high T_C oxides from the mixture of component oxides has been tried by several investigators. Single copper layer oxides such as $La_{2-x}Sr_xCuO_4$ and $Tl_2Ba_2CuO_6$ were successfully synthesized by shock loading [1-3]. The synthesis is probably favored by their relatively simple crystal structures. In contrast, formation of high T_C oxides with a large c-dimension was found to be difficult by shock compression technique, because short duration of shock pulse would be insufficient to build a completely ordered structure for the complex components. However, there still remain possibilities for formation of new metastable phases via non-equilibrium process of the shock synthesis, and persistent efforts to identify the trace amount of shock-induced phases are to be encouraged.

Irreversible chemical changes were observed in heavily shocked high T_C oxides [26-28]. Decomposition pressures were about 60, 40 and 20 GPa for LSCO, YBCO and BSCCO respectively. XPD analysis revealed reduction to Cu metal in these intensely shocked materials. In the case of BSCCO, metallic Bi was also observed above about 20 GPa. Such shock reduction has been known in many transition metal oxides [29,30], and liability of shock decomposition of high T_C oxides may be ascribed to easier reduction of Cu oxides (and also Bi oxides in BSCCO). Particular attention for preventing decomposition is needed in shock compaction and processing.

3.2 Shock Compaction

Shock compaction of YBCO has been carried out from the very beginning of research activities in high T_C oxides to form a large piece of monolithic discs or coils which can be applied for practical application. Shock compacted materials generally show a density close to that of the bulk, indicating complete collapse of voids in the starting powder materials. Shock compaction of YBCO powders was shown to be completed at as low as 3 GPa [12, 19-22]. Considerable fracturing with decreasing grain size was reported. Addition of Ag serving as a binder agent was found to help the ductility of shock compacts greatly.

Shock loading effects at macroscopic level have been examined by SEM observation of shocked and subsequently annealed BSCCO in comparison with the starting material (Fig. 2) [28]. Sintered pellets of BSCCO used as a starting material showed rather porous texture consisting of platy crystals, while the shocked specimen was found to be compact due to complete filling of voids and breakage of grains. Annealed specimen revealed crystal growth keeping the homogeneously compacted texture.

Shock consolidation of crystallographically aligned BSCCO powders was successfully carried out by Weir et al [13-15]. Prealigned micaceous powders by tapping could be consolidated by shock loading to high density bulk without appreciable voids or cracks. The degree of crystallographic alignment determined by x-ray diffraction and magnetic measurements increased with increased particle size of the starting materials.

3.3 Shock-Induced Strain

Remarkable broadening of x-ray diffraction lines has been observed in moderately shocked materials [26-28] Residual strains of LSCO and YBCO shocked to 11 GPa were estimated to be 1.0 and 0.8% respectively [24,25]. Annealing of the shocked LSCO and YBCO at 930°C

and 890°C respectively removed the induced strain almost completely. Similar shock-induced broadening of x-ray diffraction profiles was also observed in BSCCO
and TBCCO. Fig. 3 shows x-ray powder diffraction patterns of as-pelletized, shocked and subsequently annealed specimens. Shocked specimen revealed highly 00*l* oriented x-ray diffraction pattern with broadened diffraction lines, while the annealed specimen shows much sharper profiles [28].

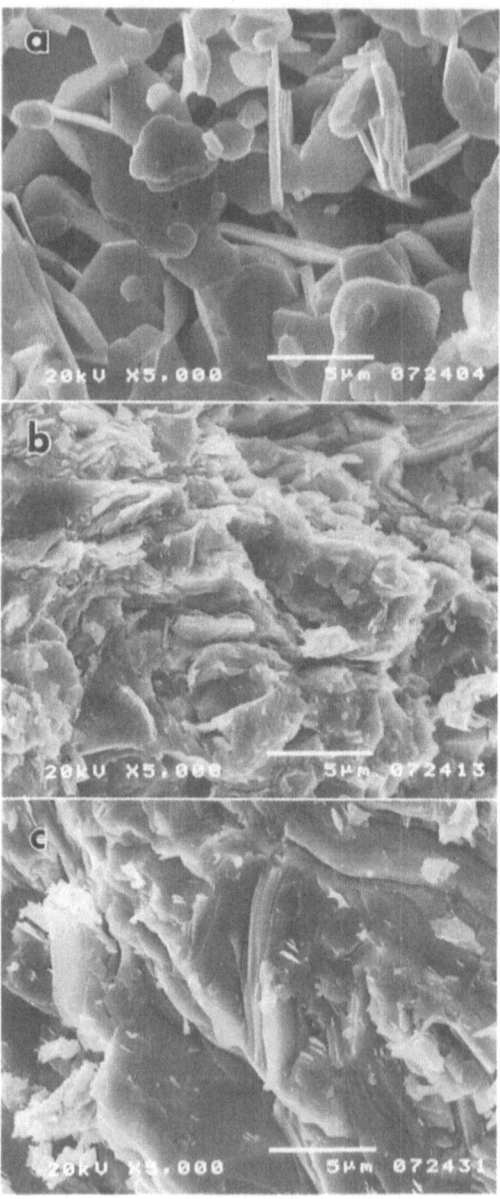

Fig. 2 SEM photographs of BSCCO; as-prepared (top), shocked to 8 GPa (middle) and subsequently annealed at 870°C (bottom).

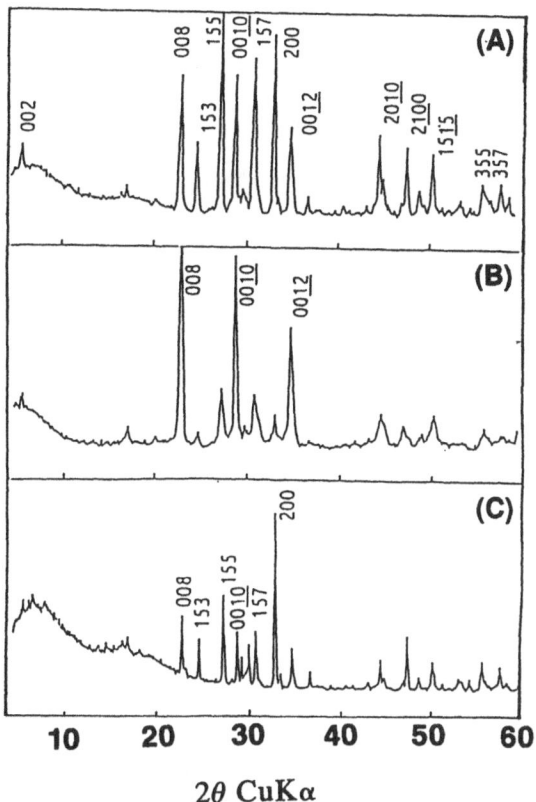

Fig. 3 XRD patterns of BSCCO; as-prepared (A), shocked to 10 GPa (B) and subsequently annealed at 870 °C (C).

The oxygen content remained nearly the same in these moderately shocked materials. Lattice parameters of high T_C oxides did not show any appreciable change by shock loading, unless significant chemical reactions took place. Apparent lattice expansion exceptionally observed in La_2CuO_4 after shock compression was ascribed to the puckered structure of the CuO_6 octahedra which resulted in an orthorhombic symmetry, since such effects were observed neither in La_2NiO_4 nor LSCO with the tetragonal symmetry.

3.4 Deformation Textures and Induced Defects (TEM Observation)

The shock deformation textures observed by TEM strongly depended on the crystal structure [27] Extensive dislocation network was observed in shocked LSCO with the K_2NiF_4 structure (Fig. 4(A)), while stacking faults and mechanical twinning prevailed in YBCO with the oxygen defective triple perovskite structure (Fig. 4(B)). Shock formation of stacking faults parallel to the basal plane of YBCO is probably related to more pronounced layer structure of YBCO compared with LSCO. Shock loading also introduced high density dislocations in YBCO, two orders of magnitude higher than conventional mechanical deformation [19-22]. These high density defects can be retained by annealing at 890°C for 1 day, while the transport properties were recovered considerably.

Fig. 4 TEM photographs of shocked LSCO(to 12 GPa) (A), YBCO (to 20 GPa) (B) and BSCCO (to 5 GPa) (C), and of shock (to 10 GPa) + Annealed (at 870 °C) BSCCO (D).

Very unique texture of kink bands was observed in BSCCO shocked to 5-10 GPa (Fig. 4(C))[14,27,28]. Kink bands were observable only for the ion-thinned specimen which was shocked parallel to the layer, and when the incident beam was also taken to be parallel to the layer. The shock-induced kinks showed a wide range in the rotation angle. Increase in shock pressure resulted in the increase in the degree of kinking and consequently in fractionation in smaller domains. In some place there was observed open fissures along which amorphous decomposed products were precipitated.

The kink bands are characteristic for the deformation in the mica-like structure with strong intralayer and very weak inter-layer bonding, which allow no dislocations. The deformation parallel to the layer is achieved only by bending or slip, so the kink is produced by a combination of a translation glide along the layer and external rotation around an axis perpendicular to the gliding direction. Easy degradation of superconductivity in shocked BSCCO is to be explained by severe deformation of kinking accompanied by local decomposition. Annealing of shocked BSCCO showed marked change in the microstructure, i. e. crystal growth along the plane, release of strain contrast and disappearance of kink bands (Fig. 4(D)).

4. Shock Effects on Superconducting Properties

4.1 Shock Effects on T_c

Shock effects on T_c of high T_c oxides were apparently variable with the material. For example, T_c of YBCO was reported to be little affected by shock loading up to 16.7 GPa, although the screening signal decreased probably due to decreased grain size [19-22]. However, the resistance data for the as-compacted specimen did not show a bulk T_c and a complete resistive transition was observed only after annealing at 890°C in O_2.

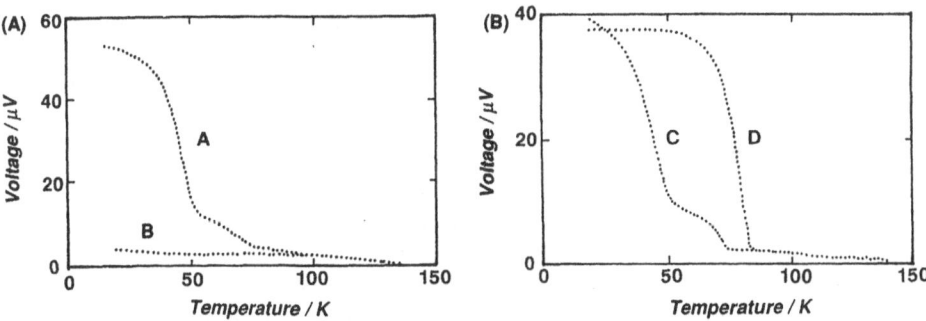

Fig. 5 Temperature variation of diamagnetic response of BSCCO; as-prepared (A) and (C), shocked to 7.8 GPa (B) and subsequently annealed at 870 °C (D).

Much pronounced decrease in T_c was noted in BSCCO, and superconductivity disappeared by shock loading to 7.8 GPa (Fig. 5(A))[27,28]. Interestingly annealing of shocked specimen at 870°C in air for 10 hours showed dramatic recovery of superconductivity with considerable improvement of T_c (Fig. 5(B)). The same heat treatment of the unshocked specimen little affected T_c.

There may be several explanations for these degradation of superconductivity by shock loading. The residual strain induced by shock loading may affect on T_c. Loss of oxygen by shock compression may also lead to decrease in hole concentration, hence decrease in T_c for the case of optimum or underdoping. Such oxygen loss may be utilized to increase the T_c, if the superconducting oxides are in an over-doping state. This was actually done for the case of Tl2201 (Fig. 6)[31]. Non-superconducting Tl2201 phase which was prepared in oxygen atmosphere became superconductive with T_c ~85 K, when shocked to 18 GPa in an open recovery fixture which allowed escape of the evolved gas.

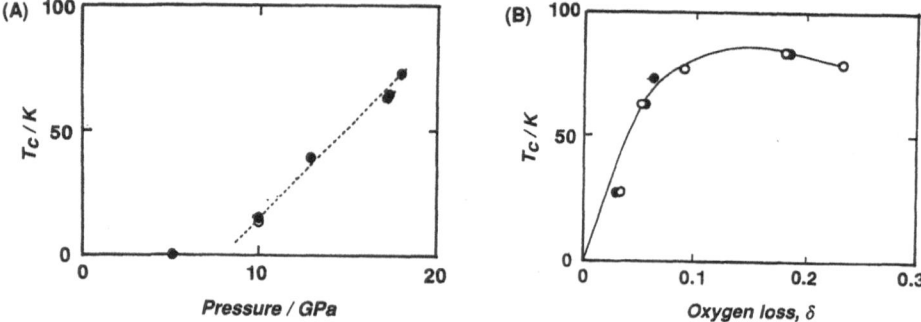

Fig. 6 (A) Dependence of T_c of the Tl2201 phase on the shock pressure. (B) Dependence of T_c of Tl 2201 on the oxygen loss: Open and solid circles correspond to the quenching in N_2 and O_2 atmosphere.

4.2 Effect on Pinning Energy

Flux-relaxation measurements were performed for the as-prepared, shocked and subsequently annealed specimens of YBCO and LSCO. Flux relaxation and magnetic hysteresis measurements of the starting powder, shock-compacted and re-sintered YBCO were made by Weir et al [19-22]. Shock processing resulted in a substantial increase in the flux pinning energy, which was most enhanced by re-sintering at 890 °C. They concluded that the intragranular critical current density of the as-shocked and resintered YBCO increased about 10 and 5 times greater than that of the starting powder.

Extensive investigation on shock effects of LSCO were carried out by Sakaguchi et al [24,25]. Logarithmic time dependence of magnetization of LSCO measured in a field of 1 T at 12 K is shown in Fig. 7. The relaxation rate was fastest in as-prepared specimen and slowest in the shock-annealed specimen. Magnetic hysteresis measured at 12 K was also compared among these three specimens, as shown in Fig. 8. Shock loading to 11 GPa considerably degraded superconductivity, but marked recovery of superconducting characteristics, even better than that of the as-prepared one, was achieved by annealing at 930°C for 12 hours in air. Flux pinning energy, E_p, estimated from the decay rate of magnetization, M, which directly measures critical current density, showed a bell-type temperature dependence with maximum at around 50 K and 15 K for YBCO and LSCO respectively (Fig. 9) [24,25]. Such temperature dependence was observed in the magnetic measurements of YBCO thin films [32].

Apparently no explicit relation was observed between pinning energy and macroscopic deformation determined by x-ray diffraction profile analysis. Remarkable development of recrystallization with 200 nm grain size observed by TEM suggests that the grain boundary may act as pinning centers in the case of LSCO.

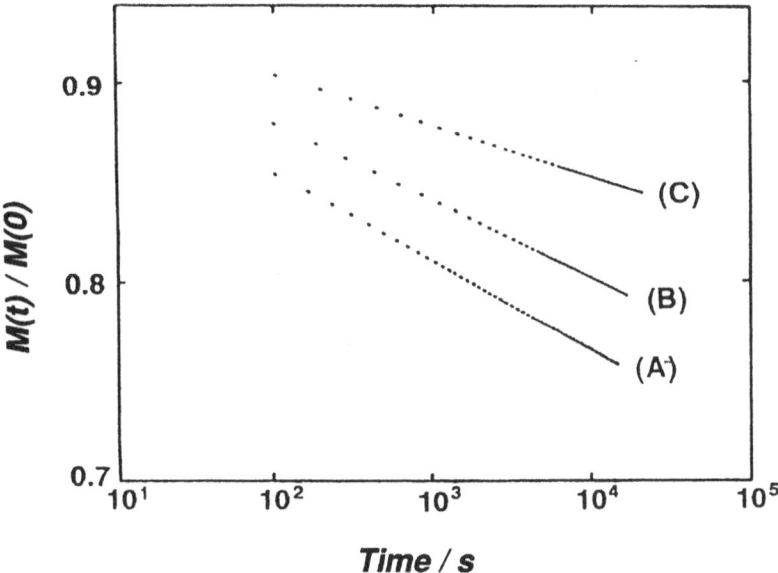

Fig. 7 Magnetization relaxation of LSCO measured at 12 K in a field of 1 T versus logarithmic time; as-prepared (A), shocked to 11 GPa (B) and shock + annealed at 930 °C (C) (from [25] with permission).

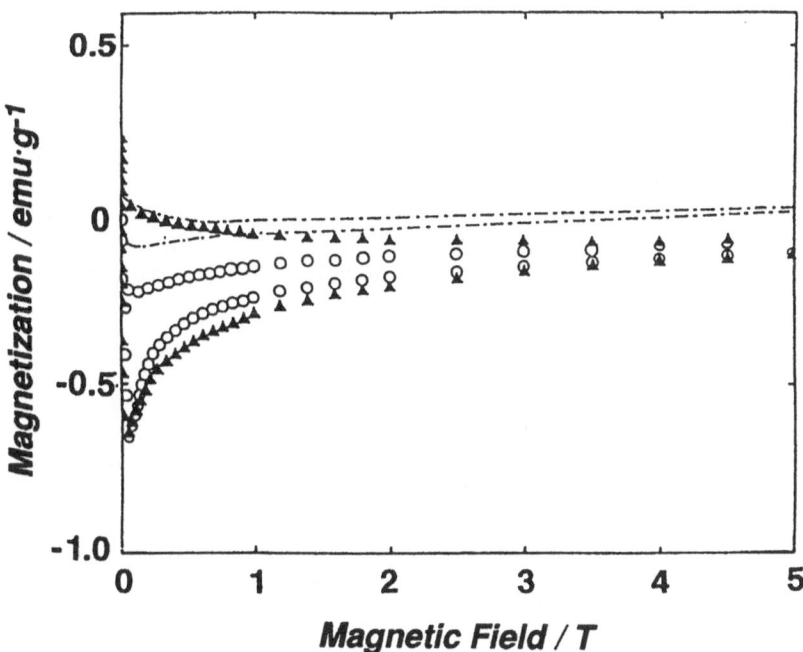

Fig. 8 Magnetization hysteresis of LSCO; as-prepared (open circle), shocked to 11 GPa (chain line) and shock + annealed at 930°C (closed triangle) (from [25] with permission).

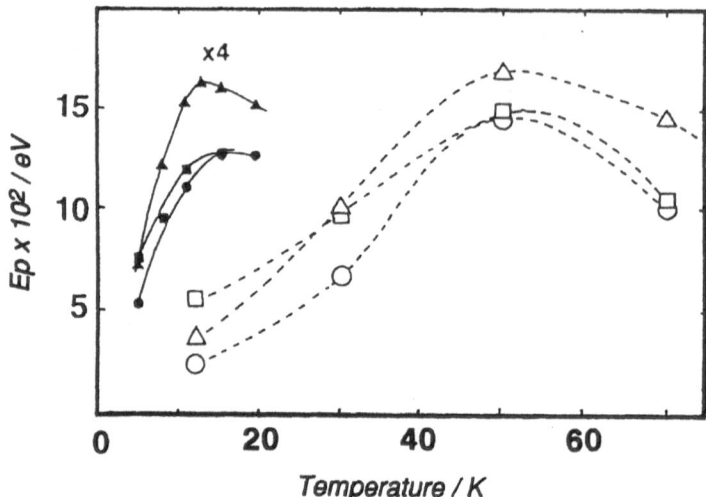

Fig. 9 Temperature dependence of pinning energy evaluated from magnetization relaxation measurements of LSCO (closed symbols) and YBCO (open symbols). Circle: as-prepared. Square: shocked to 11 GPa. Triangle: shock + annealed at 930°C for LSCO and at 890°C for YBCO.

5. Concluding Remarks

Shock-loading effects may be positively utilized for processing high T_C oxides. Firstly structure defects induced by shock loading may work as a pinning center for flux line, hence contribute to enhance J_C. Connectivity achieved by shock compaction may also improve J_C. However, there are several drawbacks in shock processing. Structure defects would be unfavorable, if short coherence length is taken into considerations. Oxygen loss also leads to degradation in T_C. Therefore some optimization is indispensable to make use of shock wave technique for processing of high T_C oxides. As revealed in magnetization relaxaion measurements, appropriate annealing of shocked materials may be promising to overcome this difficulty.

References

1. Graham RA, Venturini EL, Morosin B, Ginley DS (1987) *Phys. Lett. A 123*: 87.
2. Morosin B, Graham RA, Venturini EL, Ginley DS, Hammetter WF (1988) *Shock Waves in Condensed Matter - 1987*: edited by Schmidt SC, Holmes NC, North Holland, Amsterdam, 439.
3. Iqbal Z, Thadhani NN, Chawla N, Ramakrishna BL, Sharma R, Rao KV, Kumryev S, Reidinger F, Eckhardt H (1989) *Appl. Phys. Lett. 55*: 2339.
4. Nesterenko VF (1990) *Shock Compression of Condensed Matter-1989*: edited by Schmidt SC, Johnson JN, Davison LW, North Holland, Amsterdam, 553.

5. Iqbal Z, Thadhani NN, Chawla N, Rao KV, Skumryev S, Ramakrishna BL, Sharma R, Eckhardt H, Owens FJ (1990) *Shock Compression of Condensed Matter-1989*: edited by Schmidt SC, Johnson JN, Davison LW, North Holland, 575.

6. Toda Y, Ogura T, Masumoto T, Fukuoka K, Syono Y (1985) *Sci. Rep. Res. Inst. Tohoku Univ., A32*: 267.

7. Murr LE, Hare AW, Eror NG (1987) *Nature 329*: 37.

8. Murr LE, Monson T, Javadpour J,. Strasik M, Sudrasan U. Eror NG, Hare AW, Brasher DG, Butler DL (1988) *J. Metals, 40*: 19.

9. Matizen EV, Deribas AA, Nesterenko VF, Pershin SA, Bezverkhii PP, Voronin AN, Yefremova RI, Starikov MA (1989) *Int. J. Mod. Phys. B3*: 97.

10. Deribas A (1990) *Shock Compression of Condensed Matter-1989:* edited by Schmidt SC, Johnson JN, Davison LW, North Holland, 549.

11. Takashima K, Tonda H, Nishida M, Hagino S, Suzuki M, Takeshita T (1990) *Shock Compression of Condensed Matter-1989*: edited by Schmidt SC, Johnson JN, Davison LW, North Holland, Amsterdam, 591

12. Nellis WJ, Seaman CL, Maple MB, Early EA, Holt JB, Kamegai M, Smith GS, Hinks DG, Dabrowski DD (1989) *High Temperature Superconducting Compounds: Processing and Related Properties*: edited by Whang S, DasGupta, TMS Publ., Warrendale, PA. 249.

13. Seaman CL, Weir ST, Early EA, Nellis WJ, Maple MB, McCandless PC, Brocious WF (1990) *Appl. Phys. Lett. 57*: 93.

14. Weir ST, Nellis WJ, Seaman CL, Early EA, Maple MB, Kikuchi M, Syono Y (1991) *Physica C 184*: 1.

15. Nellis WJ, Woolf LD (1989) *MRS Bull. 63*: Weir ST, Nellis WJ, Seaman CL, Early EA, Maple MB, Kramer MJ, Syono Y, Kikuchi M, McCandless PC, Brocious WF (1992) *Shock-Wave and High-Strain-Rate Phenomena in Materials*: edited by Meyers MA, Murr LE, Staudhammer KP, Marcel Dekker, New York, 795.

16. Murr LE, Niou CS, Jin S, Tiefel TH, James ACWP, Sherwood RC, Siegrist T (1989) *Appl. Phys. Lett. 55*: 1575.

17. Sharma R, Ramakrishna BL, Iqbal Z, Thadhani NN, Chawla N (1990) *Proc. MRS Symp. HREM of Defects in Materials.*

18. Morosin B, Venturini BL, Graham RA, Ginley DS (1989) *Synth. Metals, 33*: 185.

19. Weir ST, Nellis WJ, Early EA, Seaman CL, Maple MB (1989) *Physica C 162-164*: 1263.

20. Weir ST, Nellis WJ, Kramer MJ, Seaman CL, Early EA, Maple MB (1990) *Appl. Phys. Lett., 56*: 2042.

21. Seaman CL, Early EA, Maple MB, Nellis WJ, Holt JB, Kamegai M, Smith GS (1990) *Shock Compression of Condensed Matter-1989*: North Holland, Amsterdam, 571.

22. Weir ST, Nellis WJ, Kramer MJ, Seaman CL, Early EA, Maple MB (1990) *Shock Compression of Condensed Matter-1989*: North Holland, Amsterdam,563.

23. Nellis WJ, M. B. Maple MB, Geballe TH (1988) *SPIE Multifunc. Mater. 878*: 2.

24. Sakaguchi Y, Kikuchi M, Kobayashi N, Kusaba K, Fukuoka K, Minagawa Y, Syono Y (1990) *Physica C 185-189*: 2517.

25. Sakaguchi Y, Kikuchi M, Kobayashi N, Fukuoka K, Syono Y (1992) *Physica C. 201*:183

26. Takeya H, Takei H, Jang WJ, Syono Y, Kikuchi M, Kusaba K (1990) *Jpn. J. Appl. Phys. 29*: 1252.

27. Syono Y,Nagoshi M, Kikuchi M, Tokiwa A, Aoyagi E, Suzuki T, Kusaba K, Fukuoka K (1990) *Shock Compression of Condensed Matter-1989*: North Holland, Amsterdam, 579.

28. Kikuchi M, Syono Y, Nagoshi M, Tokiwa A, Aoyagi E, Suzuki T, Kusaba K, Fukuoka K (1990) *Advances in Superconductivity II*: edited by Ishiguro T, Kajimura K, Springer, Tokyo, 603.

29. Syono Y, Kikuchi M, Goto T, Fukuoka K (1983) *J. Solid State Chem. 50*: 133.

30. Kikuchi M, Kusaba K, Fukuoka K, Syono Y (1986) *J. Solid State Chem. 63*: 386.
31. Kikuchi M, Syono Y, Kobayashi N, Oku T, Aoyagi E, Hiraga K, Kusaba K, Atou T, Tokiwa A, Fukuoka K (1990) *Appl. Phys. Lett. 57*: (1990) 813.
32. Kobayashi N, Miyoshi K, Kawabe H, Watanabe K, Yamane H, Kurosawa, H, Hirai T, Muto Y (1990) *Physica B 165-166*: 1133.

Chapter 6
Shock Compression Studies on Ceramic Materials

T. Mashimo

1. Introduction

Although shock-wave propagation in solid is a momentary phenomenon whose time interval is within several microseconds, such waves can generate ultra-high pressure conditions, to inducing changes in crystal or electric structure of condensed matter, and even nuclear reaction. The high-pressure equation of state, phase transition, dynamic mechanical behavior, etc. of solid matter are explored through by the measurement of shock-wave properties (Hugoniot) parameters (shock velocity, particle velocity, stress, etc.). The Hugoniot parameters of many materials, including elementary and compound materials, have been measured by the discreet-type methods (pin-contactor methods, flash gap method etc.,) and the continuous-type methods (condenser method, electromagnetic-gauge method, quartz-gauge method, manganin-gauge method, inclined-mirror method, laser interferometer method (VISAR), etc.) over the past 40 years by scientists chiefly in the USA and the USSR. In particular, many elementary metals and oxide minerals have been widely investigated in relation to the high-pressure physics of matter and the earth and planetary science [1-6]. However, in these early studies, the measurement methods were, in many cases, discreet type, and the specimen qualities were poor compared with more recent studies. For ceramics, the reported number of shock-compression research studies has not been many, and the ceramic specimens used in the early studies were, in many cases, sintered ceramics with large porosity.

Ceramics have a number of beneficial properties: high hardness, high melting points, high chemical stability, etc., but, ceramics are brittle. These properties are consequence of the chemical bonding states (covalent or ionic), which differ significantly from metals. Yielding properties of solids under shock compression depend on both their physical and microstructural material characteristics (crystal-chemical properties, thermodynamic properties, microstructure etc.). Brittle materials are particularly sensitive to these characteristics. In addition, recent high-performance ceramics have been developed such as microstructure-controlled ceramics, composite- or doping-type ceramics, etc. have been developed. The toughness or strength of some of these new ceramics are much improved by the formation of fracture process zones (process-zone toughening). These ceramics have potential uses in many fields, such as automobile industry, space and aeronautics industry, production system, etc. However, their performance under dynamic loading is still largely unknown.

In shock-wave measurements on solids, it is important to use controlled, good quality specimens and to use high-capability measurement facilities. We have been developing or refining the measurement facilities combined with the keyed-powder gun, and have been measuring the Hugoniot-compression curves, particle-velocity histories, stress histories and compression-shear waves on various kinds of ceramics, minerals, and semiconductors. In this report the present shock-wave measurement facilities is first described. Next, the measurement studies on selected ceramics are reviewed. The materials include aluminum oxide (Al_2O_3), zirconium dioxide (ZrO_2) and silicon nitride (Si_3N_4). The yielding properties and phase transitions are investigated. Furthermore, the yielding mechanisms and correlations with the material characteristics are phenomenologically considered on the basis of the experimental results.

2. Experimental Facilities Combined with the Keyed-Powder Gun

2.1 Keyed-Powder Gun

Experiments using combined compression-shear shock waves in solid, by which the shear moduli and material strength can be directly determined, are expected to bring remarkable advances in the studies of high-pressure equation of state, high-pressure polymorphism, melting, mechanical properties, crater problems, earth quake problems and so on [7-9]. In 1981, the author et al [10]. developed the keyed-powder gun (1SKG-KM1), which was capable of accelerating a projectile, without rotation, to velocities of over 2 km/s and of generating the combined compression-shear shock waves by oblique parallel-plate impact primarily for the shear-wave measurement in the several 10s of GPa region.

Several shock-wave measurement methods were combined with the gun: 1. the inclined-mirror method to measure the steady-state Hugoniot parameters, 2. the manganin-gauge method to measure the stress histories, 3. the electromagnetic-gauge method to measure the particle-velocity histories.

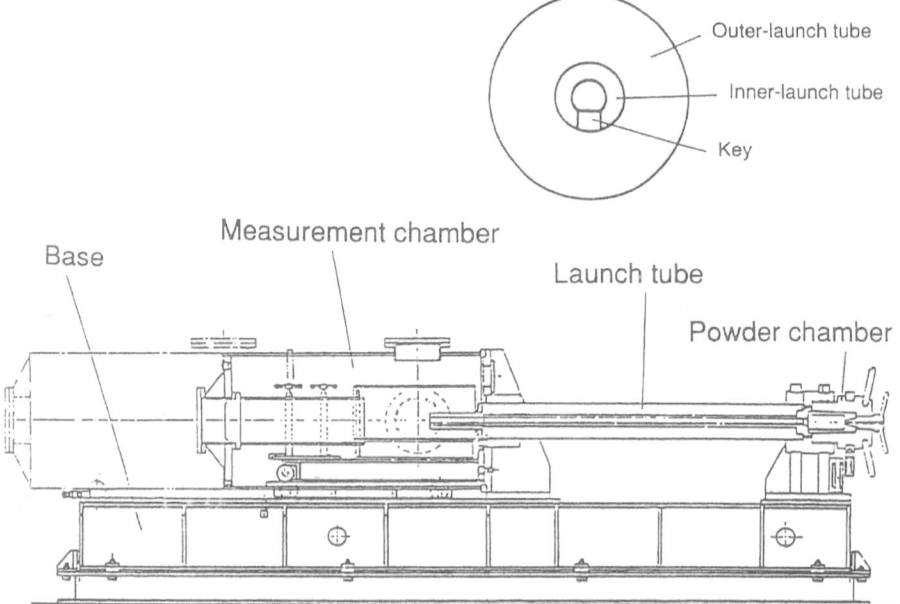

Fig. 1 Sketch of the keyed-powder gun and the cross section of the keyed launcher (from [10] with permission) .

We have been measuring the Hugoniot-compression curves, the particle- velocity histories, the stress histories, the electrical properties etc., on such materials as ceramics, minerals, and semiconductors. Furthermore, the measurement of the combined compression-shear shock waves of over 20 GPa shock amplitude in sapphire was performed by means of the electromagnetic-gauge method.

The sketch of the keyed-powder gun (1SKG-KM1) and the cross section of the keyed launcher are shown in Fig. 1. The double-wall-structure launcher was 27 mm in basic bore diameter and 1.75 m in length. The important difference of this gun from more conventional guns was the design of the launcher. The launcher was originally designed as a double-wall structure whose muzzle shape resembled an old moon. It could thus fire a projectile accelerating to a velocity of over 2 km/s without rotation due to the tenacious keyed launcher. Two types of methods for measuring of projectile velocity were combined with the gun, the reflected-light method and the electromagnetic method. The measurement facility for projectile velocity and a typical velocity signal are shown in Fig. 2.

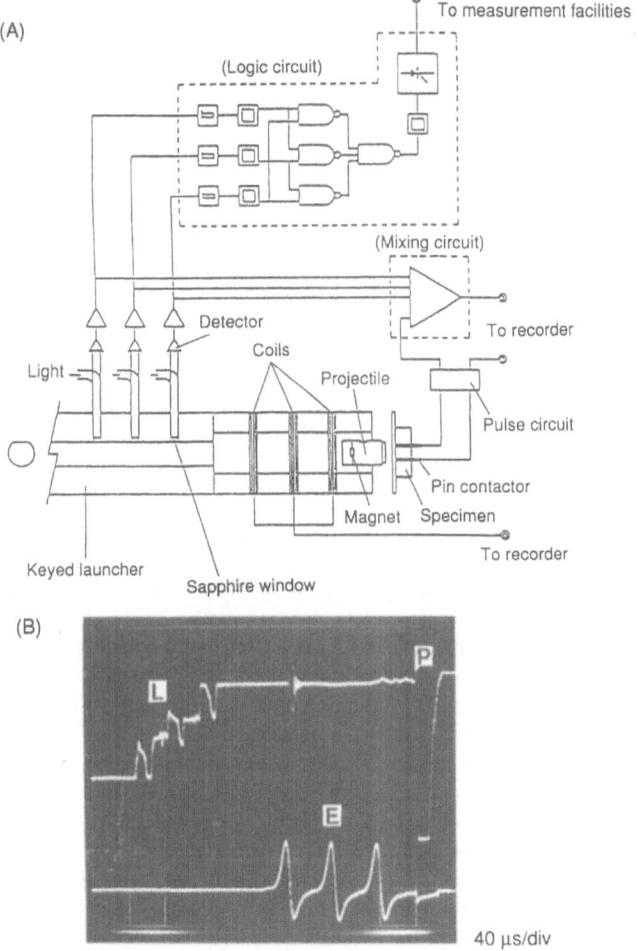

Fig. 2 (A) measurement facility of projectile velocity and (B) typical velocity signals (recorded by the 30 MHz digital memory).

In the reflected-light method, the arrival times of the projectile at three points in the extension launcher were detected by using three cross fibers, with which lights were introduced into the launcher through sapphire windows. In addition, this apparatus was equipped with the logic circuit to generate a reliable trigger signal for shock-wave measurements. In the electromagnetic method, the arrival times of a projectile which contains a small magnet at three points after the muzzle were detected by using three coils. The accuracy of this method was within 0.2%, and the tilt angle between the parallel impact plates was less than 0.2 degrees at the muzzle.

2.2 Inclined-Mirror Method

Figure 3 shows an illustration of the inclined-mirror method. This method is based on the phenomenon that the evaporated surfaces of the plane mirrors or inclined mirror on the driver plate and specimen are disturbed by the arrivals of shock wave or free surface motion. The light reflection intensity abruptly decrease, and the shock-wave arrivals or free-surface motions can be determined [11]. The steady-state Hugoniot parameters can be measured by this method to pressures over 100 GPa [12].

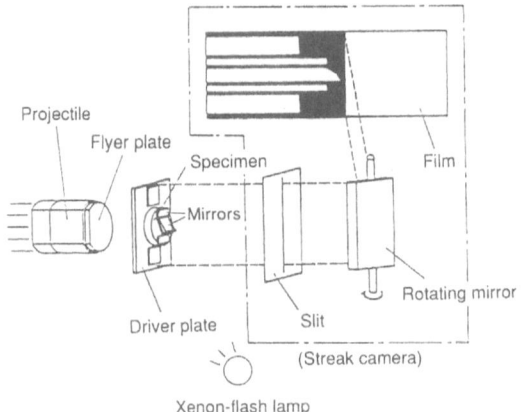

Fig. 3 Illustration of the inclined-mirror method.

In our earlier work a streak camera and Xenon flash lamp with streak rate of less than 4 mm/ms was used. In 1991, we developed a new compact high-speed streak camera whose maximum streak rate was faster than 10 mm/μs, and the new Xenon-flash lamp [13]. Fig. 4 shows a photograph of an improved streak camera and flash lamp. The streak camera consisted of a square-shape mirror, a high-frequency motor with an air bearing of faster than 120,000 rpm in maximum rotating rate, and a film-mount case of 830 mm in inner diameter. The width of slit was 15 or 30 μm. The flash light source consisted of two parallel Xenon-lamp tubes of 40 mm in discharge length and 10 mm in diameter, which were stimulated with two condenser of 100 mF and 2 kV, and were triggered by a pulse of 6 kV. Fig. 5 shows the measured streak photograph [14], for an alumina (Al_2O_3) polycrystal specimen, and an impact velocity (tungsten (W) impactor) of 1.293 km/s. The elastic shock wave arrived at the rear surfaces of the driver plate and specimen at point 1 and 2, respectively, and the plastic wave arrived at the specimen rear surface at point 3. The accuracy of this method depends on the capability of the streak camera, the dimension of the specimen, the shock velocity, etc. As a result of the present refinements of the measurement facilities, the time resolution was improved by a factor of three to four, and was estimated to be less than 7 ns.

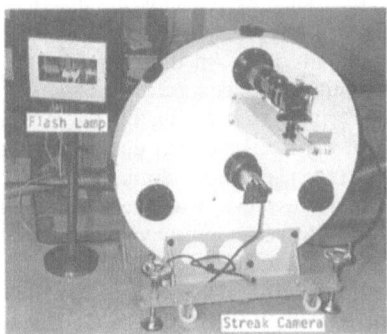

Fig. 4 Photograph of the new high-speed streak camera and Xenon flash lamp [13].

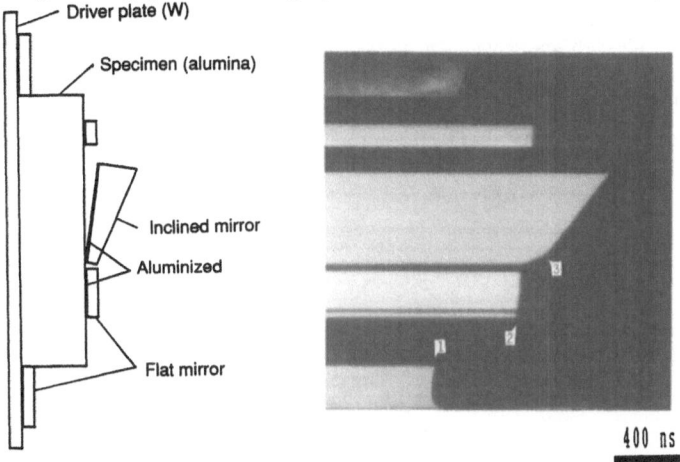

Fig. 5 Streak photograph by the inclined-mirror method, when the specimen was alumina polycrystal and the impact velocity (W) was 1.293 km/s, the inclined angle (a) was 5.012°, the slit width was 15 μm, and the streak rate at film was 8.479 mm/μs (from [13] with permission).

2.3 Manganin-Gauge Method

In order to study the kinetics of elasto-plastic transitions or phase transitions, it is necessary to measure the particle-velocity histories or stress histories at points within material. In-material gauge methods are suitable for this purpose. The manganin-gauge method is based on the piezo-resistance effect of manganin, whose piezo-resistance coefficient is almost constant up to the pressures over 50 GPa [15-17]. By using in-material manganin-gauges, stress parallel and normal to the shock propagation direction can be measured [18].

Previously a low-impedance gauge (0.3-0.6 Ω), which was originally designed by us, and provided by the Kyowa Electric Instrument Co., Ltd., was used for pressure profile measurements in the higher stress region of about several 10s of GPa [19]. The gauge consisted of a 6 μm thickness manganin foil (83.5 wt% Cu, 11.5 wt% Mn, 4.4 wt% Ni) of 0.5-1.0 mm in width, and 10-μm thickness four terminal copper foils which were produced on a

12.5 µm thickness polyimid film by the photo-etching method. The length between the first and fourth terminals was 8.5 mm. The manganin foil was attached to the terminals by spot soldering and rolling. The maximum thickness at the junction of the earlier gauge was about 50-60 µm. In 1991 we refined the gauge by reducing the thickness to 30-35 µm. During assembly, the thickness could be further decreased to 25-30 µm, by pressing the gauge between the specimens using a small vice. An illustration of the assembly is shown in Fig. 6(A). The first and fourth terminals were connected to a high-voltage pulse source, and a constant current of 5-10 A, whose ripple was less than 0.5%, was applied to the manganin gauge. The resistance changes of the manganin foil due to shock stress were detected by measuring the voltage between the second and third terminals.

Figure 6(B) shows the measured stress history to a peak stress at approximately 33.2 GPa for an alumina polycrystal specimen [20]. The impact velocity (tungsten impactor) was 1.359 km/s, and the recorder was a dual-beam oscilloscope, DS-8122, of the Iwatsu electric co., Ltd. whose maximum frequency was 100 MHz. The diffused two-step structure during stress loading is due to an elastoplastic transition. Discussion of the stress history is provided in a later section. The response time to shock stress of this method depends chiefly on the thickness of the gauge. As a result of the present refinements to the gauge, the time resolutions at rise and release of the new gauge were improved, and estimated to be less than 20 and 10 ns, respectively. The present high-impedance gauge is suitable for the measurements in the pressure range of over 10 GPa.

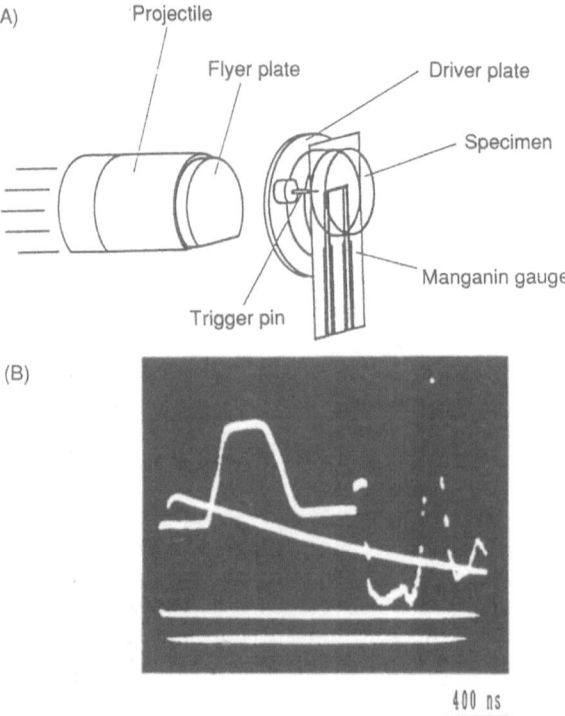

400 ns

Fig. 6 (A) Illustration of the manganin-gauge method and (B) Stress history profile, when the specimen was alumina polycrystal, the impact velocity (W) was 1.359 km/s (recorded by the 100 MHz dual-beam oscilloscope).

2.4 Electromagnetic-Gauge Method

The electromagnetic-gauge method is useful for the measurement of in-material particle-velocity histories in a non-conductive materials [21]. The principle of the method is based on Faraday's law of electromagnetic induction of a π-shaped metal loop. The measurement system of the electromagnetic-gauge method is shown in Fig. 7. The differential-type gauge used in this study consisted of a three terminal shaped copper foil of 10 μm in thickness produced on a 12.5 μm thick-polyimid film by the photo-etching method, and provided by the Kyowa Electric Instrument Co., Ltd. [22]. The symmetrical outputs were connected to a differential amplifier of the recorder to minimize the noises. To measure the shock profile, a magnetic field normal to the shock propagation direction was applied by a Helmholz coil. The pulsed current source for this coil consists of 4 μF condenser, 8 kV direct current source, gap switch, etc., and was triggered by a logic trigger signal from the projectile-velocity measurement system by the reflected-light method [10], through a delay circuit. The generated magnetic field was approximately 530 gauss at the center of the coil, and the scattering was less than 0.1% within 10 mm cubic area at the center. The details of the system was described in Ref. 22.

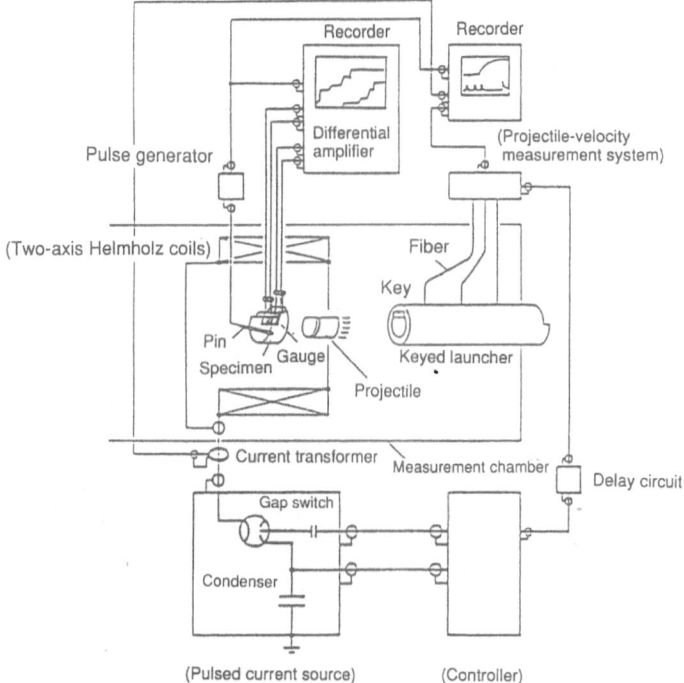

Fig. 7 Measurement system by the electromagnetic-gauge method combined with the keyed-powder gun (from [22] with permission).

Figure 8 shows the measured particle-velocity profile, for an alumina single crystal (sapphire) specimen. The impact velocity (sapphire impactor) was 1.386 km/s and the recorder was an oscilloscope of type 465 from the Sony/Textronix Co., Ltd., whose maximum frequency was 100 MHz. A two step structure due to the elastoplastic transition and a decrease behind the

elastic shock front due to stress relaxation are clearly seen in this signal. The response time of
the gauge was estimated to be less than 15 ns. It was difficult to record noise-free signals of
particle-velocity history, particularly in ionic single crystal, by this method. The velocity-
induced voltage from the gauge was usually less than 0.5 V, and easy disturbed by noise from
shock-induced polarization, electrical conduction, or pin-contactor noise. To reduce these
noise sources, stronger magnetic field and use the trigger signals which do not stimulate large
noises are beneficial. This method can be used for the measurement of combined
compression-shear shock waves. To measure the shear-wave particle velocity, a magnetic
field obliquely oriented to the shock-propagation direction is used [23]. The oblique magnetic
field was generated by a coil which was aligned to the oblique direction to the launcher axis
[24]. The magnetic field parallel to the axis was over 2500 gauss. Experimental results on
sapphire is described in a later section.

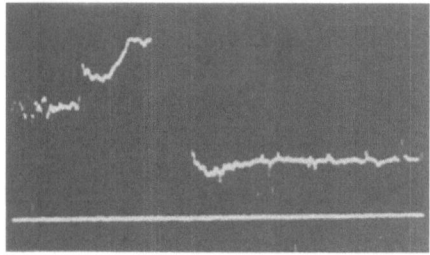

0.1 V/div
200 ns/div

Fig. 8 Particle-velocity profile, when the specimen was alumina single crystal and the impact velocity (W) was
1.386 km/s (recovered by the 100 MHz oscilloscope) (from [22] with permission).

3. Shock Compression Studies on Selected Ceramics

We have been engaged in shock compression research of ceramics in the several 10s of GPa
region. In this section, our shock compression studies on aluminum oxide (alumina),
zirconium dioxide (zirconia), silicon nitride are reviewed. Particularly, the results for alumina
and zirconia are described in detail. In these materials are included, the transformation-
toughening ceramics, grain-controlled ceramics, and whisker-doped ceramics. Properties such
a dynamic yielding, phase transitions, and high-pressure compressibility are investigated
under shock compression. The influence on the shock yielding properties of crystal-
chemistry, thermophysical properties, microstructures, and so forth are discussed within the
context of the experimental results.

3.1 Alumina (Al_2O_3)

Alumina (Al_2O_3) ceramics have been widely used due to their good mechanical, optical and
electrical properties. Sapphire, which is the alumina single crystal form, has a rhombohedral
hexagonal structure with close-packed oxygen ions and is a typical structures of the earth's
interior materials. The Hugoniot of single-crystals and polycrystals has been measured by the
flash-gap method, [25] the pin-contactor method, [26] the wire-image method [27] and the
inclined-mirror method [22,28]. We have investigated the shock compression behaviors of
alumina single crystal and a alumina polycrystal (polycrystal-I), by the measurement of

Hugoniots, stress histories, particle-velocity histories, and combined compression-shear shock waves. Recently, we measured the Hugoniots and stress histories of the another alumina polycrystal (polycrystal-II) by using the recent measurement facilities described previously (Figs. 5 and 6). The bulk densities of the polycrystal specimens (I and II) were approximately 3.875 and 3.854 g/cm^3, respectively, while that of single crystal is 3.984 g/cm^3. The porosities (Po.) at the polycrystalline specimens were about 2.3 and 3.1%, respectively. The impurities of the single crystal and the polycrystal-I materials were less than 0.1wt%, and those of the polycrystal-II were MgO (0.2-0.3 wt%) and SiO$_2$ (0.05-0.15 wt%).

Figure 9 shows the Hugoniot compression curve up to 50 GPa for the single crystal (shocked to <1210> or <0001> axis direction) and the polycrystal-I material measured by the inclined-mirror method using the earlier streak camera facility [19]. The average Hugoniot-elastic limit (HEL) determined from by the inclined-mirror method was 17 GPa. Here, the stress reduction after the HEL shock front was neglected, because it was difficult to analyze the relaxation by the inclined-mirror method. In the figure, the chained line was the extrapolated isothermal static compression curve using K_0=226 GPa and K'_0=4.0, which are derived from the static data up to 12 GPa by the x-ray diffraction method [29]. At a given density, the Hugoniot data of the single crystal exceed the extrapolated static curve by 6-8 GPa up to over 40 GPa. The stress increase due to the average shock-temperature increase of less than 50 degree C at 30 GPa was estimated to be less than 0.1 GPa by using the Mie-Grüneisen equation. If we ignore the stress relaxation, the offset of the single-crystal Hugoniot data from the static data is chiefly a consequence of the shear strength. However, this author assumes that the stress relaxation can not be ignored for single crystal.

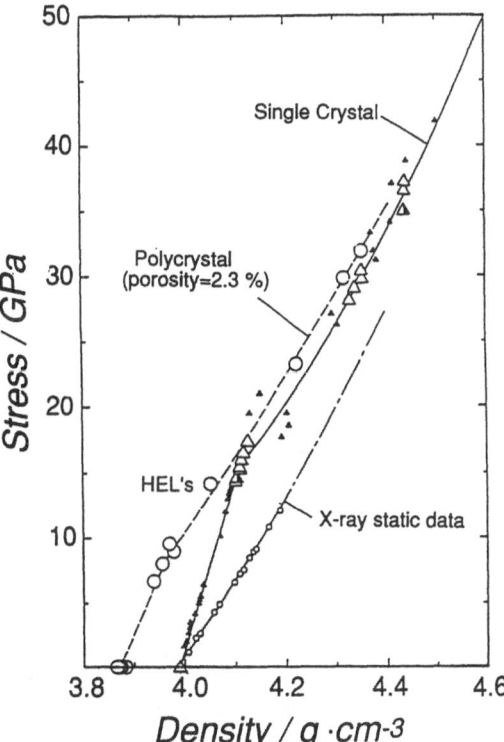

Fig. 9 Hugoniot-compression curves of alumina single crystal and polycrystal (from [19] with permission).

In contrast, the HEL stress of the polycrystal-I (Po.=2.3%) was determined to be in the range of 6-10 GPa. From consideration of these data and previous data, it was noted that the larger porosity leads to smaller the HEL stress. The strength-offset data, however, of the polycrystal-I material exceeded those of the single-crystal material by over 1 GPa in the plastic region to shock stress in excess of 30 GPa, and the data of the material with 3.5-4.3% porosity [11] exhibited even larger strength differences. The streak photograph of the polycrystal-II material (Po.=3.1%) using the inclined-mirror method with the more recent streak camera was shown in Fig. 5 [14]. The HEL stress was determined to be 9.5-13 GPa, which was slightly higher than that of the polycrystal-I material. This difference may be a consequence by binder chemistry. The Hugoniot-compression curve of the polycrystal-II material was also situated above that of single crystal in the plastic region. It was concluded that the Hugoniot offset of the polycrystal specimens from that of single crystal was chiefly caused, not by the temperature increase, but, by the remaining porosity due to the local shear strength.

Figure 10 shows the stress history profile in single crystal sapphire using the manganin-gauge method with the older gauge technique, when the peak stress was 36.5 GPa [19]. A two-step structure could not be resolved at shock front due to the small thickness of the forward specimen (approximately 1 mm). On release, a very sharp stress drop due to the elastic rarefaction wave could be observed, which is similar to a rarefaction shock. The shear strength estimated from the stress drop were 4-6 GPa in the stress region up to 40 GPa. This value was much smaller than expected from the Hugoniot offset. This may also be due to the stress relaxation and the expansion process on release. Fig. 8 showed the particle-velocity history in sapphire by the electromagnetic-gauge method when the impact velocity (sapphire impactor) was 1.386 km/s [22]. In this profile, a clear two step structure and relaxation of the particle velocity behind the elastic shock front could be clearly observed.

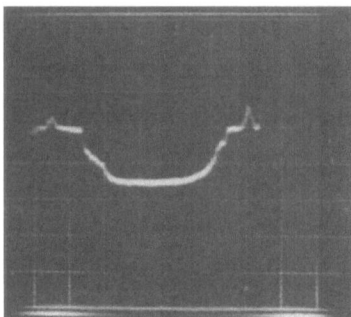

1 V/div
200 ns/div

Fig. 10 Stress history profile in alumina single crystal, when the peak stress was 36.5 GPa (recorded by the 100 MHz oscilloscope) (from [19] with permission).

These wave features were resolved because the electromagnetic gauge was thin (22.5 μm thick) and the thickness of the forward specimen was comparatively large (approximately 2 mm) allowing substantial wave separation. However, the profiles of particle-velocity history can not be directly compared with that of stress history, because the phase velocities for particle velocity and stress are not always equal. In addition, a steep increase in the particle velocity due to the elastic rarefaction wave could be also seen, which corresponded to the sharp stress drop in the manganin-gauge stress-history profile. On the other hand, in the stress history profile of polycrystal-II measured by using the newer manganin-gauge technique (Fig. 6(B)) [20], when the impact velocity (tungsten impactor) was 1.359 km/s, a diffused two-step structure and a stress drop due to the elastic rarefaction wave could be observed. The stress relaxation behind the elastic shock front observed in the single crystal material was not seen in the polycrystalline sample by the measurement of stress. Our results on polycrystals are consistent with those of Rosenberg et al [30].

These differences in Hugoniot or shock-wave profile between single-crystal and polycrystal material are thought to be cause by the differences in shock-yielding property. It has been pointed out that single crystal material may catastrophically loose shear strength behind the elastic shock front. In contrast, polycrystalline material maintains considerable shear strength. The author suggests that this tendency can be seen for many brittle materials. However, the author also suspects that all single crystal materials, particularly those with high-density structures may not totally loose shear strength behind the elastic shock front. Because the HEL stresses of single crystals are in many cases much larger than those of polycrystals, the reduction is conspicuous. The Hugoniot data of single crystal should be analyzed considering the stress relaxation. These differences are discussed in the later section.

The particle-velocity profiles of the combined compression-shear shock waves in the several 10s of GPa region were measured by the electromagnetic-gauge method. The basic measurement system was described in Sec. 2 [22]. The concept illustration of the electromagnetic-gauge method to measure both of longitudinal and shear waves and the particle-velocity histories at two points in sapphire [24], when the impact velocity (sapphire impactor) was 1.237 km/s are shown in Fig. 11(A) and (B), respectively.

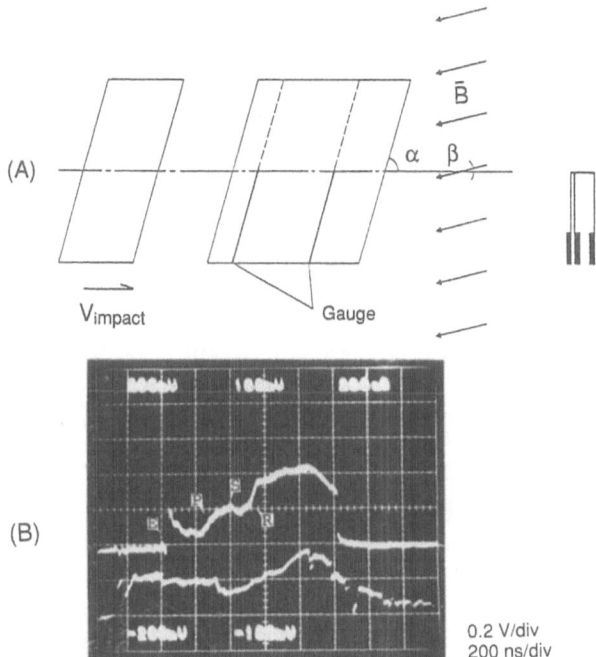

Fig. 11 (A) Concept illustration of the electromagnetic-gauge method to measure the combined compression-shear shock waves and (B) Particle-velocity histories of the combined compression-shear shock waves in alumina single crystal, when the impact velocity (W) was 1.237 km/s (recorded by the 100 MHz dual-beam oscilloscope) (from [24] with permission).

In this study, the oblique magnetic field was generated by a coil which was oriented at an oblique direction to the launcher axis. In the Fig. 11(B), the longitudinal elastic wave and plastic wave arrive at the second gauge at point E and P, respectively. The elastic wave stress was very large (>24 GPa), and the large reduction was seen in this test. The amplitude of elastic wave and the reduction decreased with the driving shock. The minus signal was clearly seen at point S, which could not be seen for the usual shock wave. This signal just corresponded to the shear wave arrival. At point R, the elastic rarefaction wave from the specimen rear surface arrived at the gauge. As a result, the shear wave velocities under shock compression of 23-27 GPa peak stresses were considerably small compared with the boldly extrapolated ones from the ultrasonic data along the a-axis direction up to 1 GPa [31]. The shear strength should be directly discussed by using the shear wave amplitude. The detailed results including the material strength will be reported, elsewhere.

3. 2 Zirconia (ZrO$_2$)

3.2.1 Cubic Zirconia
Zirconia (ZrO$_2$) exhibits uniquely different properties depending on the crystal structure, microstructure, doping material, etc. Zirconia in its pure form exists in the stable monoclinic crystal structure. Through doping with other metal oxides such as calcia (CaO), magnesia (MgO) of yitteria (Y$_2$O$_3$), zirconia can be stabilized in the cubic or tetragonal structure. The cubic zirconia are used as ionic materials with the advantage of high electrical conductivity at high temperature, which arises from the vacancies. On the other hand, the polymorphism of ZrO$_2$ provides a useful analogy to AX$_2$ compounds, which are important for the earth-interior science. So far, the Hugoniot-compression curves of pure ZrO$_2$ (monoclinic phase) have been measured for sintered polycrystal [1] and twin crystal [32].

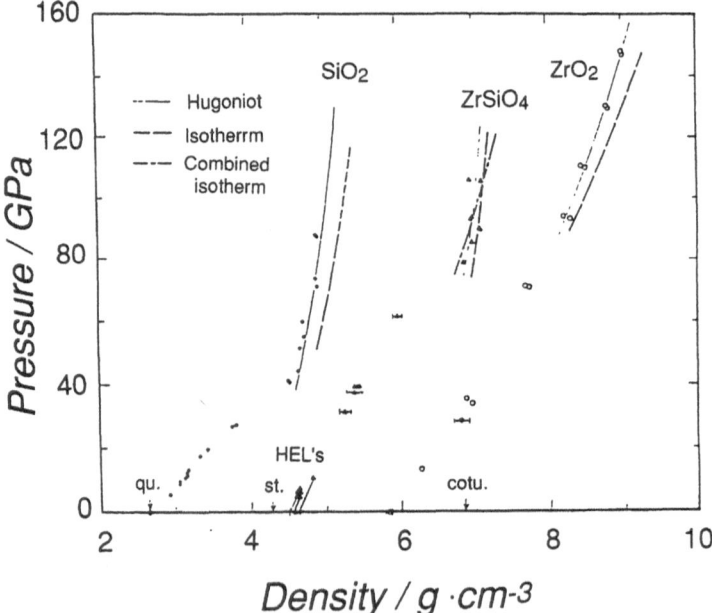

Fig. 12 Hugoniot-compression curves of pure zirconia, zircon and quartz (from [32] with permission).

The Hugoniot compression curve of pure zirconia (monoclinic phase) measured with the pin-contactor method is shown in Fig. 12, together with those of zircon [32] and quartz [33]. It was assumed that zircon transforms to a high-pressure phase, or that zircon decomposes to ZrO_2 and SiO_2, by about 70 GPa. Stabilized cubic and tetragonal zirconias should show different shock compression behaviors for each crystal condition.

We have been measuring the Hugoniot-compression curves of the Y_2O_3 and CaO-doped cubic zirconia in the pressure range up to 125 GPa using polycrystal [34,35]. and single-crystal material [36]. The Hugoniot parameters were measured by means of the inclined-mirror method using the powder gun of Kumamoto University and the two-stage light gas gun of Tohoku University [37]. The approximately 2.5 or 5 mm thick plate-shaped specimens (13x10 or 19x19 mm^2) of the Y_2O_3-doped cubic (8.0 mol%) zirconia (YCZ), and the CaO-doped (11.4 mol%) cubic zirconia (CCZ) polycrystals, and the Y_2O_3 doped cubic (9.6 mol%) zirconia (YCZ) single crystal were used. The bulk densities of the polycrystals and single crystal were measured to be approximately 5.879 (YCZ polycrystal), 5.419 g/cm^3 (CCZ polycrystal) and 5.948 g/cm^3 (YCZ single crystal), respectively. The impurity values were less than 0.1 %. The porosity of the YCZ and CCZ polycrystals were estimated considering the containing rates of HfO_2 to be about 1.5 and 4.8%, respectively. These porycrystal specimens were prepared by the normal sintering method.

The shock velocity (U_S) versus particle velocity (U_p) Hugoniot results are shown in Fig. 13 [35]. The principal results are as follows: The HEL stresses of the CCZ and YCZ polycrystals were determined to be 6-10 and 11-15 GPa, respectively, although the HEL stress is observed to depend on the porosity, amplitude of driving shock, etc. The CCZ and YCZ polycrystal behavior was elasto-isotropic solid and an elasto-plastic, respectively. This might be related to the differences in oxygen-vacancy number, ignoring the porosity effects. Furthermore, the CCZ polycrystal transformed to a high-pressure phase at 0.87 km/s in particle velocity.

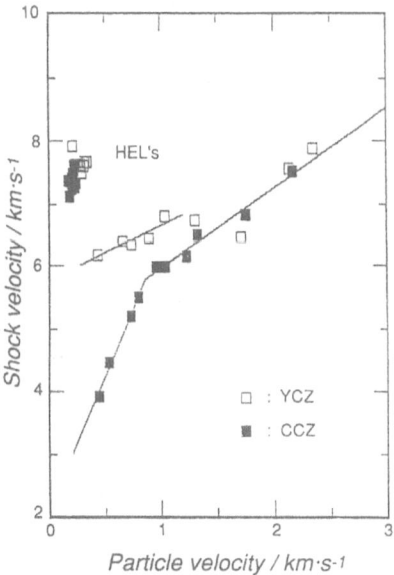

Fig. 13 U_S-U_p Hugoniot results of the YCZ and CCZ polycrystals (from [35] with permission) .

The profile characteristics of this transition was unique due to the profile like a second order transition. Recently, it was reported that the same material also transformed to an orthorhombic phase at almost the same pressure under static compression [38]. It can not be concluded at this time that these transitions are the same.

The Hugoniot compression curves parallel to the <100> and <110> axes of the YCZ single crystal are shown in Fig. 14 [36], together with those of the YCZ polycrystal. In the plastic region, the Hugoniot data converged, although the HEL data showed significant differences. The HEL stresses parallel to <100> and <110> axes were approximately 14 and 25 GPa, respectively, while that of the polycrystal was approximately 13 GPa. These anisotropic HEL's can be reasonably well understood in terms of the primary (111) cleavage plane in cubic zirconia. Just above the HEL, the shock velocities were much slower than the bulk sound velocity, although those of the polycrystal approach the bulk sound velocity. The single-crystal Hugoniot curves showed relaxation toward an isotropic compression state, whereas the polycrystal preserved considerable shear stress offset. These results clearly showed the difference in yielding property between single crystal and polycrystal zirconia, comparable to the result of alumina. The phase transition for loading parallel to <100> axis began at approximately 53 GPa, and finished by about 70 GPa. In order to discuss the equation of state of the high-pressure phase, the thermal contribution in the Hugoniot data should be analyzed. Experiments using the manganin-gauge method are now under way. The detailed results and analyses of the equation of state of the high pressure phase will be reported elsewhere.

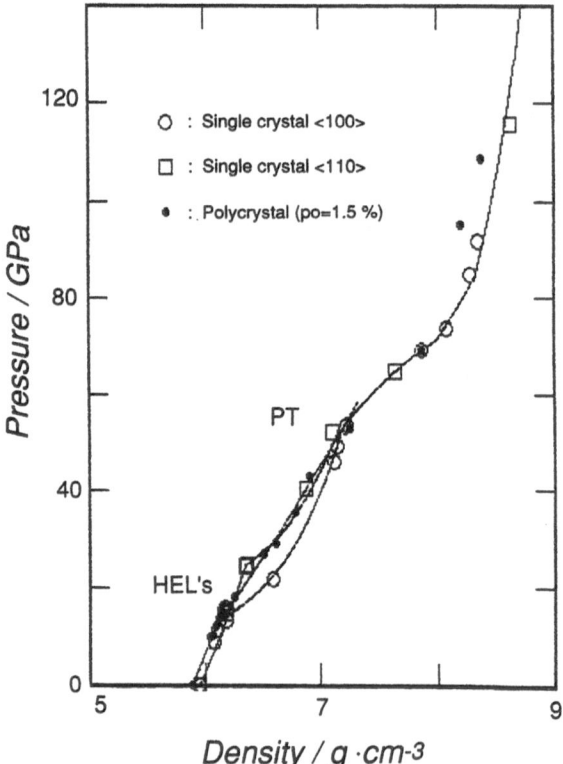

Fig. 14 Hugoniot-compression curve of the YCZ single crystal and polycrystal (from [36] with permission).

3.2.2 Tetragonal Zirconia

The stabilized or partially stabilized tetragonal zirconia has an extremely large fracture toughness for ceramics and is comparable to metals due to transformation toughening under static loading conditions [39]. The transition point (probably a different transition than in the fracture process zone) under shock compression in excess of 30 GPa for the Y_2O_3-doped (3.0 mol %) tetragonal zirconia (YTZ) measured by the inclined-mirror method had been reported previously by us [40]. Recently, another transition point with a fine step-structure at 13-17 GPa, and anomalous spall behavior were observed with the VISAR method by Grady and Mashimo [41]. The anomalous behavior of this material is interesting, but, difficult to understood. In this report, the measurement results by the inclined-mirror method and the VISAR method, and also the recovery experiment on the YTZ specimens are reviewed. The bulk densities of the YTZ test specimens were approximately 6.047 g/cm^3 for the inclined-mirror method, and were 5.954 and 6.028 g/cm^3 for the VISAR method. These specimens were produced by the hot-isotropic press method (HIP) method and provided by Sumitomo Electric Industries Ltd. The porosities were estimated considering the contained HfO_2 to be about 0, 0.3 and 1.5 %, respectively. The impurity values were less than 0.1%.

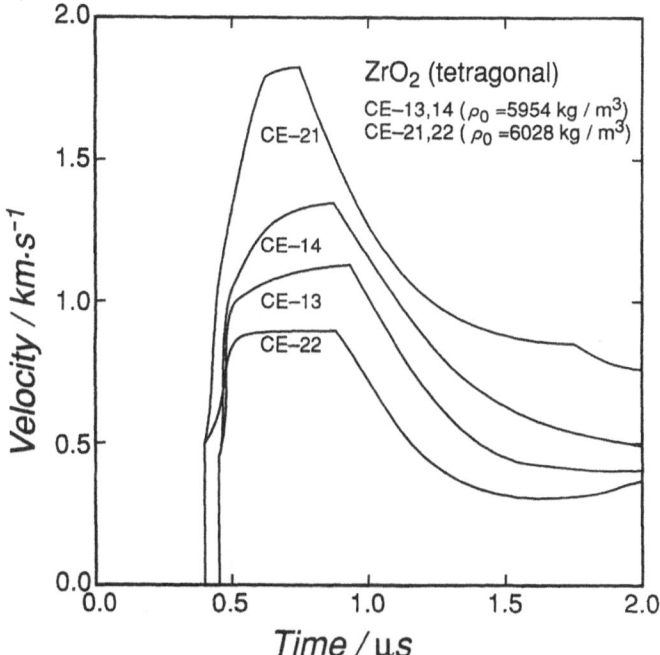

Fig. 15 Particle-velocity histories of the YTZ polycrystal by the VISAR method, when the peak stress were 29-59 GPa (from [41] with permission).

The VISAR instrumentation is capable of resolving details in the wave motion down to one or two nanoseconds and of measuring release profiles [42]. The shock-wave experiments were conducted using the single-stage powder gun of the Sandia National Laboratory. The approximately 4 mm thick (diameter=about 30 mm) plate-shaped specimen and the LiF single crystal optical window were used. Fig. 15 shows the particle-velocity histories by the VISAR method in the YTZ specimen for the peak stresses in the range of 29-59 GPa [41]. Tests CE-13 and 14 were low-density specimens (density=5.954 g/cm^3), whereas CE-21 and 22 were high- density specimens (density=6.028 g/cm^3). A three-step shock-front structure was observed for the test at a peak stress of 59 GPa. The first step was quite sharp with a delay time of about 20 ns. The transition pressure was 13-17 GPa. The second step was quite diffused, with a transition pressure of over 30 GPa. The latter transition probably corresponded to that measured by the inclined-mirror method. The phase velocities of the second wave were 7.09-7.15 km/s, which were anomalously large compared with the bulk sound velocity of 5.74 km/s. The lower yield occurred suspiciously close to the 14-16 GPa transition for tetragonal-orthorhombic (II) phase observed at static pressures, due to Ohtaka and Kume [43]. However, it was difficult to establish the nature of this lower transition— whether elastic-plastic or structural.

A typical streak photograph by the inclined-mirror method is shown in Fig. 16, for an impact velocity (tungsten impactor) of 2.620 km/s [35]. The inclined-mirror image shows a dull kink and a clear kink. The Hugoniot compression curve of the YTZ polycrystal (density=6.047 g/cm^3) by the inclined-mirror method, together with those of the YCZ polycrystal [35], is shown in Fig. 17 [44]. The first and second transition points and one Hugoniot point of the third wave measured by the VISAR method are also plotted in the figure (black circle). The shock velocity of the first shock wave was determined by the inclined-mirror method, and the shock and particle velocities of the second shock wave were determined by the VISAR method, in this analysis. The Hugoniot data of the third wave by the inclined-mirror method were analyzed considering the second transition point. The VISAR data of the third wave were in good accord with those by the inclined-mirror method. The pressures of the first and second transition points were approximately 15 and 31 GPa, respectively. Finally, this material may completely transform to the high-pressure phase by about 70 GPa, considering that the Hugoniot densities are comparable to those of the YCZ final phase [36]. The detailed analyses and discussion about the yielding or phase transitions will be reported elsewhere.

400 ns

Fig. 16 Streak photograph by the inclined-mirror method, when the specimen was the YTZ polycrystal and the impact velocity (W) was 2.620 km/s, and the streak rate was 9.472 mm/μs (from [35] with permission).

Figure 18 shows the particle-velocity histories by the VISAR method when the peak stresses were 11 and 13 GPa—well within the elastic region [41]. The rise of the shock front was very sharp. The steep release profiles at the beginning of release could be observed. This might correspond to a rarefaction shock, which would suggest a phase transition on release.

The author suggests that this transition may correspond to the tetragonal-monoclinic phase transition. In addition, we could observe a pullback signal was observed at the end of the release wave, indicating a tensile spall process. The spall strength was determined to be 1.6 GPa, which was a factor 4 to 5 higher than other high-strength ceramics and was comparable with intermediate tensile strength metals such as aluminum or copper. It was also suggested that this anomalously high value of spall strength might be related to the transformation-toughening of YTZ. This explanation is not inconsistent with the rarefaction-shock on release and the recovery experiments discussed below. These transitions and spall behaviors are discussed in relation to the process-zone toughening, in the later section.

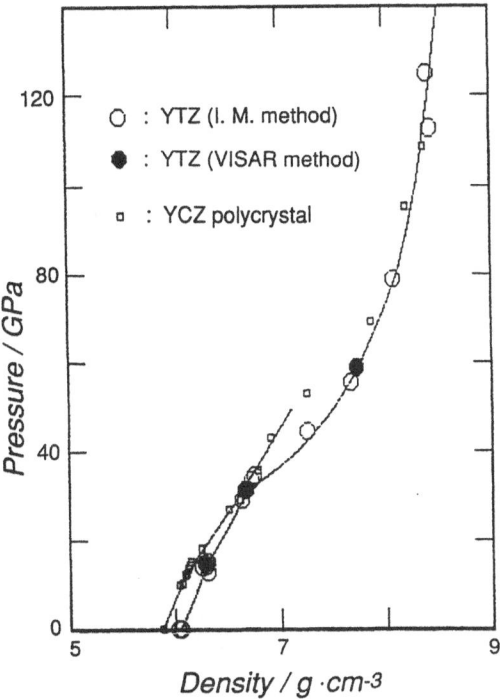

Fig. 17 Hugoniot-compression curve of the YTZ polycrystal (from [44] with permission).

We performed the recovery experiments on the YTZ polycrystalline material to get information about the shock-induced phase transition. Recovery of specimens was conducted using the momentum-trap method. Visible feature of the recovered YTZ specimens were that the specimens remained intact but with cracked surfaces. In contrast, recovered samples of the alumina and cubic zirconias (CCZ, YCZ) were crushed and looked like a powder. Microscopic observations of the recovered specimens were conducted using powder x-ray diffraction, electron microscopy, and Raman scattering spectroscopy [45]. The Raman spectrum from the broken surfaces of the YTZ specimens are shown in Fig. 19 [45], where the experimental conditions were as follows: a) before impact, b) after the bending test, c) after the shock compression (31-33 GPa), d) after the shock compression (42-45 GPa). The crystal structure before shock compression was determined by powder x-ray diffraction to be almost all tetragonal (100 %). Although the Raman band peaks (197, 181 cm^{-1}) of the monoclinic phase [46] were not observed in the specimen before impact, these band were observed in specimens after bending tests and also after shock compression. The amplitudes of the band peaks after shock compressions were higher than after the bending test. We confirmed that the Raman spectrum of the recovered specimen from the shock compression below 15 GPa also showed the band peaks of the monoclinic phase. In addition, TEM observation also showed the existence of the monoclinic phases which formed twin crystals in the recovered specimen. These results supported the supposition that a certain amount of monoclinic phase was formed under shock compression and/or at stress release. It was suggested that this transition is related to the anomalous compression and release profiles at front and release measured by the VISAR method. The detailed results will be reported, elsewhere.

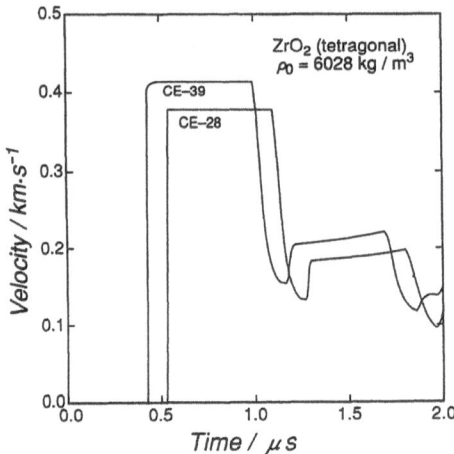

Fig. 18 Particle-velocity histories of the YTZ polycrystal by the VISAR method, when the peak stresses were 11-13 GPa (from [41] with permission).

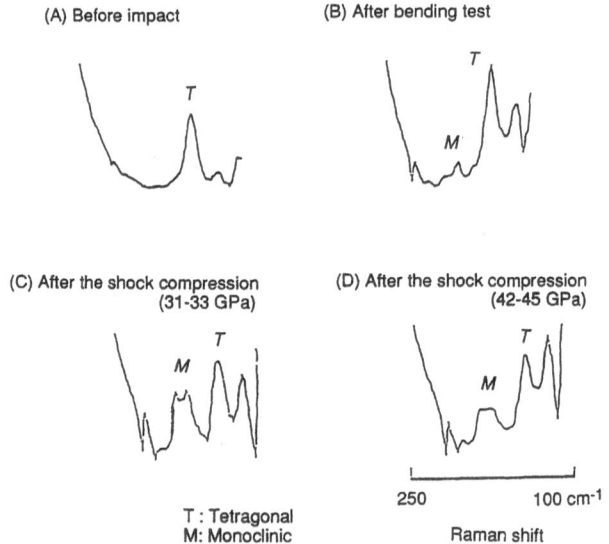

Fig. 19 Raman spectrum of the YTZ specimens before and after shock compression.

3.3 Silicon Nitride (Si_3N_4)

Silicon nitride (Si_3N_4) ceramics have good mechanical properties even under high temperatures in spite of their low density. Their use is desired in many fields, such as the automobile, space and aeronautics industries. A number of high-performance ceramics of silicon nitride have been developed. Examples include grain-controlled ceramics, whisker-doped ceramics, and nano-composite ceramics. The dynamic properties of these new ceramics have not been investigated, however, and are well understood. Recently, we started measurements of the Hugoniot properties of several Si_3N_4 ceramics, to examine the effects of the differences in microstructure, doping material, etc., on the shock-yielding properties and equation-of-state.

The Si_3N_4 specimens investigated were a normal type ceramic [47], a grain-controlled ceramic, and a SiC-whisker (20 wt%) doped ceramic, which contained considerable binder material. The binders of these materials were Y_2O_3, Al_2O_3, etc., and their rates were about the same. The differences between the normal and grain-controlled ceramics was chiefly in grain dimension and porosity. The grain of the normal or grain-controlled one has a shape of pole-like of about 0.5-1.2 μm or 0.25-0.4 μm in diameter, respectively, according to the SEM observation. The densities of the normal type and grain-controlled ceramics were approximately 3.116 and 3.236 g/cm^3, respectively. The density of the whisker-doped ceramic was approximately 3.268 g/cm^3. These specimens were provided by the Sumitomo Electric Industries Ltd.

Figure 20 shows the Hugoniot-compression curves of the normal type and grain-controlled Si_3N_4 ceramics measured by the inclined-mirror method using the earlier streak camera [48]. The elastic wave velocities and the HEL stresses of these specimens were different, while the Hugoniot-compression curves converged toward each other in the plastic region.

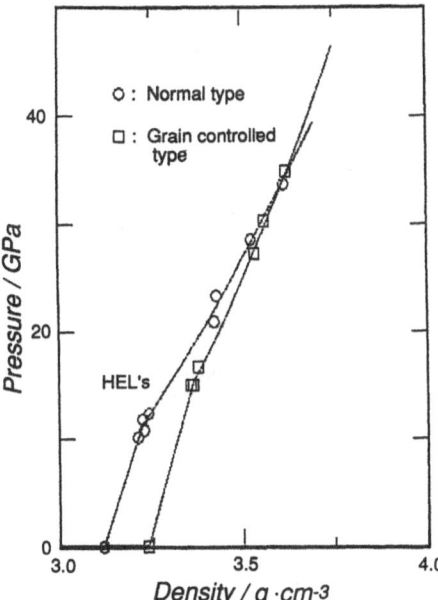

Fig. 20 Hugoniot-compression curves of the normal-type and grain controlled Si3N4 ceramics (from [48] with permission).

The HEL stress values of the normal and grain-controlled ceramics were 10-12.5 and 14-16.5 GPa, respectively, although in case of the low driving shock experiments, HEL stresses were slightly higher. The difference may be a consequence of the grain-controlling or the porosity. In addition, we recently confirmed that the another grain-controlled Si_3N_4 ceramics whose grain dimension was further finer had the higher HEL stress. It is assumed that the grain dimension of polycrystal also influences the strength under shock compression.

In Fig. 21 a streak photograph is shown of the SiC-whisker doped Si_3N_4 ceramic recently measured with the inclined-mirror method using the newer streak camera, when the impact velocity (tungsten impactor) was 1.139 km/s [49]. A very diffused kink due to elastoplastic transition can be seen in the inclined-mirror image, while those of the normal and grain-controlled ones were comparatively sharp. The HEL stress was determined to be over 17 GPa. The author assumes that this diffused kink and the high HEL stress is related to the doping of SiC whiskers. The effects of the differences in microstructure or doping on the yielding property is discussed in the later section.

400 ns

Fig. 21 A streak photograph by the inclined-mirror method, when the specimen was the whisker-doped Si_3N_4 ceramics and the impact velocity (W) was 1.139 km/s, and the streak rate was 10.107 mm/μs (from [49] with permission).

4. Phenomenological Discussion on the Shock-Yielding Phenomena of Brittle Materials

A plane shock compression state of solids is macroscopically uniaxial, while the microscopic state is unknown. When the shear stress resulting from the uniaxial deformation exceeds a certain value which is related to the yield strength, the relaxation from the uniaxial toward the isotropic state arises (yielding phenomenon), and as a result, a dissipation process is induced. The dissipation process appears both at the shock front and behind the shock front. The shock front profile is believed to be determined by a balance of the nonlinearity of matter and dissipation properties of the material. But, the dissipation process at shock front are very complicated due to the complex interplay with the stress increase due to shock-wave propagation and nonlinearity of compressibility. Here, we consider the dissipation processes behind shock front.

According to the general conservation relation of energy of a continuum in Lagrangian coordinates, the flow equation which must hold along a particle path also under shock compression is,

$$\rho DE/Dt = \tau_{ij}e_{ij} - \partial Q_i/\partial x_i ,$$

where E, Q_i, τ_{ij}, e_{ij}, ρ, x_i and t are the internal energy, the external flow of thermal energy, the stress tensor, the strain tensor rate, density, distance and time, respectively. D/Dt denotes the Lagrangeian (material) derivative. This equation is an extended form of the first law of thermodynamics. The first term of the right-hand side of the above equation is related to the dissipation of energy. At shock front, the internal energy immediately increases by the increase in the stress-strain term due to shock wave propagation, but, the process is complicated. If the internal energy by shock wave propagation has been thrown at rise wave and the heat flow is negligible, the change in internal energy behind shock front is equal to the stress-strain term. The strain rate tensor consists of the mean strain rates and the shear strain rates. The mean strain rates are thought to be very small behind the shock front, because the volume change due to the relaxation to isotropic state is small. The shear strain rates, which chiefly determine the shear relaxation, depend on the slip system, that is, the yielding property or viscosity of matter. If we can determine the shear strain rates behind the shock front, we can directly discuss the dissipation process. However, it is very difficult to measure directly the shear strain rates under shock compression. Consequently we must indirectly discuss the yielding phenomena by using the measured mean stresses or longitudinal particle velocities.

On the other hand, ceramics are hard and brittle. This is generally understood to be due to: 1. the dislocation density is low, 2. the dislocation mobility is small, 3. the slip modes are few, and, 4. the activation energy for nucreation or growth of dislocation is small, etc. These properties are thought to be a consequence of the chemical bonding state of covalent or ionic solids, which are much different from those of metals. The yielding phenomena of ceramics are complicated even under static loading conditions. The yielding phenomena under shock compression (shock-yielding phenomena) are further complicated, by the ill-defined slip system. The shock-yielding properties of brittle materials strongly depend on their chemical and microstructural characteristics. Many high-performance ceramics, with unusual crystal structure, and controlled microstructure or composition have recently been developed. The mechanical properties of these ceramics are sometimes improved by formation of the process zones (process-zone toughening). It is thought to be an interesting problem as to whether the process-zone toughening is effective under shock compression.

In this section, the shock-yielding phenomena of ceramics including some of the recently developed ceramics are considered on the basis of the experimental results. Before discussing the shock-yielding phenomena, we shall consider some problems in the experimental and analytical procedures of shock compression study.

4.1 Some Problems in Experimental and Analysis of Shock Compression of Solids

Some basic problems exist in the experimental and analytical procedures of shock compression of solids. The conservation equations of shock compression, which are used in the analysis of the Hugoniot data, represent only one-dimensional motions of particles of continuum. However, in the plastic region, atoms must also move normal to the shock propagation direction during relaxation toward an isotropic state, while the atomic motions coincide with the continuum particle in the elastic region. In addition, the Hugoniot data are the average values, while the shock compression states of solids in the plastic region and mixed phase region are thought to be nonuniform. So, we can not get the direct informations about crystal lattice or heterogeneous state from the Hugoniot data. In addition, it is also difficult to directly determine the shear strain under shock compression, as discussed earlier. Furthermore, it is currently very difficult to perform an in-situ microscopic observation of the dynamic lattice state, and post-shock recovered specimens have, in many cases, changed due to pressure release or by annealing after shock compression.

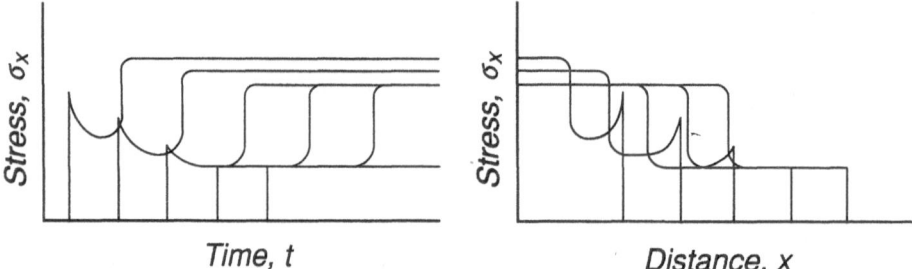

Fig. 22 Model diagram of the typical relaxation patterns of two-step structure shock wave.

Profiles of the shock waves with dynamic transitions (including yielding) are often unsteady and complex, because these transitions are usually heterogeneous time-dependent phenomena. So, it is difficult to know whether states at the shock front or the behind shock front chiefly effect the shock-wave profile and velocity, or what changes they may have on the Hugoniot data. Model diagram of the typical relaxation patterns of two-step structure shock-wave is shown in Fig. 22 (shock front). As a result, the Hugoniot data and shock-wave profiles may depend on the thickness of the specimen, the driving stress, the materials of driver plate or impactor plate, etc. Effects of the specimen thickness and the driving stress are particularly important. Furthermore, it is difficult to measure the phase velocity change and to analyze the stress change, particularly in the measurement of particle velocity, while particle velocity is more essential than stress for the shock wave analyses. In addition, the time resolutions of many stress measurement technique are not good, although high-time resolved measurement methods for particle velocity, such as the VISAR, has been developed.

4.2 Classification of the Shock-Yielding Phenomena of Solids

Figure 23 shows a model diagram of the stress (s)-density (r) Hugoniots, stress profiles, and shock velocity (U_s)-particle velocity (U_p) Hugoniots for toughening solid (1), perfect elasto-plastic solid (2), quasi-elasto-plastic solid (3), quasi-elasto-isotropic solid (4) and perfect elasto-isotropic solid (5). Here the direction of stress vector is parallel to the shock-wave propagation. In the figure, the perfect elasto-plastic solid does not loose shear strength above the HEL, and maintains almost constant offset from an isotropic curve. For the toughening solid, the toughness increases, and the material may show nonlinear deformation and high strength. On the other hand, the perfect elasto-isotropic solid catastrophically loses all shear strength above the HEL, and the stress converges to an isotropic state. The small offset from the isotropic curve for a perfect elastoplastic solid is the thermal pressure due to temperature increase. Strictly speaking, the perfect elasto-plastic solid and the perfect elasto-isotropic solid are not expected to exist, because the strength under plastic deformation is usually less than before plastic deformation, and under dynamic conditions even a liquid preserves some shear strength due to the viscosity. The quasi-elasto-plastic solid and quasi-elasto-isotropic solid are situated between the perfect elasto-plastic solid and the perfect elasto-isotropic solid, which loose some differing degrees of shear strength behind the HEL front. The difference between them is a comparative problem. Almost all materials may be included in these types of solid. The strength at the HEL, or in the plastic region, does not always correlate with the strength on release, or with the spall strength, because the mechanisms of the compressive strength and the tensile strength of materials are not necessarily the same. Particularly, many ceramics have high compressive strength, but, low tensile strength under static conditions. As a result, release profile of stress histories of many ceramics do not show the large elastic stress drop which is predicted by a simple model. Consequently, it is difficult to directly determine the shear strength from the elastic stress drop at release.

The differences in yielding property are thought to be caused by the difference in slip mechanisms. Possible slip mechanisms of the shock-yielding solid are cracks (including cleavages, microcracks), dislocations (including dislocation (slip) band, extended dislocation, kink band, stacking fault, etc.) and complex slip zones of cracks and dislocations. The principal differences among them are in the bonding strength (stored deformation energy) and dimension of clusters in the support of shear strength. The preserved shear strength of a crack should be much smaller compared with that of a dislocation. On the other hand, it is known that the stacking faults around impurities sometimes increase the strength of metals. Creep phenomena can not occur under shock compression due to the short duration. In this report, it is suggested that the mechanisms of the shock-yielding phenomena are classified into the following six types: 1. plastic deformation, 2. plastic-brittle zone deformation, 3. brittle destruction deformation, 4. partial-melting deformation, 5. lattice destruction deformation. The author further suggests the additional special mechanism: 6. toughening deformation, which is based on the process zone. This type deformation is discussed later. Fig. 24 shows the model illustrations for the typical slip systems by crack (a), dislocation band (b), complex slip zone (c) and process zone (d). The tentative classification of the mechanisms, slip systems, characterizations, materials, etc. for shock-yielding phenomena, are shown in Table 1.

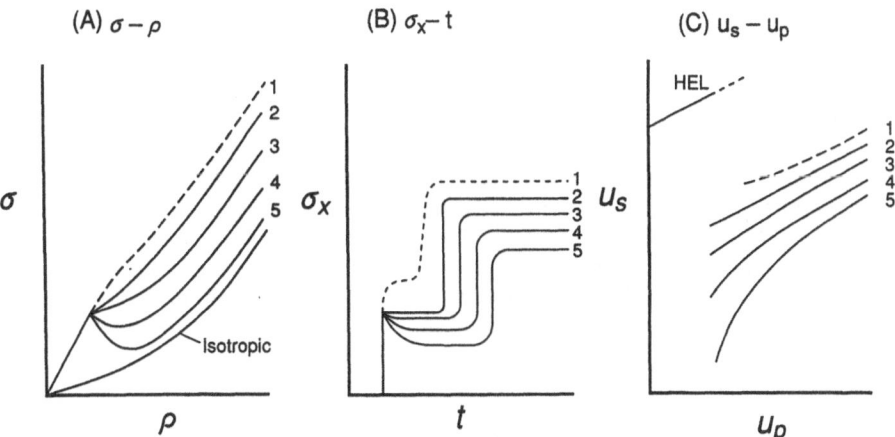

Fig. 23 Model diagram of the σ-ρ Hugoniots (A), stress profiles (B) and Us-Up Hugoniots (C) for toughening solid (1), perfect elasto-plastic solid (2), quasi-elasto-plastic solid (3), quasi-elasto-isotropic solid (4) and perfect elasto-isotropic solid (5). Here the direction of stress vector is parallel to the shock-wave propagation.

Perfect plastic deformation is usually based on dislocations which uniformly multiply and grow. Many metals are classified as elasto-plastic solid. However, all metals do not behave as the perfect elasto-plastic solid, because the deformation is not always uniform. In fact, stress relaxation is also observed in some metals. It has been found that even some polymers and ionic materials behave as elasto-plastic solid in low stress region [18,50]. However, for such low-strength materials, the shear strength can be ignored in sufficiently high pressure region. For comparison, it is also known that most ceramics shows plastic deformation at sufficiently high temperature.

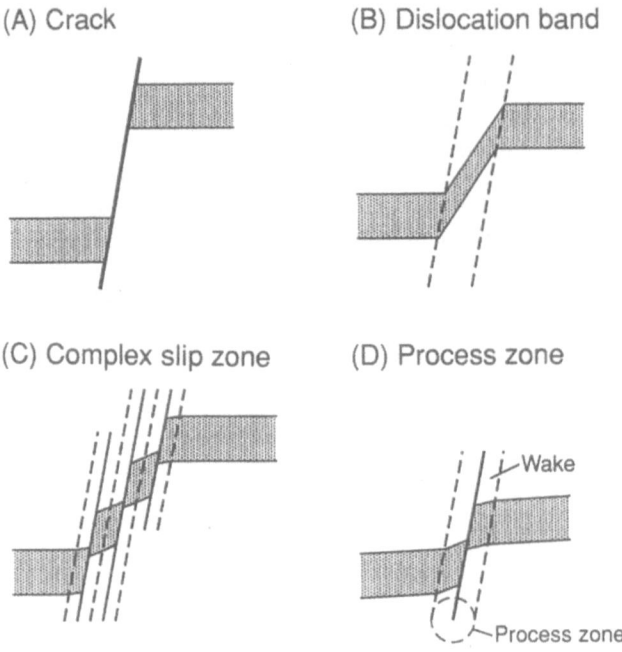

Fig. 24   Model illustrations of the slip systems by crack (A), dislocation band (B), complex slip zone (C), and process zone (D).

Brittle destruction deformation is caused by cracks and cleavages. The plastic-brittle zone deformation is suggested to be caused by the complex slip zones which consist of combined cracks (including microcracks) and dislocations. The complex slip zones are assumed to be formed at the tips of cracks. The cracks or complex slip zones are apt to grow along cleavage plane or the maximum shear stress planes (usually about 45 degrees to the shock propagation direction in continuum material). These types of deformations are attended by a wide range friction resistance between the clusters, which are surrounded by slip zones or cracks, while their interior remains in almostly elastic state. The materials may behave as the quasi-elasto-plastic solid or quasi-elasto-isotropic solid. Many brittle materials may be included into these types. Actually, in the low stress region, cracks may play the dominant role, while in the high stress region the dislocation or complex slip zones may become more significant.

The planar slip structures of the cracks or complex slip zones (including lamellae structure) have been observed in recovered specimens of quartz [51,52], olivine [53,54], periclase [53,55]. In certain minerals, it was pointed out that above the HEL many cracks were observed along certain crystal planes, while below the HEL only irregular cracks occurred in the recovered specimens [56,57]. In quartz, cross lined-shaped luminescences, with directions parallel to the cleavage planes, were observed under shock compression by in-situ observation with an electric framing camera [58]. Similar luminescences was not observed for fused silica [58]. It has been pointed out for some minerals that the greater the driving stress, the higher the temperature increases, the less the deformation cluster size, the less the macroscopic shear strength becomes, and, as a result, a uniform state may be approached [53,59]. This is consistent with the fact that above a certain stress the luminescence of quartz was not observed [58]. In addition, the author expects that at the tips of cracks or slip zones, a process zone can be formed in some brittle materials, such as the transforming-toughening or composite type ceramics. If a process zone is formed as part of the slip mechanism, the toughening behavior may work even under shock compression. Such materials are expected to behave as the toughening solid.

Table 1   Tentative classification of the shock yielding phenomena
(mechanisms, slip systems, characterizations, materials, etc.)

deformation mechanism	slip system	preserved strength	yeilding type	characterization, materials
1. plastic deformation	dislocations (including slip band, extended dislocation, stacking faults, etc.)	large	elasto-plastic	metal
2. plastic-brittle zone deformation	complex slip zones (cracks, micro-cracks, dislocations)	middle	elasto-plastic elasto-isotropic	brittle material (high stress region)
3. brittle destruction deformation	cracks, cleavages	small	elastro-isotropic	brittle material (low stress region)
4. partial-melting deformation	melting zones (and cracks, dislocations)	very small	(perfect ?) elasto-isotropic	low-melting temp. and, low-thermal conductivity material
5. lattice destruction deformation	melting, lattice instability ? vacancies ?	very small	(perfect?) elasto-isotropic	low dense structure material, low-melting temp. material, unstable phase material? material with vacancies?
6. toughening deformation	process zones, microcracks ? stacking faults ?	very large	toughening	composite-type material? transformation-toughening materials

Among the former five deformation mechanisms, the partial-melting deformation and lattice destruction deformation are the most effective in causing the loss of shear strength. In the partial-melting deformation, the slip mechanism is the partial melting of zones along cracks or slip planes, which can not support large shear stresses [59, 60]. The lattice destruction deformation suggested here is caused by a lattice instability or lattice collapse due to melting or structure transition, or by large vacancy densities. The author believes that the partial-melting deformation occurs in low-melting temperature and low-thermal conductivity solids, whereas the lattice destruction deformation occurs in the low-melting, low-density structure or unstable-structure solids. It has been pointed out that silica, some silicates, and some semiconductors, which have comparatively low thermal conductivities, low melting temperatures or low-density structures, behave as elasto-isotropic solids [33, 61, 62]. Amorphous phase regions were observed along slip zones in recovered specimen of silica, olivine, feldspar, etc. [63-65], although it was suggested that these amorphous phases were the reversion products of high-density, short-range-order solids [66]. The CaO-doped cubic zirconia also appears to behave as an elasto-isotropic solid [34]. The author believes that this is related to the large initial vacancy density due to the doping of CaO. It is, however, very difficult to uniquely confirm the slip mechanisms from the experimental results. Probably, several slip mechanisms work together in many cases. Further measurements and observational data are needed to clarify the yielding mechanisms in these high-performance materials.

4.3 Correlation with Some Crystal Sate and Thermal Property

It has been pointed out [59,67] that the thermal conductivity strongly affects the yielding behavior of brittle materials through the thermal shear-banding phenomenon. The author suggested that the chemical-crystal properties should be ranked as a large factor in determining the shock-yielding property [19,68]. However, the important correlations among them have not been well understood. We shall consider correlations among the shock yielding characteristics and the chemical-crystal properties, on the basis of the experimental results. The important properties have been suggested to be chemical-crystal, crystal state (single, polycrystal or amorphous), thermal conductivity, microstructure, porosity, phase transition, etc.

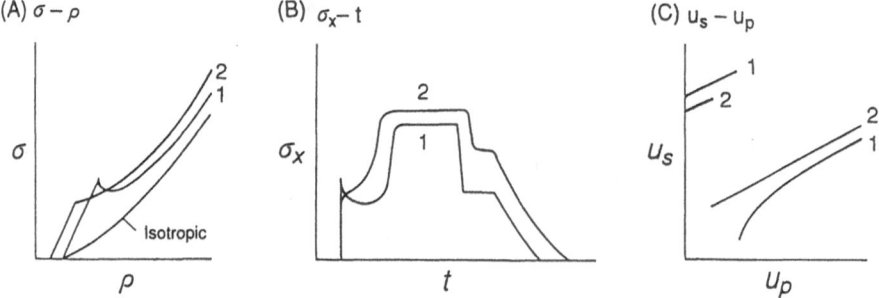

Fig. 25 Model diagram of the σ-ρ Hugoniots (A), stress profile (B) and Us-Up Hugoniot (C) of single crystal (1) and polycrystal (2). Here the direction of stress vector is parallel to the shock-wave propgation.

Some experimental results showed the differences in σ-ρ Hugoniots, or stress history even for the same compound. The crystal state can be identified as an influential factor to shock yielding. Single crystals show high HEL stress and large stress relaxation in contrast to polycrystals. The porosity effects are significant in polycrystal behavior, and usually reduces the HEL stress. Fig. 25 shows the model diagram of the σ-ρ Hugoniots, stress profiles and U_s-U_p Hugoniots of single crystal and polycrystal of high-strength ceramics, respectively. This behavior has been observed in both alumina and zirconia (YCZ). Fig. 26 shows a model illustrating the deformation in single crystal and polycrystal material under shock compression. It is assumed that cleavages or cracks can easily appear or grow along cleavage plane or the maximum shear stress planes in single crystal. However, it is believed that, in polycrystals, the growth, multiplication and motion of cracks or dislocations are disturbed by the grain boundaries, stacking faults and impurities, or by the interactions with each other, and so many microcracks or stacking faults are formed. As a result, the material may become toughened, and the stress behind the elastic shock front increases, as shown in the figure. This mechanism is thought to be similar to the process-zone toughening, which is discussed later. Furthermore, it is suggested that the grain dimension of polycrystal influences even the HEL stress of some ceramics, considering the result that the finer grain-controlled Si_3N_4 ceramics had the higher HEL stress as shown in Section 3.3.

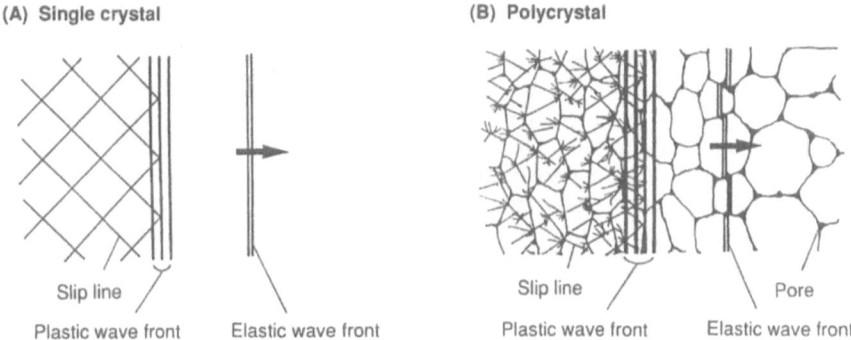

Fig. 26 Model illustrations of the deformation images in single crystal (A) and polycrystal (B) material under shock compression.

For single crystal, the yielding property depends on the shock propagation direction, because of the orientations of cleavage planes. This behavior was observed in the anisotropic HEL stresses of the YTZ single crystal (<100> and <110> axes). However, the author believes that all single crystals of high-strength ceramics do not totally loose shear strength behind the elastic shock front. Because the HEL stress of single crystal is, in many cases, larger than polycrystals, the reduction is more conspicuous. The Hugoniot- compression curves should be analyzed considering the stress relaxation, to more accurately access the Hugoniot offset from the isotropic curve. In contrast, the elastic stress drop on release of single crystal is larger than that of polycrystal. This tendency was observed in alumina, as shown in Figs. 6 and 10. On release, single crystals can macroscopically maintain a considerable shear strength, while polycrystals loose more shear strength in the expansion process. This may be due to the remaining large elastic regions for single crystals, or due to the larger crack or slip zone density for polycrystals.

Crystalline and amorphous materials at the same compound may also show difference in shock-yielding property. It is believed that in amorphous material, dislocations or cracks can not easily move or grow, because amorphous structure has no long-range lattice ordering or cleavage plane. It has been pointed out that fused silica or polymer preserves a degree of shear strength, whereas quartz immediately undergoes collapse to an isotropic state behind the elastic shock front [18,33,69]. Amorphous metals have very high strength and high hardness under static conditions. A useful study would be to investigate the shock-yielding properties of amorphous metals. In addition, impurities or solute atoms may also affects the shock-yielding phenomena, although, at present, the correlation of these features with the strength is confusing. It is generally known that the perfect crystals have high strength. On the other hand, the existence of impurities sometimes increases the strength through stacking faults in metals. Even under shock compression, it is expected that impurities influence the strength of brittle materials, as well as in metals. Vacancies, on the other hand, may reduce the strength, as seen on the CaO-doped stabilized zirconia.

It had been suggested that the heterogeneous shock yielding of brittle materials was mainly caused by high local temperatures due principally to their low thermal conductivities. Recently, Rosenberg et al. measured the stress histories normal and parallel to the shock propagation direction, and showed the pure AlN ceramics behave as elasto-plastic solid [70]. We also measured the Hugoniot properties of several AlN ceramics and confirmed similar result. We have also confirmed phase transformation in pure AlN polycrystal by the inclined-mirror method using the newer streak camera system. As a result, it can be concluded that the heat conductivity is not large factor in determining the shock-yielding property of AlN, because the heat conductivities of AlN ceramics are much greater than those of the other ceramics, and are as large as those of metals [71,72]. However, low thermal conductivity may affect partial melting deformation. It has been pointed out that tungsten, a metal, looses shear strength above the HEL, but, preserves a degree of shear strength in the plastic region [73]. Tungsten is thought to be a critical material, since tungsten is slightly brittle despite of metal, and has high hardness and high thermal conductivity. Diamond has the highest hardness, and has the highest thermal conductivity. It would be useful to investigate the shock-yielding properties of diamond.

Effects of Microstructure and Phase Transition: Process-Zone Toughening?

The yielding under shock compression of brittle materials is assumed to be caused by cracks and dislocations. It is expected that polycrystal materials, whose grain particles are fine, will support a larger shear strength, because the growth or motion of cracks or dislocations is inhibited by the many grain boundaries. For example, the finer grain and denser Si_3N_4 ceramics had higher HEL stresses. Furthermore, it was pointed out that a deformation process zone increases the stored deformation energy, and is responsible for the nonlinear deformation and high toughness under static conditions. Process zones can be caused by phase transformation, forming of microcracks or stacking faults, or the existence of whiskers or fibers, etc., as is illustrated in Fig. 24(D). It is also expected that if process zones are formed under shock compression, the compression profile of the shock wave will become nonlinear in shape, and the HEL stress will increase, as shown in Fig. 23(A) and (B) (type 1). In fact, anomalous free-surface motion and a high HEL stress were observed on SiC-whisker doped Si_3N_4 ceramics (Fig. 21). The author believes that this is caused by the process zones formed at the SiC whiskers, or is related to the phase transition of SiC [74].

The Y_2O_3-doped tetragonal zirconia showed very different shock-wave behaviors from the cubic zirconia, despite being the same basic compound. It was found that the tetragonal zirconia showed an anomalous behaviors within shock front and on release. The shock front indicated a fine three-step structure. It is believed that the tetra-monoclinic, or tetra-orthorhombic (II) transformation probably plays an important role in this behavior. However, at present, it is difficult to determine what types of transition (including yielding) is responsible for the observed wave structure. The author would like to suggest two possibilities: 1. the low transition point corresponds to the beginning of elastoplastic transition (while the yielding is disturbed by the transformation toughening, and as a result, the second shock wave velocity is anomalously high), and the high transition point corresponds to a phase transition, or 2. the low transition point corresponds to the beginning of the tetragonal-orthorhombic (II) phase transition, and the high transition point corresponds to regular elasto-plastic transition with phase transition, respectively. As mentioned in Sec. 3, the recovered YTZ specimens from the shock compression to over 30 GPa were intact with crack surfaces whereas those of alumina, CCZ and YCZ were crushed and looked like a powder. It was also found that a certain amount of monoclinic phase was observed in the recovered specimens from shock compression. In addition, upon release from a relatively low stress shock state of about 13 GPa, the unusually large spall strength was observed. The author suggested that this release behavior was due to the tetragonal-monoclinic transformation, considering the rarefaction shock profile upon release and the recovery experiments from shock stresses below 15 GPa. These facts supported the transformation toughening in YTZ under shock compression. In the transformation-toughened materials, it is believed that process zones are formed at the tips of cracks, slip zone, or dislocations, and they often disappear or their growth is disturbed by the energy absorption or volume increase due to the transformation, as shown in Fig. 24(D).

The author suggests that process-zone toughening under shock compression can be examined by high-resolution measurements of the shock-wave profile, by the observation of process zone wakes (trace) in the recovered specimen, or by measurement of residual strain energy. We have confirmed the effects of the difference in doping or microstructure on the Hugoniot or shock wave profile for the ZrO_2 ceramics, the Si_3N_4 ceramics and also the AlN ceramics. The recovery experiments are now in progress on Si_3N_4 and AlN ceramics. Further experimental results are needed. Additional investigation on ZrO_2 is needed to clarify the yielding mechanism and transitions. Further work is also necessary on other high-performance ceramics, including the nano-composite ceramics and the fiber-doped ceramics to examine other toughening mechanisms.

5. Concluding Remarks

The purpose of our study was to investigate the shock yielding properties and phase transition properties of ceramics including the recently developed high-performance ceramics, and to understand the deformation mechanisms and correlate these mechanisms with physical characteristics. In this report, earlier and more recent shock-wave measurement facilities were described. Next, the studies on several types of alumina, zirconia and silicon nitride were reviewed, primarily putting the yielding behavior into focus. In particular, the crystal state and porosity of alumina, the doping, crystal state and transformation toughening of zirconia, and the grain controlling, doping of whiskers and porosity of silicon nitride were investigated. A tentative classification of shock-yielding phenomena of solids was suggested, and the mechanisms and correlations with material charactristics, including transitions and microstructures were phenomenologically discussed on the basis of the experimental results. The possibility of the process-zone toughening under shock compression was also suggested. In addition, some problems in the experimental and analytical procedures of shock compression study of solids were pointed out.

We have not yet acquired sufficient Hugoniot or observation data to fully understand the shock-yielding phenomena in ceramics. It is expected that shock-wave measurements on newly developed materials and also on the already studied materials using high-quality specimens are to be performed and will yield valuable insight. In particular, ultra-hard ceramics such as diamond, carbides (B_4C, SiC, TaC ...), borides (TiB_2, ZrB_2 ...), and nitride (BN, TiN ...) are expected to be investigated in detail. The study of some of these ceramics have already been started [74-76]. Special metals such as amorphous metals or special intermetallic compounds such as aluminides and titanides are also expected to be investigated. It is important that recently-developed high-performance ceramics, such as the composite- or doping-type ceramics, be investigated to examine process-zone toughening under shock compression.

The measurement of the strain rates is necessary in the investigation of the shock-yielding phenomena. We can directly discuss the slip mechanism by using the shear strain rates as already mentioned, although the measurement of them is not previously possible. The investigation of mean strain rate or stress relaxation has not been sufficiently performed on brittle materials, although significant advances on metals have been made [77,78]. For this propose, high-resolution time-dependent measurements of stress and particle velocity are needed in shock-wave testing of ceramics. In addition, the special dynamic-compression methods, the combined compression-shear shock wave method [79], and the quasi-isentropic plane wave method [80] will provide new and interesting informations on the yielding properties, equation-of-state and phase-transition of ceramics. In particular, the measurement of shear wave, is believed to provide the direct information on yielding. At the same time, parallel investigation of detailed microscopic observations on the recovered specimens will be performed using the recently advanced technologies or facilities, to clarify microscopical yielding mechanisms. Theoretical studies are also expected to advance. The analytic theories of shock-yielding phenomena of brittle materials have been recently suggested [81-83]. The shock yielding phenomena of brittle materials are very complicated due to many factors such as slip systems, nonuniformity, relaxation, thermophysical properties, chemical and microstructural characteristics, and phase transitions including melting. It will be difficult to accurately simulate shock-wave profiles and compression curves of brittle materials, even after addressing these factors.

References

1. *LASL Shock Hugoniot Data* (1980) edited by Marsh SP, University of California, Berkeley.
2. *Compendium of Shock Wave Data 1* (1966) ed. by van Thiel M, University of California, Livermore, California, (NTIS).
3. McQueen RG, Marsh SP, Taylor JW, Fritz JN, Carter WJ (1970) *High-Velocity Impact Phenomena:* ed. by Kinslow R, Academic Press, New York, 244.
4. Al'tshuler LV (1965) *Sov. Phys. Uspekhi. 8*: 52.
5. Ahrens TJ, Gregson VG Jr (1964) *J. Geophys. Res. 69*: 4839.
6. McQueen RG, Marsh SP, Fritz JN (1967) *J. Geophys. Res. 72*: 4999.
7. Abou-Seyed AS, Clifton RJ, Herman L (1976) *Exp. Mech. 6*: 127.
8. Gupta YM (1976) *Appl. Phys. Lett. 29*: 694.
9. Chhabildas LC, Swegle JW (1980) *J. Appl. Phys. 51*: 4799.
10. Mashimo T, Ozaki S, Nagayama K (1984) *Rev. Sci. Instr. 55*: 226.
11. Ahrens TJ, Gust WH, Royce EB (1968) *J. Appl. Phys. 39*: 4610.
12. Goto T, Syono Y (1980) *SCI. REP. RITU. A-Vol 29*: 32.
13. Mashimo T, Nakamura A, Hamada Y (1992) *Proc. 20th Internat. Congress on High Speed Photography and Photonics* (to be publiushed).
14. unpublished data.
15. Keough DD, Wong TY (1970) *J. Appl. Phys. 41*: 3608.
16. Gupta YM (1983) *J. Appl. phys. 54*: 6265
17. Vantine HC, Erickson LM, Janzen JA (1980) *J. Appl. Phys. 51*: 1957.
18. Chartagnac PF (1982) *J. Appl. Phys. 53*: 948.
19. Mashimo T, Hanaoka Y, Nagayama K (1988) *J. Appl. Phys. 63*: 327.
20. unpublished data.
21. Dremin AN, and Adadurov GA (1964) *Sov. Phys. Solid State 6*: 1397.
22. Mashimo T, Nagayama K (1986) *Jpn. J. Appl. Phys. 25, Suppl. 25-1*: 103.

23. Gupta YM, Keough DD, Walter DF, Dao KC, Henly D, Urtiew A (1980) *Rev. Sci. Instr. 51:* 183.
24. Mashimo T (1988) *Shock Wave in Condensed Matter 1987*: ed. Schmidt SC, Holmes NC, North-Holland, 285.
25. McQueen RG, Marsh SP (1960) in *Handbook of Physical Constants*: ed. Clerk SP Jr, Geophysical Society of America, New York, Chap. 7.
26. Pavlovskii MN (1971) *Sov. Phys. Solid State 12*: 1736.
27. Graham RA, Brooks WP (1971) *Phys. Chem. Solids 32*: 2311.
28. Gust WH, Royce EB (1971) *J. Appl. Phys. 42*: 276.
29. Sato T, Akimoto S (1979) *J. Appl. Phys. 50*: 5285.
30. Rosenberg Z, Yaziv D, Yeshurun Y, Bless SJ (1987) *J. Appl. Phys. 62*: 1120.
31. Gieske JH, Barsch GR (1968) *Phys. Stat. Sol. 29:* 121.
32. Mashimo T, Nagayama K, Sawaoka A (1983) *Phys. Chem. Minerals, 9*: 237.
33. Wackerle J (1962) J. *Appl. Phys. 33*: 922.
34. Mashimo T, Kodama M, Nagayama K (1988) *Advans in Ceramics 24*: 329.
35. Mashimo T, Kodama M, Kusaba K, Fukuoka K, Syono Y (1990) *Shock Compression of Condensed Matter 1989*: edited by Schmidt SC, Johnson JN, Davison LW, North-Holland, 469.
36. Mashimo T, Kodama M, Kusaba K, Fukuokja K, Syono Y (1992) *Proc. 18th Internat. Symp. Shock Waves*: ed. Takayama K, Springer-Verlag, 441.
37. Syono Y, Goto T (1980) *SCI. REP. RITU. A-Vol 29*: 17.
38. Arashi H, Private Communication.
39. Garvie RC, Hannink RN, Pasoe RT (1975) *Nature 256*: 713.
40. Mashimo T (1988) *J. Appl. Phys. 63*: 4747.
41. Grady DE, Mashimo T (1992) *J. Appl. Phys. 71*: 4868.
42. Barker LM, Hollenbach RE (1972) *J. Appl. Phys. 43*: 4669.
43. Ohtaka O, Kume S, Ito E (1988) *J. Am. Ceram. Soc. 71*: C-448.
44. unpublished data
45. Nakamura A, Mashimo T, Nishida M, Matsuzaki S (1990) *Proc. 1989 Nationat. Symp. Shock Wave Phenomena*: 145
46. Ogata T, Kihara M, Nakamura K, Kobayashi K (1988) *J. Ceram. Soci. Jpn. 96*: 310.
47. Kamijo E, Honda M, Higuchi M, Yamakawa H, Komura O (1983) *Sumitomodenki 123*: 139 (in Japanese).
48. Mashimo T, Nakamura A, Wakamori K, Miyake M (1990) *J. Soc. Mat. Sci. 39*: 1615 (in Japanese).
49. unpublished data
50. Asay JR, Hicks DL, Holdridge DB (1975) *J. Appl. Phys. 46*: 4316.
51. Stöffler D (1972) *Fortschr. Mineral, 49*: 50.
52. Ananin AV, Brevson ON, Dremin AN, Pershin SV, Tatsii VF (1974) *Combustion Expros. Shock Waves 10*: 426.
53. Muller WF, Hornemann U (1969) *Earth. Planet. Sci. Lett. 7*: 251.
54. Reimold WU, Stoffler D (1978) *Proc. Lunner. Planet. Sci. Conf. 9th.*: 2805.
55. Klein MJ (1965) *Phil. Mag. 12*: 735.
56. Bauer JF (1979) *Proc. Lunar. Planet. Sci. Conf. 10th*: 2573.
57. Mori H (1985) *J. Jpn. Crys. Soc. 27*: 179.
58. Brannon PJ, Konrad CH, Morris RW, Jones ED, Asay JR (1983) *SAND82-2469.*
59. Grady DE (1980) *J Geophys J Res. 85*: 913.
60. Granz AJ (1988) *Phys. Chem. Minerals, 16*: 221.
61. Goto T, Syono Y (1985) J. *Appl. Phys. 58*: 2548.

62. Goto T, Sato T, Syono Y (1982) *Jpn. J. Appl. Phys. 21*: L369.
63. Engelhardt W, Stoffler D (1968) *Shock Metamorphism of Natural Minerals*: edited by French B, Short N, Mono. Press. Baltimore, 159.
64. Jeanloz R, Ahrens TJ, Lally JS, Nord GL Jr, Christie JM, Heuer AH (1972) *Science 197*: 457.
65. Bogdanov AG, Popov S AS, Rundenko VS (1971) *Engl. Transl. Acad. Sci. USSR Proc. Chem. Sect. 201*: 1011.
66. DeCarli PS, Milton DJ (1965) *Science 147*: 144.)
67. Davison L, Graham RA (1979) *Phys. Rep. 55*: 255.
68. Mashimo T (1988) *Shock Waves in Condensed Matter. 1987*: edited by Schmidt SC, Holmes NC, North-Holland, 289.
69. Bless SJ, Brar NS, Rozenberg A (1988) *Shock Waves in Condesed Matter 1987*: edited by Schmidt SC, Holmes NC, North-Holland, 309.
70. Rosenberg Z, Brar NS, Bless SJ (1991) *J. Appl. Phys. 70:* 167.
71. Sumitomo Electric Industri Co. Ltd., Private Communicatins.
72. Horiguchi A, Ueno F, Tsuge A (1986) *Toshiba Review, 44*: 616.
73. Asay JR, Chhabildas LC, Dandekar DP (1980) *J. Appl. Phys. 51*: 4774.
74. Gust WH, Holt AC, Royce EB (1973) *J. Appl. Phys. 44*: 550
75. Gust WH, Royce EB (1971) *J. Appl. Phyts. 42*: 276.
76. Kipp ME, Grady DE (1990) *Shock Compression of Condensed Matter 1989:* edited by Schmidt SC, Johnson JN, Davison LW, North-Holland, 469.
77. Swegle JW, Grady DE (1985) *J. Appl. Phys. 58*: 692.
78. Gilman JJ (1979) *J. Appl. Phys. 50*: 4059.
79. Abou-Seyed AS, Clifton RJ, Herman L (1976) *Exp. Mech. 6*: 127.
80. Steinberg D, Cochran S, Guinan M (1980) *J. Appl. Phys. 51*: 1498.
81. Barker LM, Scott DD (1984) *Sandia Report SAND84-0432*.
82. Sternberg J (1988) *J. Appl. Phys. 65*: 3417.
83. Addessio FL, Johnson TN (1990) *J. Appl. Phys. 67*: 3275.

Chapter 7
The Role of Thermal Energy in Shock Consolidation

M. A. Meyers, S. S. Shang and K. Hokamoto

1. Introduction

Dynamic consolidation has considerable potential for densifying high strength materials which are very difficult to sinter by conventional techniques. Formation of dense compacts requires the collapse of the gaps between the particles as well as considerable amount of energy deposited at the particle surfaces for interparticle bonding. The ultra rapid deformation and energy deposition in shock consolidation produces partial melting at the particle surfaces followed by a rapid solidification via heat conduction into the interior of the particles. A series of attempts have been made by a number of investigators to consolidate these difficult-to-consolidate powders [1-6]. However, there exist two major problems. One is cracking of the compacts at both the microscopic and macroscopic level. The other is a lack of uniformity in microstructure and mechanical properties within the resulting compacts. Three novel approaches have been implemented: (1) shock consolidation of pre-heated specimens; (2) shock densification at a low pressure (just above threshold for pore collapse) followed by hot isostatic pressing (hipping); (3) use of local shock-induced reactions to increase temperatures of particle interfaces and enhance bonding. Fig. 1 shows, in a schematic fashion, these three approaches.

In this chapter we will present an analysis of the energy deposition processes at the shock front (Section 2). The experimental techniques developed by the authors and co-workers to carry out the explosive consolidation and additional thermal energy application are described in Section 3. In Section 4 the experimental results obtained in shock consolidation experiments of nickel-base superalloys [9], titanium aluminides [7, 11-12], silicides [13], and, silicon carbide, diamond and boron nitride [8] are discussed.

2. Energy Deposition During Shock Processing

It is very important to estimate the total energy needed to consolidate a material and to determine the shock parameters required to effect such a consolidation by shock waves. This predictive capability has been obtained, for soft materials, through the energy flux models of Gourdin [14] and Schwarz *et al* [15]. For harder materials, the energy expended in plastic deformation becomes an important component of the overall equation and energy predictions incorporating plastic work have been made by Nesterenko [16] and Ferreira and Meyers [20].

In this section the total energy requirement will be calculated for some typical "hard" materials: SiC, c-BN, diamond, and Ti_3Al. This estimate of the overall energy enables us to establish the minimum shock energy required as well as the energy partitioning, when other sources of energy other than shock energy are used.

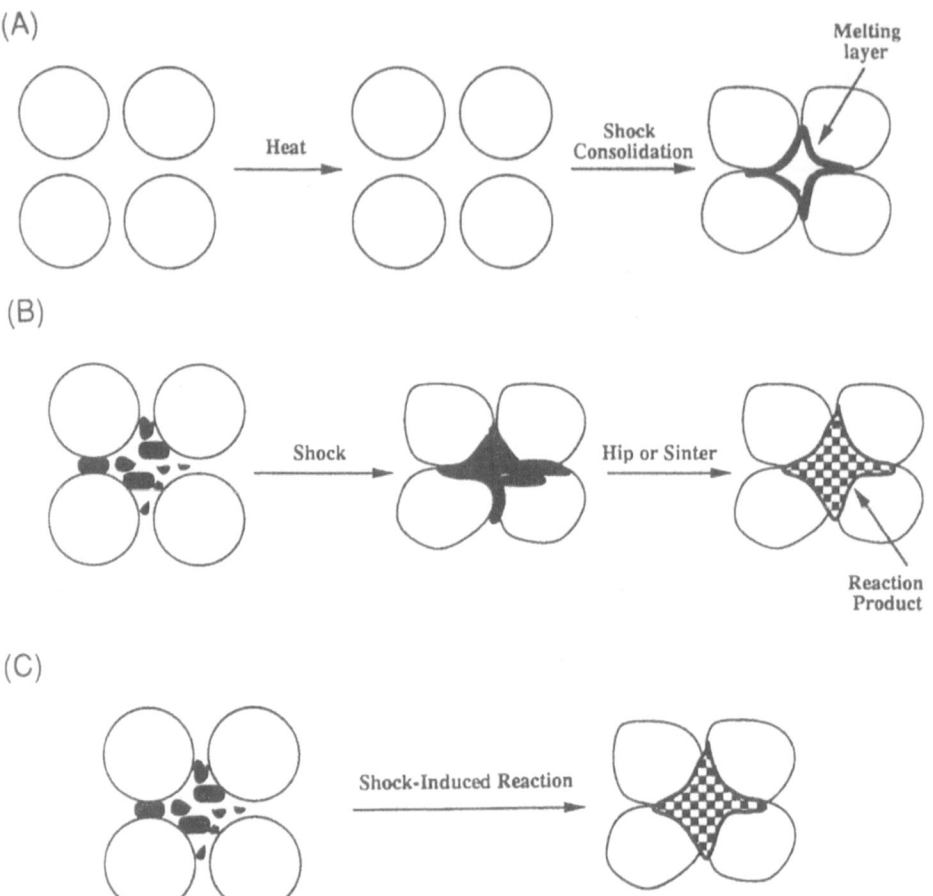

Fig. 1   Schematic representation of three approaches used to improve performance of shock consolidation and minimize cracking: (A) pre-heating of powders; (B) post-shock hip or sinter processing; (C) use of shock-induced reaction to assist bonding.

The discussion presented below is an extension of the work by Schwarz *et al.* [15] and Ferreira and Meyers [20]. Schwarz *et al* [15] only considered the energy required to produce a melting layer of a specified thickness around the particles. This melting energy was equated to the shock energy, and a required shock pressure could be in such a manner established. The expression developed by Schwarz *et al* [15] is:

$$E_m = \left[\overline{C}_p \left(T_m - T_0\right) + H_m\right] L \tag{1}$$

where $\overline{C}_p$ is the average specific heat, H_m is the latent heat of fusion, L is the fraction melted, T_m is the melting temperature of the solid material and T_0 its initial temperature.

However, there are other energy dissipation mechanisms at the shock front other than melting at the particle surfaces, and Fig. 2 shows schematically the various phenomena occurring during shock compaction. These are:

(1) The material is plastically deformed; the collapse of the voids requires plastic flow. A *plastic deformation energy* has to be computed.

(2) The plastic flow of the material is a dynamic process, leading to interparticle impact, friction, and plastic flow beyond the flow geometrically necessary to collapse the voids. We will call this component *"microkinetic energy"*; Nesterenko [16] was the first investigator to consider it.

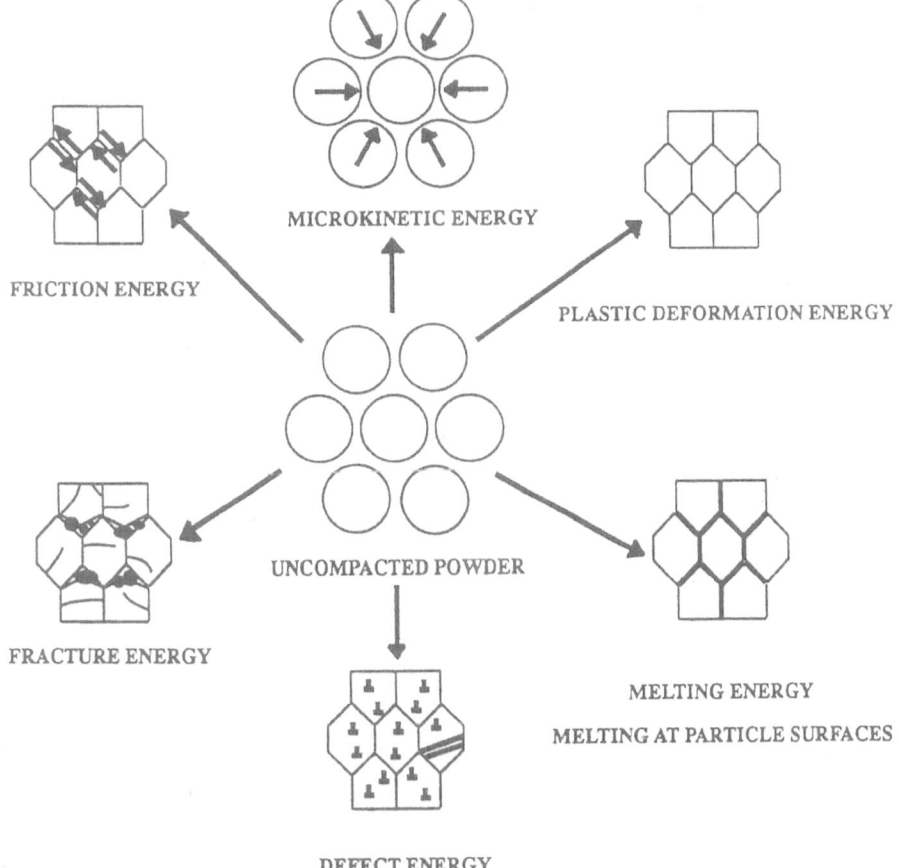

Fig. 2 Various modes of energy dissipation in shock compression of powders.

(3) *Melting at interparticle regions*: It is known that energy is preferentially deposited at the particle surface, leading eventually to their melting. This is the main component of Schwarz *et al* [15] 's model.

(4) *Defect energy*: Point, line, and interfacial defects are produced by the passage of the shock wave; Meyers and Murr [17] provide quantitative assessments of these defect concentrations.

(5) *Friction energy*: The rearrangement of the powders at the shock front requires relative motions, under the applied stress. Thus, friction may play a role in energy deposition at the shock front.

(6) *Fracture energy*: Brittle materials may consolidate by fracturing. The comminuted particles can more efficiently fill the voids.

(7) *Gas compression*: Compaction is most often conducted with the powder being initially at atmospheric pressure. Thus, the gaps between the powders are filled with gas. Shock compaction of the powders compresses and heats these gases. This effect was considered first by Lotrich, Akashi, and Sawaoka [18]. Elliott and Staudhammer [19] demonstrated, by conducting compaction experiments at different gas pressures, the importance of the interparticle gas on shock consolidation.

Some of the energetic contributions given above were used by Ferreira and Meyers [20] to calculate an overall energy expenditure term in shock consolidation. They proposed the simple equation:

$$E_T = E_{v.c.} + E_m + E_d \tag{2}$$

where $E_{v.c.}$, E_m, and E_d are the void collapse, melting, and defect energies. They used the following expressions for the three energy components above:

$$E_{V.C.} = \frac{2}{3} YV_s \{[\alpha_0 \ln \alpha_0 - (\alpha_0 - 1) \ln (\alpha_0 - 1)] - [\alpha \ln \alpha - (\alpha - 1) \ln (\alpha - 1)]\} \tag{3}$$

where a_0 and a are the initial and final distention of the material, Y is the yield stress and V_s is the solid volume. The distention is defined as the ratio between the specific volumes of the porous and the densified materials. This expression is due to Carroll and Holt [21]. For the melting energy, Ferreira and Meyers [20] subtracted the energy involved in the plastic deformation of the material, i.e., part of the energy is expended in plastic deformation and it should not be computed twice. Therefore, Schwarz *et al*'s [15] equation (Eq. 1) was changed to:

$$E_m = \left[\overline{C}_p \left(T_m - T_0 - \frac{E_{v.c.}}{\overline{C}_p}\right) + H_m\right] L \tag{4}$$

The third component of energy considered by Ferreira and Meyers [20] was the dislocation energy:

$$E_d = \left(\frac{Gb^2}{10} + \frac{Gb^2}{4\pi} \ln \frac{\rho_d^{-1/2}}{5b}\right) \frac{\rho_p}{\rho_0} \tag{4A}$$

where G is the shear modulus, b is the Burgers vector, ρ_d is the dislocation density, ρ_p is the powder density, and ρ_0 is the density of the consolidated material. According to the literature [17] one finds that the dislocation density for shock loaded materials in the pressure range where shock consolidation occurs is approximately equal to 5×10^{10} cm^{-2}. The computed value of E_d was found to be negligible.

This energy E_T was equated to the shock energy, E_S, which is equal to:

$$E_S = \frac{P}{2} (V_{00} - V) \tag{5}$$

V_{00} and V are the initial and shock compressed specific volumes, respectively. The relationship between P and V is obtained, for powders, through the equation of state for the solid material and the Mie-Grüneisen equation formulation.

By setting:

$$E_S = E_T$$

it is possible to obtain the pressure required to shock consolidate a powder. The details of these calculations are described elsewhere [20]. The required pressure for shock compaction of different powder materials was calculated at several distentions and the same particle diameter (40 μm), with the hardness values ranging from 1.2 GPa to 98 GPa; from these results one can obtain a master plot. The pressure vs hardness curves were fitted to a line, as shown in Fig. 3.

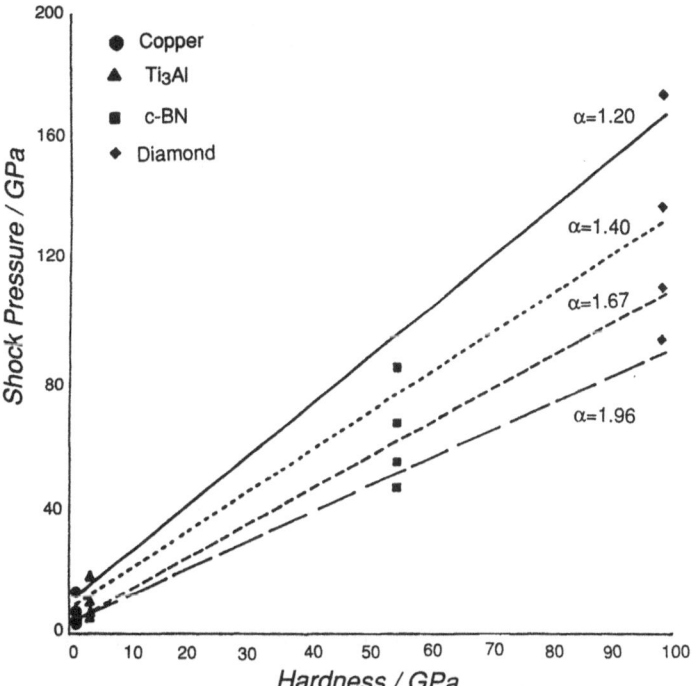

Fig. 3 Pressure required for shock consolidation vs hardness at several distentions (from Ferreira and Meyers [20]) with permission).

The results show that as distention decreases, higher pressure is required to shock consolidate the material. It is also shown that at same melting fraction (same particle diameter) and at same distention, the greater the strength, the greater is the pressure required for shock compaction. Thus, as a first approximation, these calculations can be used as a starting point for prediction of the shock consolidation pressures.

More recently, Shang and Meyers [23] have modified the model proposed by Ferreira and Meyers making it more physically realistic. If one looks at Fig. 2, one can see that the various mechanisms are not independent. Indeed, the production of a molten layer is the direct consequence of the various processes: void collapse, microkinetic energy, gas heating, interparticle friction. Shang and Meyers [23] proposed an alternative method to estimate the void collapse energy and calculated the microkinetic energy separately. In the computation of the void collapse energy they used the model developed by Helle et al [24].

The void collapse energy is strongly affected by particle geometry, particle contact areas and mechanical properties of the particle and adjacent particle. Two major steps in the development of this model are:

(1) Calculation of the forces transmitted to the contact areas.
(2) Calculation of the total contact area as a function of the relative density.
Fischmeister and Arzt [25] infer that the force transmitted to contact areas is

$$f = 4\pi R^2 P/\rho Z \tag{6}$$

where R is the particle diameter, P the applied stress, ρ the current density and Z the coordination number of the powders. When pressure is applied to a powder compact, this causes the total contact area to increase. The total average area of contact per particle is

$$ZA_c = 4\pi \rho \left(\frac{\rho - \rho_0}{1 - \rho_0}\right) R^2 \tag{7}$$

The effective pressure on each particle contact is

$$P_{eff} = f/A_c = 4\pi R^2 P/A_c Z\rho \tag{8}$$

Using equation (7), the contact pressure is given

$$P_{eff} = \frac{P(1 - \rho_0)}{\rho^2 (\rho - \rho_0)} \tag{9}$$

The effect of trapped gas is ignored. The plastic yield condition [26] is assumed to be

$$P_{eff} \geq 2.97 Y_y \tag{10}$$

Y is the flow stress of the material. The external pressure which will cause plastic yielding is

$$P_y = 2.97 \rho^2 \frac{(\rho - \rho_0)}{(1 - \rho_0)} Y_y \tag{11}$$

The total energy required to densify the material is given by

$$E_d = \int_{V_0}^{V_f} P_y \, dV = \frac{2.97 \, Y_y}{1 - \rho_0} \int_{V_0}^{V_f} V^{-2} \left(V^{-1} - \rho_0\right) dV$$

$$E_{v.c.} = \frac{2.97\ Y_y}{1-\rho_0}\left[\frac{(-\rho_f^2-\rho_0^2)}{2} + \rho_f\,\rho_0\right]$$

(12)

Figures 4 (A), (B), (C), and (D) show the external pressure vs specific volume for diamond, c-BN, SiC ad Ti3Al during the plastic yielding stage (application of Eq. 11). The overall void collapse energy done on compacting the powder is given by the area under the pressure-volume curve.

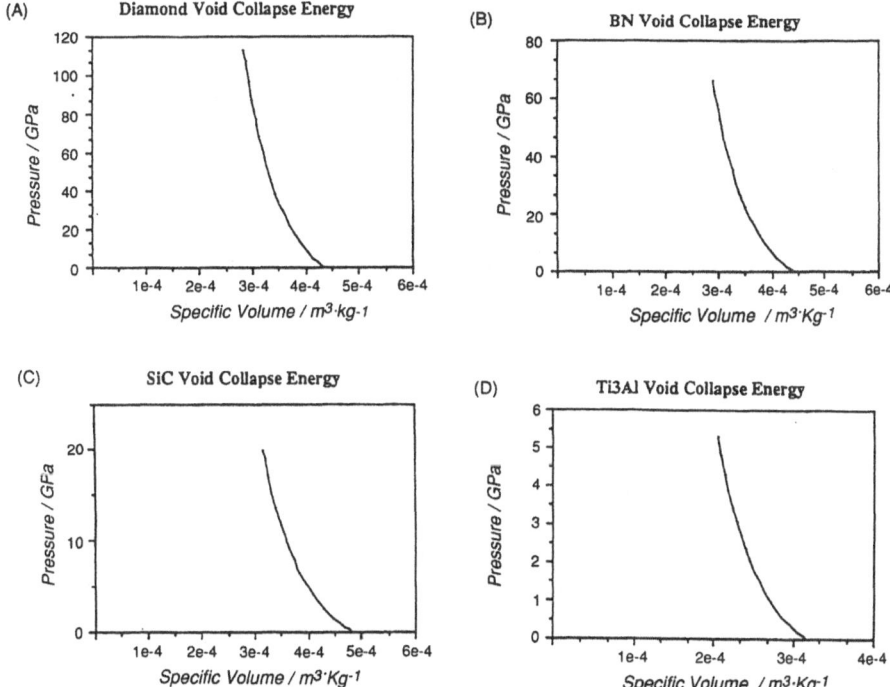

Fig. 4 The external pressure vs specific volume for diamond, c-BN, SiC and Ti3Al during the plastic yielding stage, according to Helle *et al*'s [24] model.

Nesterenko [16] developed a model which describes the relative movement of particles under dynamic compression, as shown in Fig. 5. In this model, the external shell impacts the central core (with radius c) at an impact velocity of V. This impact velocity is estimated to be

$$V = \frac{\text{average relative movement of particles}}{\text{shock rise time}}$$

(13)

For large particle size (50-150 μm), the interstitial void volume is larger than that of small particle size (10-50 μm). Relative movement of particles, upon densification, is greater for large particle size, which causes larger deformation per interstitial site. Furthermore, he defined the dimensions of central core and shell.

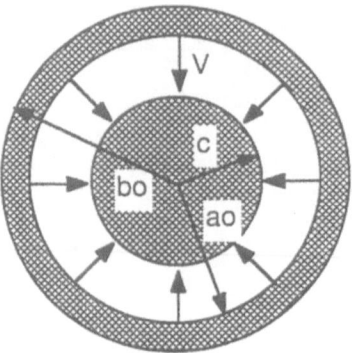

Fig. 5   Carroll-Holt [21] sphere collapse configuration modified by Nesterenko [16] to incorporate micromechanical energy and particle size.

$$\alpha_0 = \rho_s / \rho_p$$

$$a_0 = R$$

$$b_0 = R \, \alpha_0^{1/3}$$

$$c = R \, (2 - \alpha_0)^{1/3}$$

where α_0 is the initial distention and R is the average size. The shock rise time can be estimated from Gourdin's model [14]. The impact velocity is given by

$$V = \frac{a_0 - c}{t} \qquad (14)$$

therefore, the microkinetic energy is given as

$$E_k = \frac{1}{2} \, fV^2 = \frac{1}{2} \, f \left(\frac{a_0 - c}{t} \right)^2 \qquad (15)$$

$$f = \frac{b_0^3 - a_0^3}{b_0^3 - a_0^3 + c^3}$$

Staudhammer and Murr [27] have shown that interstitial void volume is strongly dependent on particle size, especially for micron or submicron powders. Fig. 6 shows particle size versus actual void volume for monosize spherical powders. For fine particles (<10 μm), larger interstitial void volumes for fine particles will result in larger energy deposition, but relative movement of constrained particle will decrease. Therefore, the effective pressure transmitted to contact areas will decrease; this will reduce shock energy deposited at the powder surfaces. At low shock pressure or for powders with higher initial densities (<50 %TD), the shock rise time is longer. From Eq. 15, it can be seen that the microkinetic energy decreases with increasing of the shock rise time. Nesterenko's model correctly describes the lack of melting in these compacts due to insufficiency of relative movement of the particles.

Shang and Meyers [33] considered the deformation, microkinetic, and defect energies and computed shock energies. The pressure can be estimated by equating the shock energy (Eq. 5) with the sum of the energies dissipated (Eq. 4A, 12, and 15):

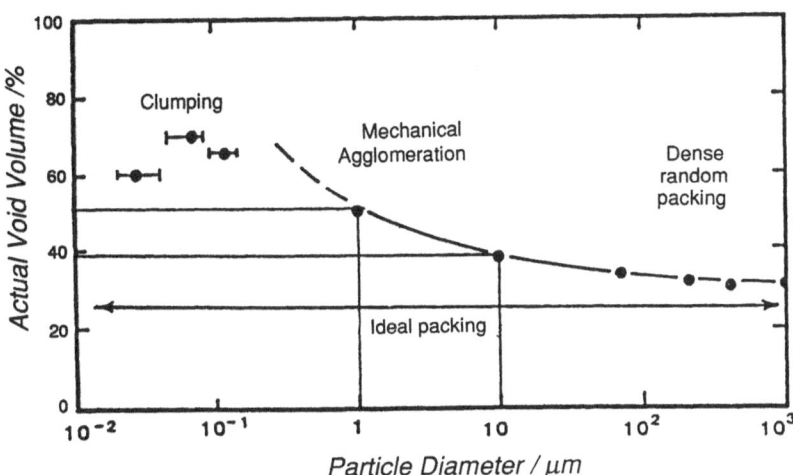

Fig. 6 Particle diameter vs void volume for monosize powders (from Staudhammer and Murr [27]).

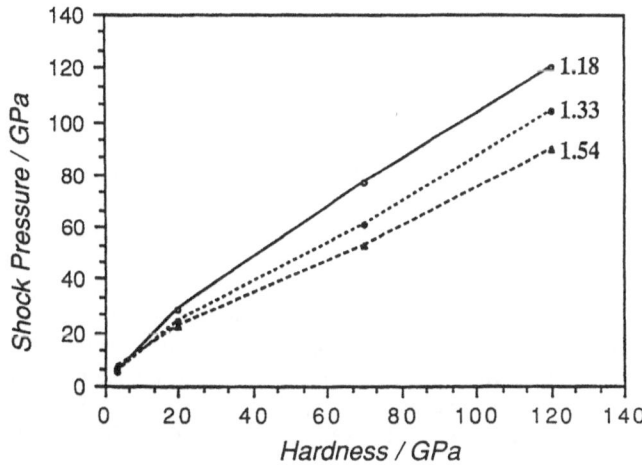

Fig. 7 Pressure required for shock consolidation vs hardness for three distentions according to improved model (from Shang and Meyers [23]).

$$P = \frac{2\rho_0 \rho}{(\rho - \rho_0)} \left[\left(\frac{Gb^2}{10} + \frac{Gb^2}{4\pi} \ln \frac{p_d^{-1/2}}{5b} \right) \frac{\rho_P}{\rho_0} + \frac{1}{2} f \left(\frac{a_0 - c}{t} \right)^2 + \frac{2.97 \, Y_y}{1 - \rho_0} \left(\frac{-\rho_f^2 - \rho_0^2}{2} + \rho_f \rho_0 \right) \right] \quad (16)$$

Fig. 7 shows the predicted pressures required to consolidate materials with different hardnesses incorporating the ideas discussed in the previous paragraphs (Shang and Meyers [23]). The calculations were conducted for four materials: Ti_3Al, SiC, BN, and diamond.

Three different distentions (1.18, 1.33, and 1.54) were used. It can be seen that the pressure required is decreased as the distention is increased. These predictions are fairly close, but somewhat lower than the ones obtained by Ferreira and Meyers [20].

In the case of pre-heating or shock-induced reactions, these energetic terms can be incorporated into the overall computation. Thus, the shock energy can be decreased by non-shock energy terms.

3. Experimental Techniques

A wide range of experimental setups have been used over the past thirty years to shock consolidate powders. These systems are grouped into two classes:
   1 - explosively driven systems
   2 - gas/propellant driven systems
The authors will restrict themselves to the description of the systems that they have used. For detailed descriptions of other systems, the works of Graham [28], Thadhani [29], Korth [30], Vreeland [31], Ahrens [32], Sawaoka [33], Prümmer [34], Staudhammer and Murr [27] should be consulted.

Explosively generated cylindrical and planar systems were used and/or developed at the Center for Explosives Technology Research, New Mexico Tech, Socorro, New Mexico, by Meyers and co-workers [7-13, 35, 36, 38].

3.1 Cylindrical system

The cylindrical system is the most common geometry for shock consolidation because of the ease with which set-ups can be developed. This system has been used by Meyers and co-workers since 1981 [35]. In Europe, it has been used since the 60's. A detailed description is given by Prümmer [34]. A modification of the cylindrical system has been developed by Meyers and Wang [36] in 1988. Fig. 8(A) shows the basic components of this system. The powder is contained in the internal tube. The external tube is surrounded by the explosive charge, which is detonated at one end; this external tube acts as a flyer tube, impacting the internal tube. This technique generates pressures in the powder that can be several time higher than the ones generated by the single tube technique. Thus, low-detonation-velocity explosives can be used to consolidate hard powders, and the formation of the Mach stem is decreased. Significant improvements in compact quality were obtained in nickel-base superalloys [9], titanium alloys [7, 11, 12], and aluminum-lithium alloys [13].

The radial cross-sectional view of the system employing the double tube configuration is shown in Fig. 8(B). By applying an analysis similar to that developed by Gurney [37] for the velocity of fragments, one can estimate the velocity at which the flyer tube is accelerated inwards by the explosive. The chemical energy of the explosive is equated to the sum of the kinetic energy of the gases and that of the tube. One obtains the following equation:

$$V_p = \sqrt{2E} \left(3 / \left[5(m/c) + 2(m/c)^2 \cdot \frac{R + r_0}{r_0} + \frac{2r_0}{R + r_0} \right] \right)^{1/2}$$

(17)

where V_p is the velocity of the flyer tube and m/c is the ratio between the mass of the flyer tube and the mass of the explosive charge. R and r_0 are shown in Fig. 8(B). The importance of this equation is that it allows the selection of a pre-established pressure in the powder. The shock pressure is directly related to the impact velocity.

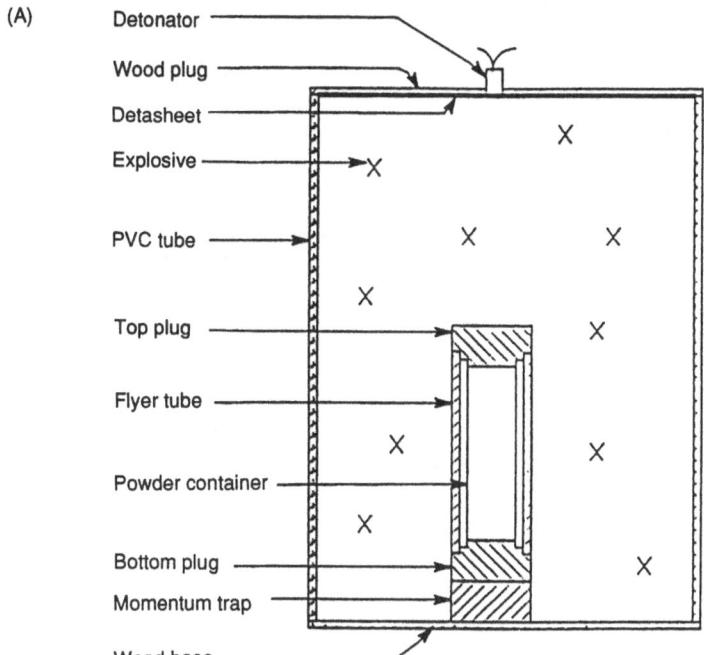

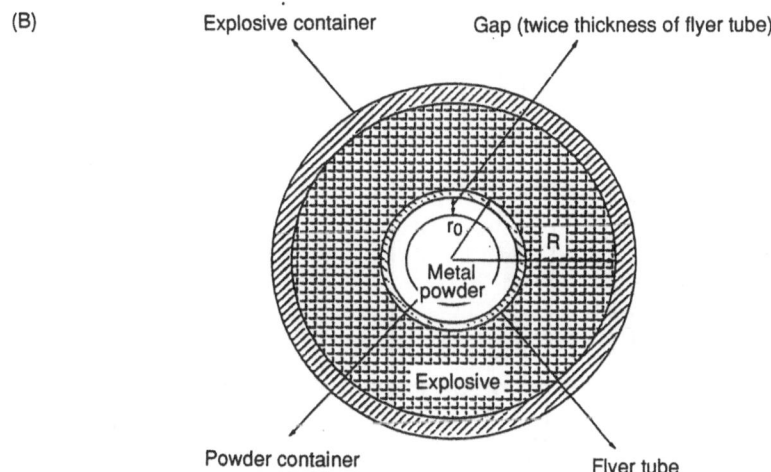

Fig. 8 (A) Cylindrical shock consolidation system using flyer tube and (B) cross-section of system.

A schematic of the high temperature test device is illustrated in Fig. 9(B). The heated powder container is dropped into a mild steel pipe (which acts as the flyer tube) by the release of a solenoid switch). The bottom portion of the mild steel flyer tube (equivalent to the height of the powder container) is surrounded by the explosive charge in a PVC pipe. The charge is initiated at the bottom using a detonator. Prior to the test, the powder container (loaded and sealed under argon atmosphere) is heated in a furnace at the desired temperature and loaded in the test device. Proper precautions are taken to minimize the temperature loss in the powder container by firing the charge in less than three minutes after the hot sample is removed from the furnace. The capsule is propelled upwards by the detonation at the bottom and has to be trapped. An armor plate standing on four support legs (Fig. 9(A)) and having the bottom surface protected by sand bags has been successfully used. This system was developed by Wang et al. [9].

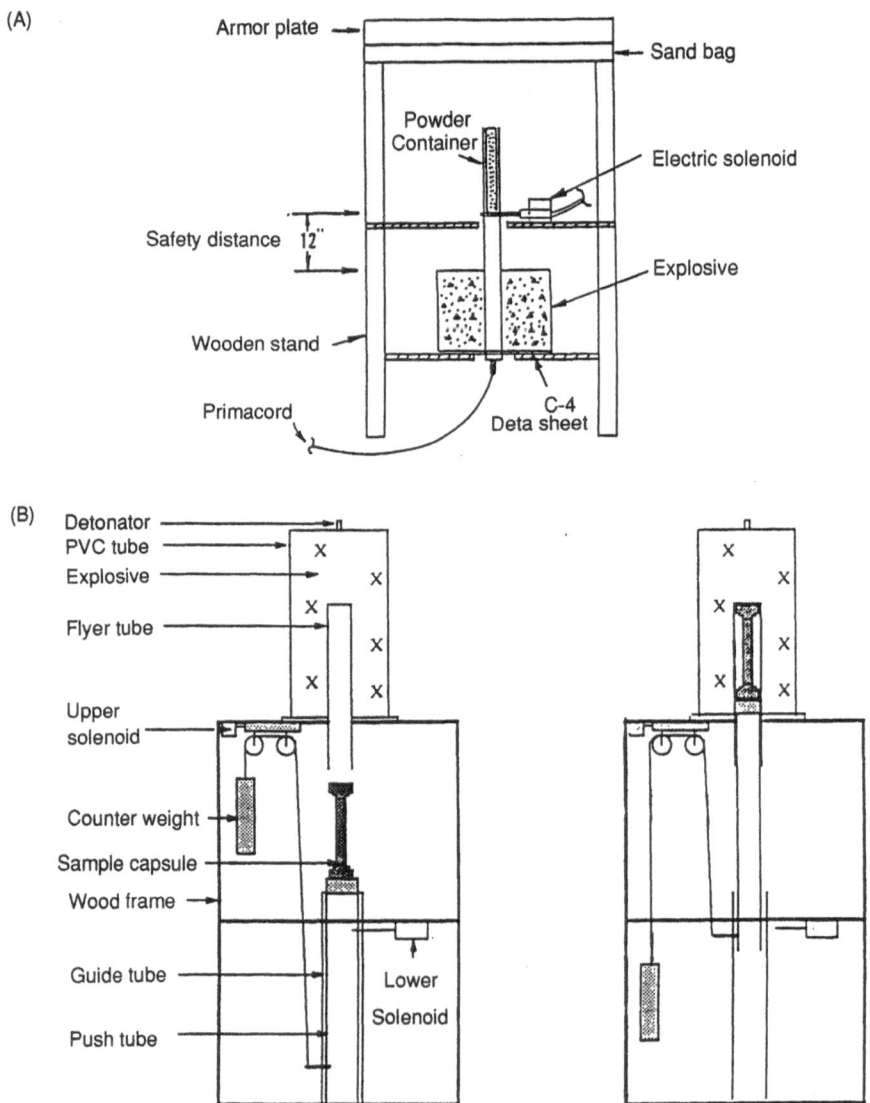

Fig. 9 High temperature cylindrical shock consolidated systems: (A) system used by Wang, Meyers, and Szecket [9] and (B) system used by Ferreira et al. [11] (from [11] with permission).

A more elaborate system, with enhanced safety features and in which detonation is initiated at the top, is shown in Fig. 9(B). In this system, the hot capsule containing the powder is placed below the explosive charge. A push-tube activated by a counterweight raises the capsule until it penetrates the explosive charge (Fig. 9(B)). The system is activated by a solenoid that releases the push-tube. In case of emergency or failure to detonate, a second safety solenoid drops the counter weight and the capsule is freed to drop from the explosive. This eliminates the possibility of accidental exposure of the explosive to high temperatures for prolonged times. This system was developed by Ferreira et al. [11].

3.2 Sawaoka system

Sawaoka and co-workers [1, 2, 33], in Japan, developed a planar impact system using a flyer plate in a planar parallel impact against a target in which capsules with the powders are placed. Very high impact velocity experiments can be conducted for consolidation of certain very hard and difficult-to-bond materials like ceramics. The Sawaoka fixtures utilize explosively driven plates for impact, to generate very high shock pressures (in the range of about 20-100 GPa) [1, 2].

The set-up employs stainless steel flyer plates impacting stainless steel capsules at velocities of 1.5-3.0 km/s. The fixture configuration and the explosive assembly is shown in Fig. 10. The flyer plate is accelerated downwards by the detonation of an explosive charge resting on its top. The explosive charge is initiated simultaneously over the top surface by an explosive system called "mousetrap assembly" or by a plane-wave lens. These set-ups create a plane wave that initiates the detonation at the top surface simultaneously. Up to 8 or 12 individual capsules containing the powder can be utilized, and are held in cavities in the recovery fixture. The shock pressures and temperatures within these capsules are quite inhomogeneous because the powder has a different shock impedance and velocity than the capsule. Computer simulations are used to obtain calculated pressures and temperatures within the capsules [28, 38].

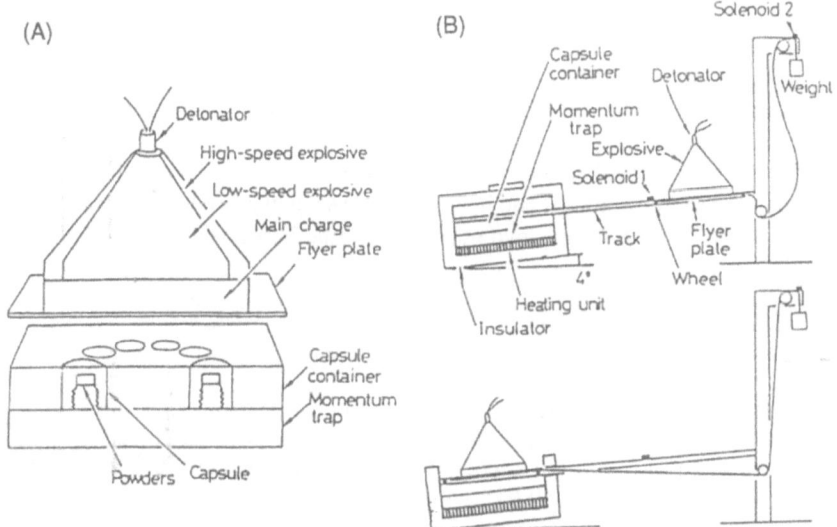

Fig. 10 Sawaoka planar shock consolidation system: (A) section of system showing flyer plate acceleration setup with explosive lens and cross-section of capsules and tooling and (B) high-temperature adaptation of system.

A modification of this system for high-temperature consolidation was developed by Yu and Meyers [38]. The set-up is shown in Fig. 10(B). The capsules and surrounding tooling are placed in a discardable furnace, which is heated to the desired temperature. The main explosive charge and plane-wave generator, resting on the flyer plate, are placed on a cart that can descend a 4° ramp, once the solenoid is activated. When the main charge arrives to its final destination it can be detonated. If any problem arises, the charge can be rapidly withdrawn by the activation of solenoid 2, which is a significant safety feature. Pre-shock temperatures as high as 970 K have been achieved by this method. For these high temperatures, the capsule material has to be a nickel-base superalloy instead of stainless steel.

4. Consolidation Experiments: Results and Discussion

Section 2 describes, in a quantitative manner, the energy expenditure in shock consolidation. The problems encountered with shock consolidation are primarily cracking and incomplete bonding. Fig. 11(A) shows plots of critical flaw size as a function of tensile stress level for materials having different fracture toughnesses. These plots were made using the well known fracture mechanics equation: $\sigma = K_{IC} / \sqrt{\pi a}$. It is difficult to conceive a shock consolidation process in which no flaws are left, and the particle size is a good indicator of the inherent flaw size in a shock consolidated material. The three fracture toughnesses given, 5, 50, and 100 MPa $\sqrt{m}$, are characteristic of brittle (ceramics), tough (steel, titanium alloys) and very ductile materials (copper, nickel), respectively. Fig. 11(B) shows the critical tensile stresses for 25 mm and 10 mm particle sizes as a function of the compressive stresses needed to consolidate the respective powders. The compressive stresses were taken from the calculations given in Section 2 (Eq. 16), at a distention corresponding to an initial density of 65% of the theoretical density (this is a typical value for powders). Tensile stresses due to reflections are always present in shock consolidation systems. Additional sources of stresses are residual stresses due to the loading inhomogeneity and the temperature gradients during cooling. The amplitude of the reflected tensile stresses can be as high as the compressive stresses. However, in well designed systems a significant portion of the tensile stresses is trapped. When the tensile stresses exceed the critical tensile stresses for the specific material, failure occurs; this is shown in Fig. 11(B) in a schematic fashion. A realistic line shows $\sigma_t = 0.1\sigma_c$, i.e., the tensile reflections have, at most, an amplitude of 10% of the compressive pulse. Thus, the hatched region is termed "Danger Zone". By reducing the shock amplitude it is possible to translate the curve downwards, and point $\boxed{A}$ (corresponding to a hypothetical material) is changed to $\boxed{B}$. Thus, there are two approaches to be implemented for improved shock consolidation.

  (a) reduction of tensile stresses
  (b) reduction of shock energy

This chapter deals exclusively with methods used to decrease the shock energy necessary for consolidation. Shock energy is replaced by thermal or chemical energy in an effort to improve compact quality. In this section, results obtained by the authors using the three approaches delineated in Fig. 1 will be discussed.

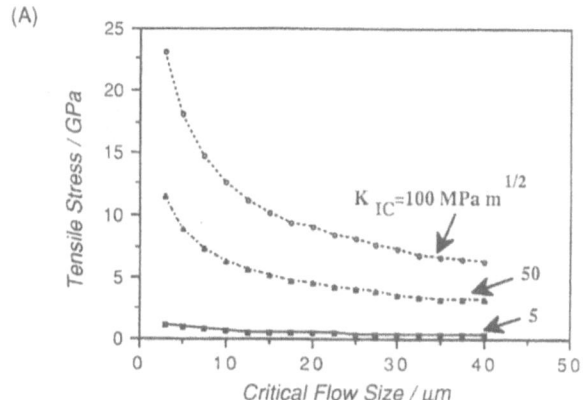

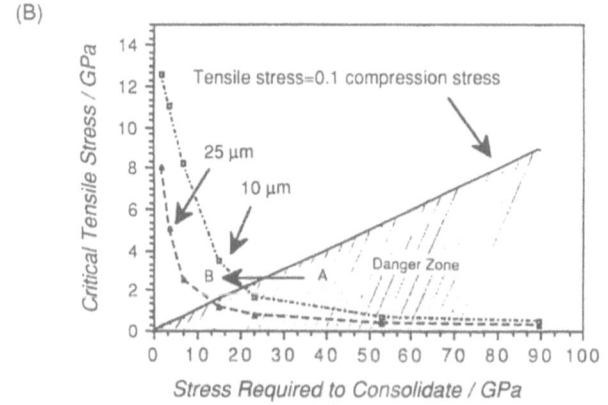

Fig. 11 (A) critical flaw size as a function of tensile stress for materials with different fracture toughnesses and (B) variation of critical tensile stress for 10 and 25 μm flaw activation with shock stress required for consolidation: notice danger zone.

4.1 Hot shock consolidation

Wang, Meyers, and Szecket [9] developed a high temperature shock consolidation system, shown in Fig. 9(A), for nickel-base superalloys. Earlier experiments by Meyers, Gupta, and Murr [35], Meyers and Pak [39], and Wang, Meyers, and Graham [40] had yielded good bonding but a great deal of cracking, using a single tube cylindrical geometry, or the Sandia calibrated fixtures Bear series [40]). High-temperature shock consolidation experiments were conducted and compared with room-temperature experimental results. As is expected from energetic considerations, the melting fraction increases with temperature, at a constant pressure. The volume fraction of interparticle molten (and resolidified) regions could be established for shock consolidation experiments conducted at different temperatures and pressures, because they etched differently. The results are shown in Fig. 12. At a fixed pressure, the molten fraction increases with initial temperature. This is due to both the decrease in flow stress of the material with temperature and to the non-shock energy component provided by pre-heating the powders. It was possible, by using preheating, to obtain good consolidation of the superalloy with absence of cracking. The best mechanical properties were obtained by pre-heating the specimen to 525°C and using a shock pressure of 18 GPa, resulting in approximately 25% apparent interparticle melting. Tensile tests were conducted and typical results are displayed in Fig. 13. Aging of the alloys subsequent to shock consolidation further increased the yield and ultimate tensile strengths.

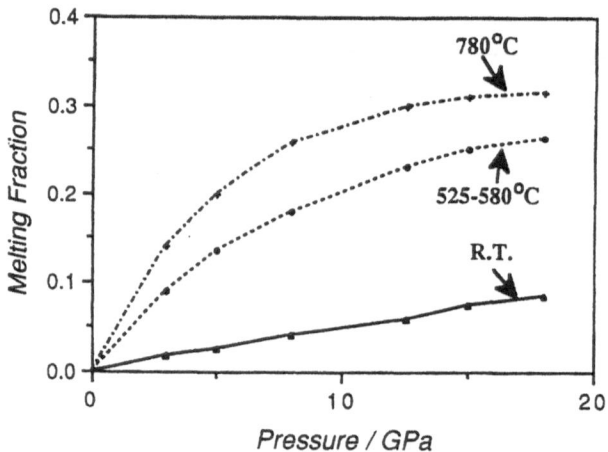

Fig. 12 Shock-induced melting (as a fraction of total volume) as a function of shock pressure for IN718 superalloy powder at three different temperatures (from Wang, Meyers, and Szecket [9] with permission).

Ferreira *et al.* [11, 12] shock consolidated titanium aluminides at ambient and high temperatures, using the experimental set-up shown in Fig. 9(B). Pre-shock temperatures were 600°C, 750°C, and 900°C. Two basic compositions produced by rapid solidification rate were used: Ti-21 wt pct Nb-14 wt pct Al and Ti-30.9 wt pct Al-14.2 wt pct Nb. These powders exhibit the basic structures of Ti_3Al and TiAl, respectively. The increase in pre-shock temperatures led to a decrease in cracking, in some cases. Fig. 14 shows the macrocrack surface area per unit volume for the various conditions. Ti_3Al undergoes a BCC-tetragonal transformation upon heating and this embrittles the original ductile microstructure retained by rapid solidification. This example illustrates the fact that by increasing the pre-shock temperature one can also increase the propensity for cracking.

Diamond, cubic boron nitride and silicon carbide powders were explosively consolidated at high temperature by using a planar impact system (Fig. 10). Hot shock consolidated materials exhibited a decrease of surface cracks as compared with specimens consolidated at room temperature. The consolidated diamond specimen showed evidence of surface melting of the particles by small crystallized grains generated during cooling of this molten layer. 4-8 mm diamond consolidated showed good bonding. Excess cracks and generation of graphite were observed by the increase of particle size. Hot consolidated diamond admixed with graphite showed the highest hardness value, but the improvement of the bonding was not confirmed by the specimen admixed with silicon powders. Hot consolidated 40-50 µm c-BN showed good bonding compared with the specimen which has small particles. The improvement of the bonding of c-BN powders by the addition of graphite could not be recognized in this investigation [8]. These results are also described by Shang, Hokamoto, and Meyers [8]. A detailed description is presented in the three following sub-sections.

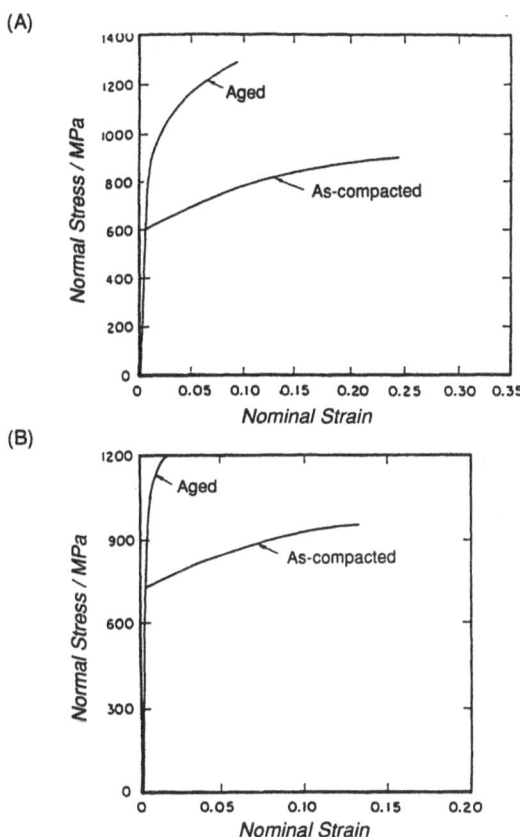

Fig. 13 Tensile stress-strain response of IN718 powders shock consolidated at (A) 740°C and (B) 525°C in the as-consolidated and consolidated plus aged conditions (from Wang, Meyers, and Szecket [9] with permission).

4.1.1 Characterization of Silicon Carbide

Two different sizes of pure SiC powders (7 and 44 μm) were used as starting materials in the high temperature shock experiments. Well consolidated samples were obtained upon shocking preheated SiC powders with 65% of the theoretical density at an impact velocity of 1.9 km/s. The shock pressure was calculated to be 12 GPa by using the one-dimensional impedance matching technique. However, cracking is not totally eliminated and circumferential cracks are noticeable. Scanning electron microscopic (SEM) analysis of recovered SiC compacts revealed that SiC particles underwent a considerable amount of deformation and the material was fully densified.

Fig. 15 shows the microstructure of a fracture surface of 7 μm SiC compact. The fractograph indicates that cracks are transgranular and this is evidence for excellent bonding of the powders. The relative densities of the 7 and 44 μm SiC compacts were 98.8% and 98.5% of theoretical, respectively. Both samples had a very close average microhardness value of 28 GPa. Transmission electron microscopic (TEM) analysis (Fig. 16) revealed the substructural features of the 44 μm SiC compacts. We can observe lattice distortion in the interior of SiC particle, indicating that the particle was heavily deformed (Fig. 16(A)). Fig. 16(B) shows the interparticle region. The dark round regions (8-80 nm) in the micrograph are nanocrystalline SiC particles, while the white surrounding material corresponds to the amorphous phase. This interfacial layer was formed due to the very rapid collapse of the gaps between the particles as well as the rapid deposition of energy upon interparticle sliding. Furthermore, this thin interfacial layer (0.2-0.3 μm) and high thermal conductivity of SiC (13 W/m°C) also promote the tendency for these nanocrystalline structure and amorphous phase. The energy deposition time is on the order of the wave transit time through the particle (a few nanoseconds).

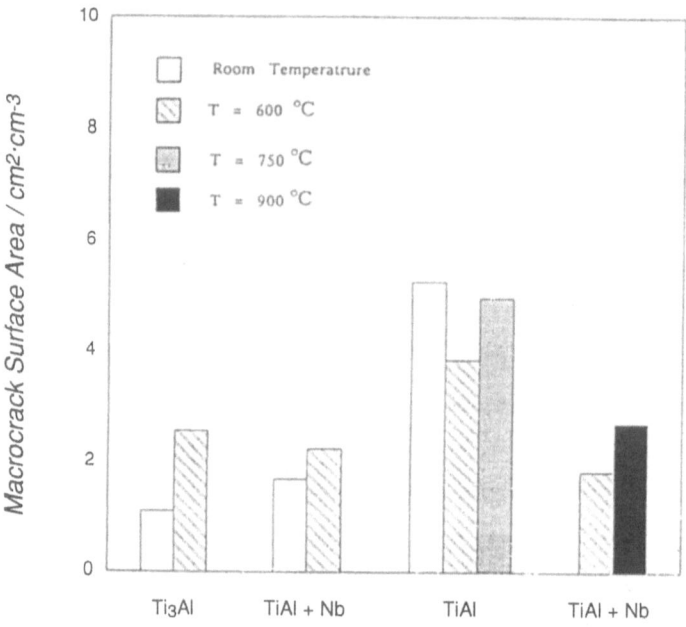

Fig. 14 Effect of initial temperature on the average crack density of compacts (from Ferreira *et al* [11]).

Fig. 15 Scanning electron micrograph of fracture surface of consolidated 7 μm SiC.

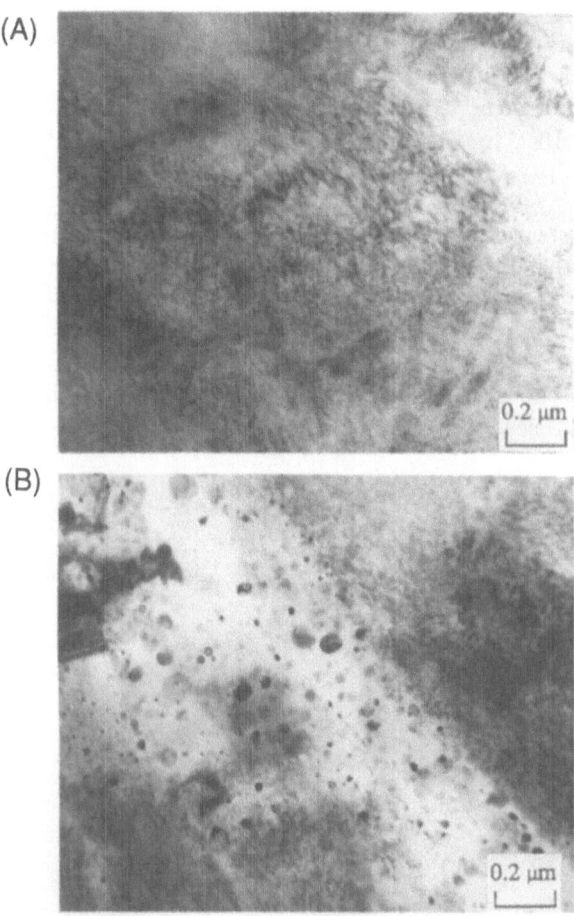

Fig. 16 Transmission electron micrograph of consolidated 325 mesh SiC showing (A) lattice distortion in the interior of SiC particle and (B) the surface layer in the interparticle region.

The melted or sublimated interparticle material resolidifies at cooling rates as high as 10^5 to 10^{10}°C/s [41]. It is believed that amorphous SiC phase forms at higher cooling rate than nanocrystalline structure and that the crystalline phase nucleates within the amorphous region while it is being cooled. However, another possibility is that melted or sublimated SiC reacted with oxygen to form $Si_xC_yO_z$ or SiO_2 which also has the tendency to form an amorphous phase.

4.1.2 Charcterization of Diamond

Three different sizes of natural diamond powders (4-8, 10-15, and 20-25 μm) were used as starting materials. The high-temperature planar impact system (Fig. 10(B)) was used for consolidation. The diamond samples were compacted at impact velocities of 1.2 and 1.9 km/s. The samples are well consolidated and the microhardness values of samples are in the range of 20-30 GPa. Diamond particles are very brittle and are subject to high bending moments due to their highly irregular shape. The number of cracks generated in the consolidated diamond powders tends to increase with an increase in the particle size of the starting diamond powders [1]. Generation of small amounts of graphite was detected by x-ray diffraction in the compacted 10-15 and 20-25 µm diamond powders. The transformation to graphite usually occurs in larger particle size specimens consolidated at room temperature. For larger particle size (10-25 µm), the interstitial void volume is larger than that of the small particle size (4-8 µm). Relative movement of constrained particles, upon densification, is greater for large particle size, which causes larger deformation per interstitial site. Therefore, larger interstitial void volumes will result in larger shock energy deposited at the powder surfaces and cause phase transformation. Furthermore, this transformation is also enhanced with the help of preheating. Scanning electron micrographs from the interstitial melting areas of the compacted 20-25 µm diamond powders is shown in Fig. 17. Small round diamond particles (0.5-1 µm) were observed in this sample. These small diamond particles are believed to be produced during the resolidification of the melting layer. Thicker surface layer (2-4 µm) were generated due to large particle size.

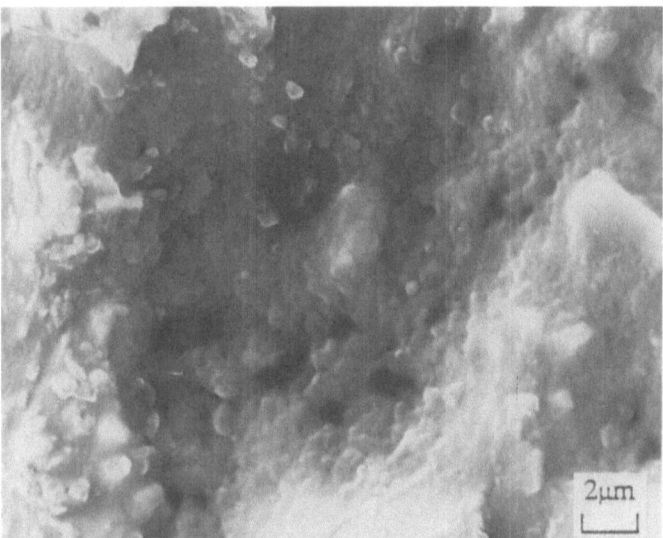

Fig. 17 Scanning electron micrograph of the melting layer after the resolidification from 20-25 µm diamond.

The recovered samples of compacted 4-8 µm diamond/C (15 wt%) and 4-8 µm diamond/Si (7.5 wt%) mixture were also examined. Deformation of the softer component graphite decreases the pressure concentration and fills the pores between diamond particles. X-ray diffraction showed that part of the graphite powders was transformed to diamond. This is consistent with theoretical prediction [32] and results obtained by Potter and Ahrens [43]. These investigations showed, in room temperature experiments, that graphite transformed into diamond at pressures of 15-22 GPa. The consolidated diamond/C sample shows an extremely high microhardness value: >55 GPa. However, the recovered diamond/Si sample is not very well consolidated. We expected the heat generated from reaction between silicon and carbon to enhance the bonding, but the reaction did not occur. Furthermore, the addition of silicon will reduce the effective pressure transmitted to contact areas between diamond powders due to poor constraint. This resulted in poor consolidation.

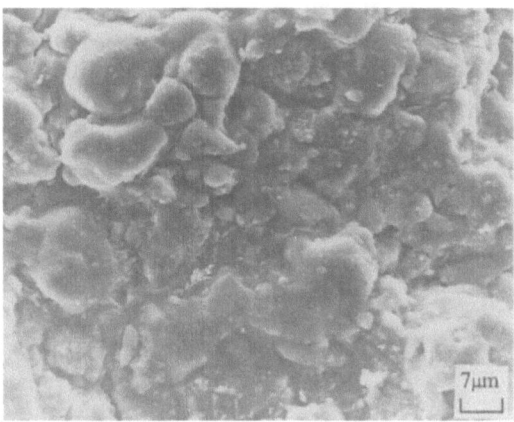

Fig. 18 Scanning electron micrograph of fracture surface from consolidated 10-20 μm c-BN.

4.1.3 Characterization of c-BN.

The recovered c-BN (10-20 and 40-50 μm) samples are very well consolidated. Fig. 18 is a fracture surface of 10-20 μm c-BN which shows the melting of particle surfaces. This transgranular fractograph suggests that the bonding of the particle is excellent. The morphology of the c-BN is polyhedral. This particle geometry enhances the constraint between the particles during void collapse stage and results in larger energy deposition at the powder surfaces. Therefore, the microhardness value of the samples can reach 61 GPa. A highly angular particle shape would tend to decrease surface temperatures due to (1) poor constraint between particles and (2) generation of microcracks in the particle reducing the energy dissipation on the surface.

C-BN admixed with C (15 wt%) and a mixture of Ti (11 wt%) and C(9 wt%) was also consolidated. The graphite powders are severely deformed and uniformly distributed on the surfaces of the c-BN powder after shock consolidation. However, the cooling rate was not sufficiently high for the transformation of the graphite into diamond. The remaining graphite causes the degradation of the consolidated c-BN sample. Recovered c-BN admixed with Ti and C composite showed low microhardness value. The exothermic reaction of Ti and C did not take place during the passage of the shock wave. The remaining Ti and C caused the degradation of the sample.

4.2 Shock Consolidation Followed by Annealing or Hot Isostatic Pressing

Coker et al [10] shock consolidated titanium alloys at room temperature and followed this treatment by HIP. This resulted in significant microstructural changes and improvement of mechanical properties. Fig. 19(A, B) shows typical microstructures after shock consolidation and hipping (Fig. 19(C)). The Ti-6%Al-4%V alloy powder had a profuse dispersion of erbia and responded in a heterogeneous way to shock-wave passage. Whereas the interiors of the particles exhibited a lath structure (resulting from a martensitic phase transformation) with a high density of dislocations and deformation twins, the interparticle regions were characterized by a small grain size and a low dislocation density. The latter features are shown in Fig. 19(B) and are the result of interparticle melting. Hot isostatic pressing annealed most of the dislocations; Fig. 19(C) shows the resultant microstructure, with the characteristic erbia stringers. Hipping has, as a primary objective, the healing of the cracks that remain after the shock process. Fig. 20 shows the yield stress and total elongation for shocked and shock + hipped titanium. The hip process enhanced the ductility of titanium considerably, while only decreasing the yield stress by approximately 10 pct. Conventionally forged material is also presented, for comparison purposes.

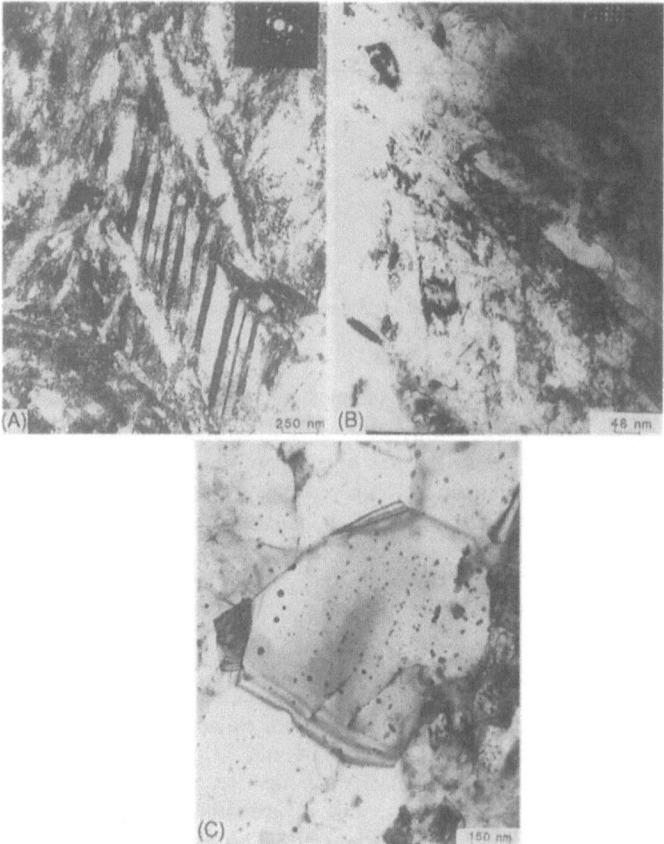

Fig. 19 Transmission electron micrograph showing (A) and (B) as-shocked and (C) shocked + hipped microstructure in a titanium alloy (from Coker *et al* [10] with permission).

In an effort to eliminate cracking in Ti₃Al, elemental Ti and Al were added to the Ti₃Al powders and the mixture was shock consolidated. This work is described in detail by Shang and Meyers [7]. Annealing and hipping treatments were conducted subsequent to this; the objective of this treatment was to produce the reaction between Ti and Al powders. A double tube setup with low detonation velocity explosive was used. The shock pressure was calculated to be 9.5 GPa. Scanning electron microscopic analysis of recovered compacts after dynamic densification revealed that Ti and Al particles underwent a considerable amount of deformation and that the material was fully densified. The Ti₃Al particles seem to retain their initial sphericity. Well densified Ti₃Al+Ti+Al compacts were obtained (~95-98% of the crystal density) and machined for subsequent hipping. Hot isostatic pressing (hipping) was conducted at 600°C and 1000°C for one hour. The shock densified Ti₃Al compacts were

(Ti + Al) and porosities were observed in specimens at different temperatures. These porosities were due to either (a) shrinkage that occurs when Ti and Al reacted; or (b) trapped gases causing expansion of voids. The density of Ti_3Al is 4.2 g/cm^3, while that of Ti and Al are 4.5 and 2.7 g/cm^3, respectively. Thus, the reaction leads to a 25% shrinkage. Since the capsules were not evacuated prior to shock densification, there is trapped air in them. Thus, when the reactants melt, the entrapped gases would expand and form spherical voids, which minimize the surface area. Consequently, hot isostatic pressing (HIP) was conducted at 600°C and 1,000°C for one hour. SEM analysis of the material recovered from hipping at 600°C revealed that Ti and Al powders reacted with each other (Fig. 21(A)). The dark regions correspond to the voids (marked in figure) and the grayish regions (marked in figure) correspond to the reaction products of Ti and Al. Analysis of the material recovered after hipping at 1,000°C (Fig. 21(B)) revealed that the reaction products bonded the Ti_3Al powders very well, as compared to the compacts produced using hipping at 600°C. One can also see the phase separation in the interiors of the particles, after 1,000°C HIP.

Compression tests also showed strong bonding between Ti_3Al powders after 1,000°C hipping. These 1,000°C-hipped specimens have an average ultimate compressive strength of 2 GPa and compressive fracture strain of 20%. In contrast, the average values of ultimate compressive strength and compressive fracture strain of dynamic-densified specimens are 1.4 GPa and 4.5%, respectively. Post-shock hipping can minimize the microcracks and tremendously increase the mechanical properties of the samples. Figs. 22(A) and (B) show the compressive property data for Ti_3Al alloys at room temperature. The data are from shock-densified specimens which were hipped or annealed at various temperatures.

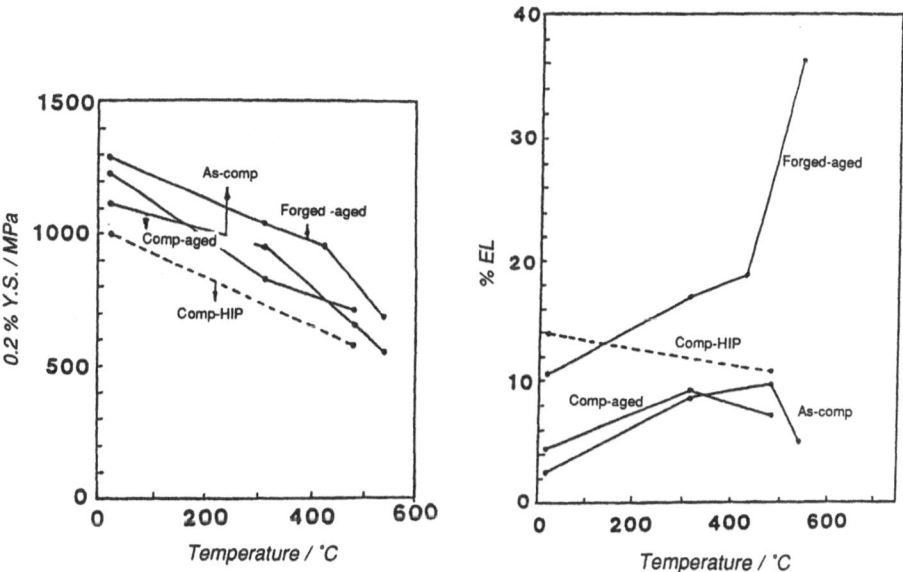

Fig. 20 (A) Yield stress and (B) elongation of commercial purity titanium conventionally forged and shock (Comp.) processed as a function of temperature (from Coker *et al* [10] with permission).

The data demonstrate that Ti_3Al exhibits an apparent increase in ductility with increasing temperature of annealing or hipping. The lower ultimate compressive stress and ductility of annealed specimens, compared to the 1000°C-hipped specimen, could be due to two reasons: (1) Ti and Al were not totally reacted in annealed specimens (2) shrinkage occurred when Ti and Al reacted, with the production of voids. Hipping at 600°C resulted in the a_2 (hcp) and orthorhombic structure and these structures exhibited a high yield stress but low ductility. The fracture surfaces from failed compressive specimens were observed by SEM. The specimen which was hipped at 600°C failed by a mixture of shear and axial splitting. Axial splitting is produced by tensile stresses at the tip of the pre-existing cracks and voids due to the compressive forces. Observation of the axial splitting region showed that the individual particles were heavily deformed and exhibited cleavage regions.

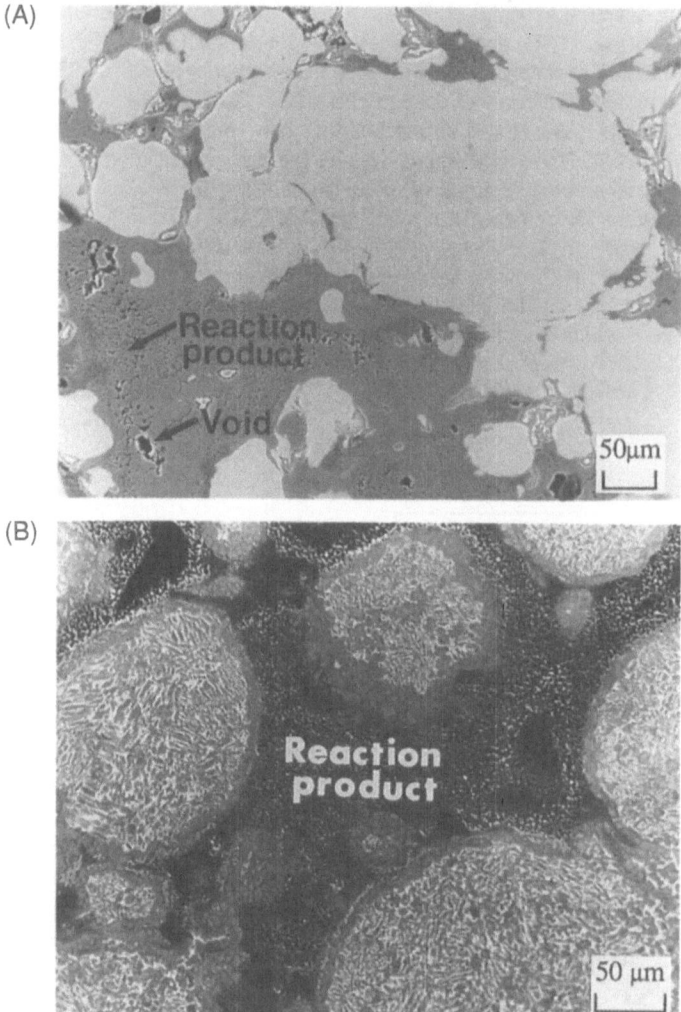

Fig. 21 Scanning electron micrograph from shock-densified Ti3Al+Ti+Al alloys (A) hipped at 600°C and (B) hipped at 1,000°C at 1,000°C in which the different regions were identified by EDS analysis (from Shang and Meyers[7]).

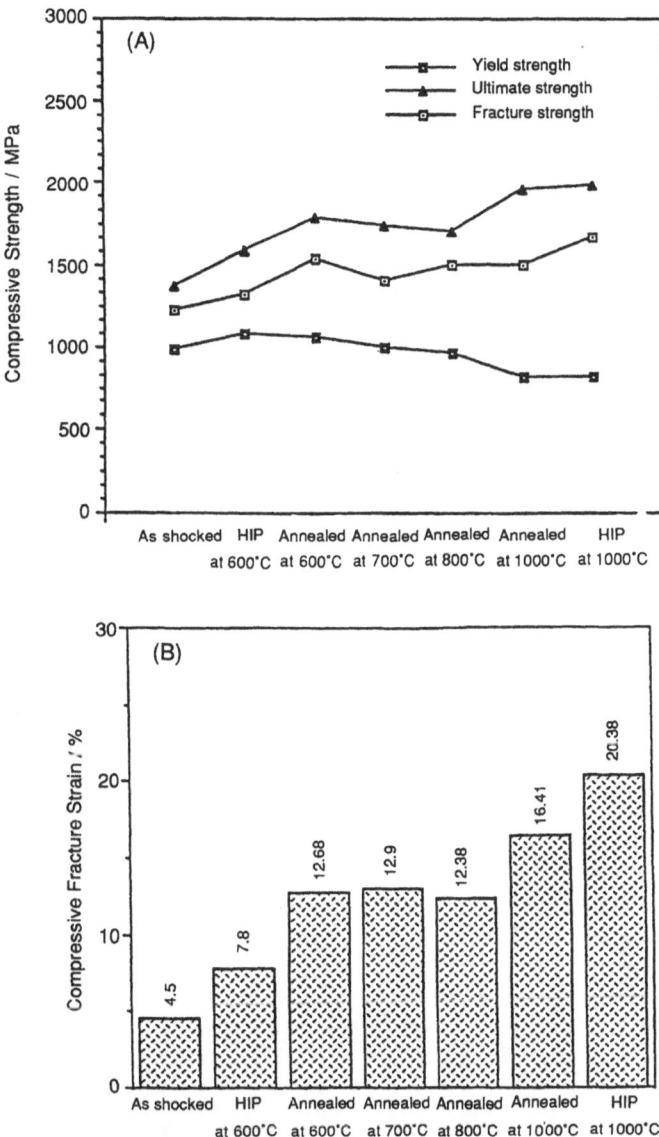

Fig. 22 Compressive properties of Ti₃Al hipped or annealed from 600°C to 1,000°C (A) yield strength, ultimate compressive strength, and fracture strength and (B) maximum compressive fracture strain (from Shang and Meyers[7]).

Figure 23 shows the reason for the enhanced ductility of the compact hipped to 1,000°C. Phase separation occurs within the particles, and a crack is shown propagating through the two-phase region. Microcracks form and propagate through the brittle a_2 phase (dark) and are arrested or retarded by the B2 phase (light).

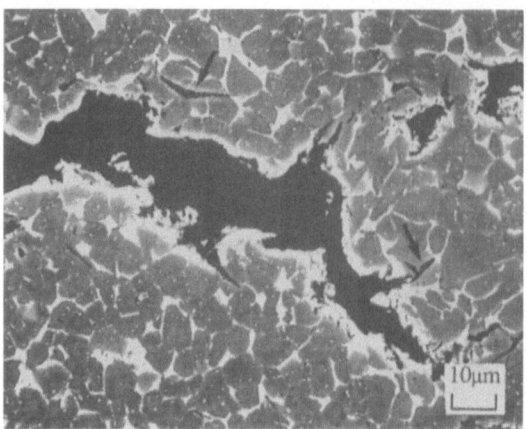

Fig. 23 Scanning electron micrograph showing intergranular and transgranular cracks in the interior of Ti3Al particle; white particles are ductile phase and dark particles are brittle phase. Arrow shows how cracks are arrested by ductile phase (from Shang and Meyers [7]).

SHOCK SYNTHESIS AND CONSOLIDATION

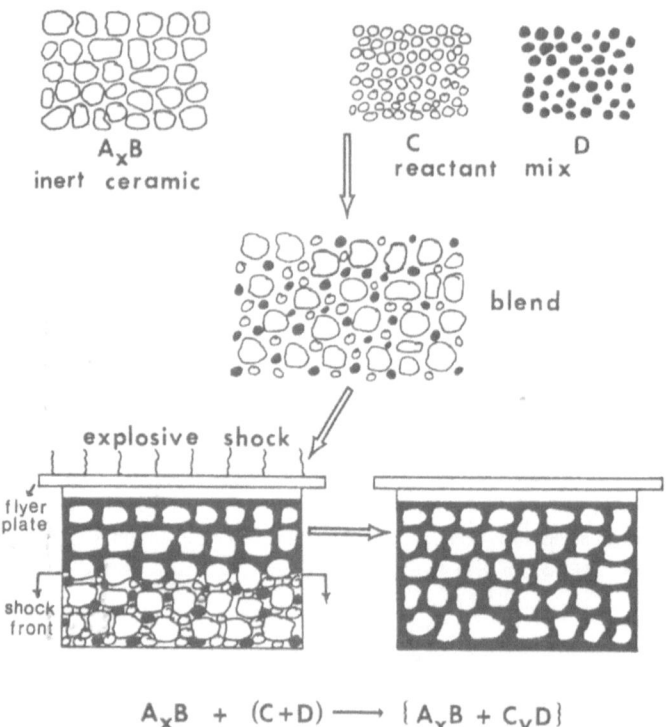

Fig. 24 Sequence of events in chemically-assisted shock consolidation.

4.3 Reaction-assisted shock consolidation

The use of shock-induced chemical reactions to help the consolidation of ceramic powders was pioneered by Sawaoka and Akashi [44] and is described in their patent. Elemental powder mixtures (e.g., Ti and C) are added to the ceramic (e.g., BN) powder. The shock wave triggers the highly exothermic Ti + C reaction, which produces the ceramic TiC and at the same time releases a great amount of heat. This heat helps to "soften" the BN surfaces and has been found to be instrumental in the bonding process. Fig. 24 shows the process schematically. Yu, Meyers, and Thadhani [13] and Yu and Meyers [43] applied this concept to aluminides and silicides, respectively. Their results are presented in the two sections that follow, in an abbreviated form.

4.3.1 Chemically-assisted shock consolidation of TiAl [13]
The hard and brittle TiAl powders are difficult to shock consolidate without additives. Thus, Nb and Al powders were added to TiAl powders with the intent of initiating a shock-induced reaction to generate heat and produce an intermetallic compound binder phase to assist in the consolidation of TiAl. Fig. 25 shows the cross section of a sample containing TiAl, Nb, and Al, shocked at 2.3 km/s impact velocity. It is clearly seen that the bottom portion of the capsule underwent better bonding. This corresponds to the region of higher mean bulk temperature. The profile of the well consolidated region is clearly evident.

Figure 26, at higher magnification, shows the optical micrograph of a region near the top of the cross section. It is clearly seen that Nb powders just surround TiAl powders; they did not react with Al powders. Fig. 26 (B) shows the quantitative analysis of Fig. 26 (A) by SEM. There is no reaction between Nb and TiAl powders. At the central region of the cross section, reaction took place with residual Nb particles remaining. Fig. 26 (C) (bottom part of cross section) shows that Nb and Al powders reacted with each other and with TiAl powders. The reaction products bonded the TiAl powders together. Fig. 26 (D) shows that all points in this region (O, P, Q, R) were reaction products which had almost the same composition.

It is interesting to notice that reaction does not take place as postulated in Fig. 24, but in a more complex mode, with involvement of the inert intermetallic compound and binders. This aluminum was found to react with the TiAl powder. They are thought to be due to a Ti-Al-Nb ternary compound. These results clearly show that chemically-induced bonding is a concept that can be applied to assist in consolidation of metallic systems.

SHOCK WAVE

Fig. 25 Typical cross-section of a shocked compact containing elemental powder reactants: notice bottom part (reacted) and top part (unreacted) (from Yu, Meyers, and Thadhani [13]).

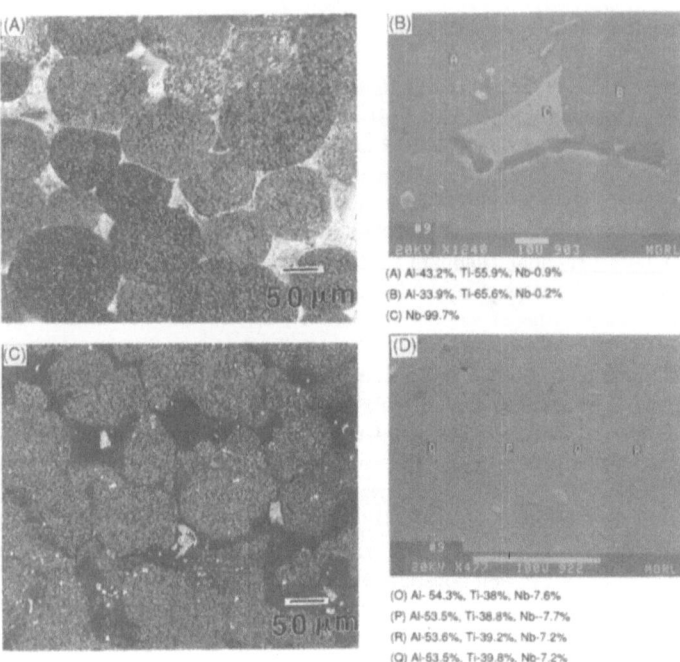

(A) Al-43.2%, Ti-55.9%, Nb-0.9%
(B) Al-33.9%, Ti-65.6%, Nb-0.2%
(C) Nb-99.7%

(O) Al- 54.3%, Ti-38%, Nb-7.6%
(P) Al-53.5%, Ti-38.8%, Nb--7.7%
(R) Al-53.6%, Ti-39.2%, Nb-7.2%
(Q) Al-53.5%, Ti-39.8%, Nb-7.2%

Fig. 26 (A) Optical micrograph of the top region of capsule containing TiAl + Nb + Al, (B) quantitative analysis of (A), (C) Optical micrograph of the bottom region of capsule, and (D) quantitative analysis of (C) (from Yu, Meyers, and Thadhani [13]).

4.3.2 Chemically-Assisted Shock Consolidation of Silicides [43]

Different quantities of reactants (niobium and silicon powders) mixed in NbSi$_2$ stoichiometry were added to NbSi$_2$ powders: 10, 30, and 50 wt%. The micrographs of the cross-sections of recovered compacts are shown in Figs. 27-29. In Fig. 27, the composition of the specimen was Nb-Si (10 wt %) and NbSi$_2$ (90 wt %), and most of the areas show that Nb-Si-NbSi$_2$ powders remained unreacted; however, the powders were well compacted. In this case, the silicon powder underwent large plastic deformation, surrounding the niobium and NbSi$_2$ powders, but the shock pressure and temperature were not sufficient to initiate the chemical reactions between reactant materials. As the quantity of Nb-Si powders was increased, the extent of the chemical reaction was also observed to increase. In the specimen containing 30 wt% Nb-Si powders, two different regions were observed and are shown in Fig. 28. The top portion corresponds to the unreacted region, and the bottom portion reveals fully reacted region, in accordance with the maximum temperature profiles. Figs. 28(B) and 28(C) show a scanning electron micrograph and composition maps for silicon and niobium. It is clear that no reaction occurred; this represents the top portion of the capsule. The morphology of the reacted region is quite different and the profuse presence of voids is noteworthy. Compositional mapping (not shown) demonstrated that reaction between niobium and silicon took place. As the quantity of the Nb-Si powders was increased to 50 wt% (Fig. 29), most areas showed full reaction and an appearance similar to that of 100% elemental powders. The higher magnification micrograph of the specimen containing 50 wt% Nb-Si powder mixture is shown in Fig. 29(B) and (C). The dot mapping shows that the composition is uniform (except for voids). From these results, it seems that both the quantity of Nb-Si powders and shock parameters have a considerable effect on the synthesis process.

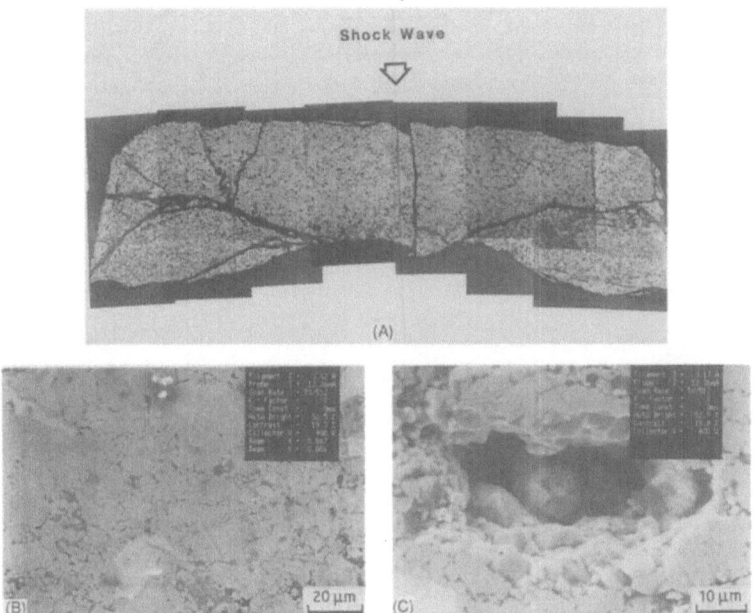

Fig. 27 (A) Optical micrograph of cross-section of compact (Nb 6.2 wt%, Si 3.8 wt%, NbSi$_2$ 90 wt% (B) Scanning electron micrograph of local region in (A), (C) Scanning electron micrograph of void of (A) (from Yu and Meyers [43] with permission).

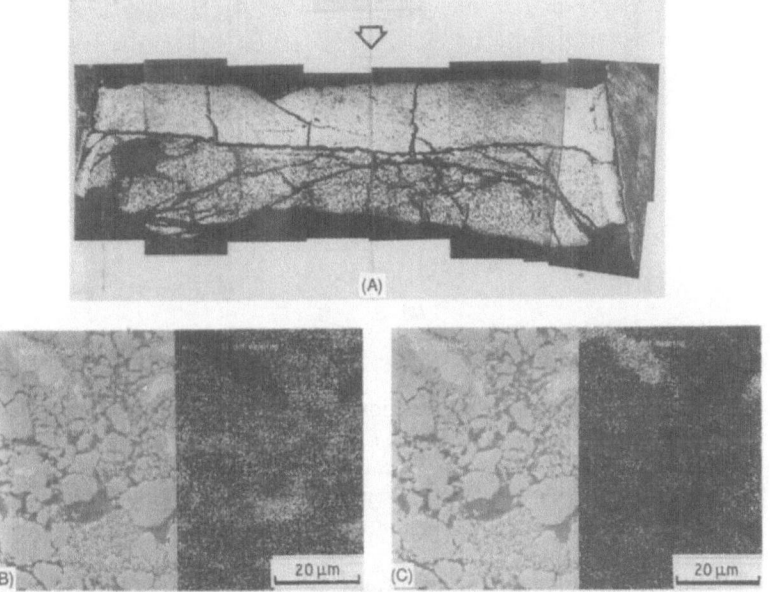

Fig. 28 (A) Optical micrograph of cross-section of compact (Nb 18.6 wt%, Si 11.4 wt%, NbSi$_2$ 70 wt%). (B) Scanning electron micrograph of unreacted region, silicon dot mapping. (C) Niobium do mapping. ((D) Scanning electron micrograph of reacted region.) (from Yu and Meyers [43] with permission).

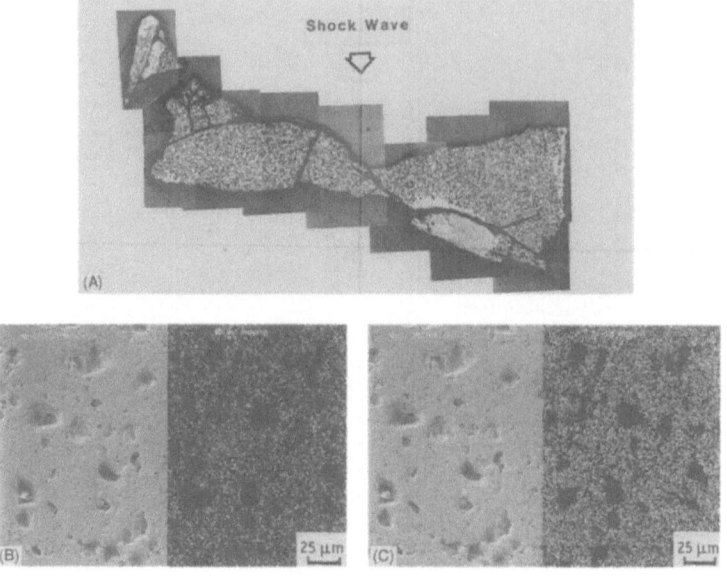

Fig. 29 (A) Optical micrograph of cross-section of compact (Nb 31 wt%, Si 19 wt%, NbSi$_2$ 50 wt%), (B) Scanning electron micrograph of unreacted region, silicon dot mapping, and (C) Niobium dot mapping and (D) Scanning electron micrograph of reacted region (from Yu and Meyers [43]).

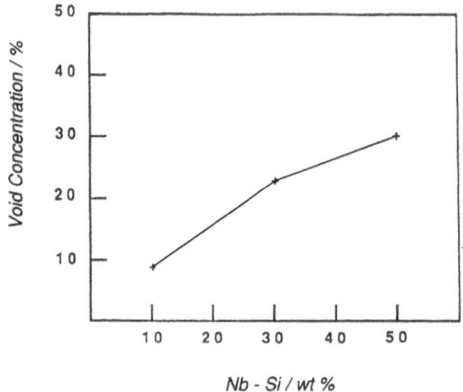

Fig. 30 Void concentration plotted against Nb-Si weight percentage (from Yu and Meyers [43]).

The void concentrations in the reacted regions were 9%, 24% and 30% for 10% Nb + Si, 30%, and 50% addition, as shown in the plot in Fig. 30. The increase in the number of voids with increased proportion of reactants indicates that more shrinkage and gas evolution occurred due to the heat generated by the exothermic reaction of Nb-Si powders. This shows that the degree of reaction depends on pressure, temperature, and the fraction of reactant materials.

The preliminary results obtained on chemically-assisted shock consolidation indicate that reactions can be helpful in shock consolidation. Careful control and optimization of experimental conditions and materials systems is necessary for successfully obtaining good compacts. Very encouraging results have recently been obtained by Potter and Ahrens [6,

42], Tan and Ahrens [5], and Ahrens *et al.* [32] by the addition of graphite, Si, SiC, and Si_3N_4 to diamond or boron nitride powders. It is felt by the co-authors that this area requires concentrated and intensive research.

5. Conclusions

It is very difficult to shock consolidate hard materials (intermetallic compounds and ceramics) because of two reasons:
(a) high shock pressures are required to generate sufficient energy for bonding of the powders
(b) the fracture toughness of these hard materials is such that they are very flaw-sensitive.
Thus, small flaws invariably present after shock densification are easily-activated by tensile reflected stresses or thermo-mechanical stresses during cooling (residual due to temperature gradients). Fig. 11 shows the shock stresses required for consolidation as well as critical stresses for activation of flaws. It is clear that ceramics and intermetallics are in the "danger zone", arbitrarily set for a tensile stress equal to 10% of the compressive shock stress.

Ongoing research efforts aimed at circumventing this major problem are described. The basic approach is to provide energy to the material through means complementary to the shock energy. Three different approaches have been implemented, with varying degrees of success:
1) Use of local shock-induced reactions to increase temperatures of particle interfaces and enhance bonding.
2) Shock densification at a low pressure (just above threshold for pore collapse) followed by hot isostatic pressing or sintering.
3)Shock consolidation of pre-heated specimens.

These approaches have been applied to a number of metals, intermetallic compounds, and ceramics. In combination with optimized fixture designs, they could lead to crack-free compacts of simple shapes (disks and cylinders).

References

1. Akashi T, Sawaoka AB (1987) *J. Mater. Sci.* 22: 3276.
2. Akashi T, Sawaoka AB (1987) *J. Mater. Sci.* 22: 1127.
3. Sawai S, Kondo K (1990) *J. Am. Ceram. Soc. 73*: 2428.
4. Sawai S, Kondo K (1988) *J. Am. Ceram. Soc. 71*: C-185
5. Tan T, Ahrens TJ (1988) *J. Mater. Res. 3:* 1010.
6. Potter DK, Ahrens TJ (1987) *Appl. Phys. Lett. 51*: 317.
7. Shang SS, Meyers MA (1991) *Metall. Trans. 22A*: 2667.
8. Shang SS, Hokamoto K, Meyers MA (1992) *J. Mater. Sci.* 27:5470.
9. Wang SL, Meyers MA, Szecket A (1988) *J. Mater. Sci . 23*: 1786.
10. Coker HL, Meyers MA, Wessels JF (1991) *J. Mater. Sci. 25*: 1277.
11. Ferreira A, Meyers MA, Thadhani NN, Chang SN, Kough JR (1991) *Metall. Trans. 22A*: 685.
12. Ferreira A, Meyers MA, Thadhani NN (1992) *Metall. Trans.*, in press.
13. Yu LH, Meyers MA, Thadhani NN (1990) *J. Mater. Res.* 5: 302.
14. Gourdin WH (1984) *J. Appl. Phys. 55*: 172.
15. Schwarz RB, Kasiraj P, Vreeland T Jr, Ahrens TJ (1984) *Acta Metall. 32*: 1243.

16. Nesterenko VF (1988) Proc. Novosibirsk-Conference on Dynamic Compaction, 100.
17. Meyers MA, Murr LE (1981) in *Shock Wave and High - Strain - Rate Phenomena in Metals*, eds. Meyers MA, Murr LE, Plenum Press, N.Y., 487.
18. Lotrich VF, Akashi T, Sawaoka A (1986) in *Metallurgical Applications of Shock Wave and High - Strain - Rate Phenomena*, eds. Murr LE, Staudhammer KP, Meyers MA,
19. Elliott NE, Staudhammer KP (1992) in *Shock Wave and High -Strain - Rate Phenomena in Materials*, eds. Meyers MA, Murr LE, Staudhammer KP, Marcel Dekker Inc., N.Y.,371.
20. Ferreira A, Meyers MA (1992) in*Shock Wave and High - Strain - Rate Phenomena in Matrials*, eds. Meyers MA, Murr LE, Staudhammer KP, Marcel Dekker   Inc., N.Y., 361.
21. Carroll MM, Holt AC (1972) *J. Appl. Phys., 43*: 1626.   .
22. Norwood FR, Graham RA (1992) in *Shock Wave and High- Strain - Rate Phenomena in Materials*, eds. Meyers MA, Murr LE, Staudhammer KP, Marcel Dekker Inc., N.Y.,989.
23. Shang SS, Meyers MA, unpublished results.
24. Helle AS, Easterling KE, Ashby MF (1985) *Acta Metall. 33*: 2163.
25. Fischmeister HF, Arzt E (1983) *Powder Metall. 26*: 82.
26. Arzt E (1982) *Acta Metall. 30*: 1883.
27. Staudhammer KP, Murr LE (1988) in *Shock Waves for Industrial Applications*, ed. Murr LE, NOYES Publishers, N.J., 237.
28. Norwood FR, Graham RA, Sawaoka A (1986) in *Shock Waves in Condensed Matter*, ed. Gupta YM, Plenum Press, 837.
29. Thadhani NN, Holman GT, Romero B, Graham RA (1991) CETR Report No. A-01-91
30. Korth GE, Flinn JE, Green RC (1986) in *Metallurgical Applications of Shock Wave and High - Strain - Rate Phenomena*, eds. Murr LE, Staudhammer KP, Meyers MA, Marcel Dekker Inc., N.Y., 129.
31. Mutz AH, Vreeland T Jr. (1992) in *Shock Wave and High- Strain - Rate Phenomena in Materials*, eds. Meyers MA, Murr LE, Staudhammer KP, Marcel Dekker Inc., N.Y., 425.
32. Ahrens TJ, Bond GM, Yang W, Liu G (1992) in *Shock Wave and High - Strain - Rate* Inc., N.Y., 339.
33. Sawaoka AB, Horie Y (1992) in *Shock Wave and High - Strain - Rate Phenomena in Materials*, eds. Meyers MA, Murr LE, Staudhammer KP Marcel Dekker   Inc., N.Y., 323.
34. Prümmer R (1987) *Explosivverdichtung Pulvriger Substanzen* , Springer-Verlag, Berlin, Germany..
35. Meyers MA, Gupta BB, Murr LE (1981) *J. of Metals 33*: 21.
36. Meyers MA, Wang SL (1988) *Acta Metall. 4*: 925.
37. Gurney RK (1943) *The Initial Velocities of Fragments From Bombs, Shells, and Grenades*, BRL Report 405.
38. Yu LH, Meyers MA (1992) in *Metallurgical Applications of Shock Wave and High-Strain-Rate Phenomena*, eds. Meyers MA, Murr LE, Staudhammer KP, Marcel Dekker Inc., N.Y., 303.
39. Meyers MA, Pak H.-r, (1985) *J. Mater. Sci.* 20: 2133.
40. Wang SL, Meyers MA, Graham RA (1986) in *Shock Waves in Condensed Matter*, ed. Gupta YM, Plenum Press, 731.
41. Morris DG (1981): *Met. Sci .* 15: 116.
42. Potter DK, Ahrens TJ (1988) *J. Appl. Phys. 63*: 910.
43. Yu LH, Meyers MA (1991) *J. Mater. Sci. 26*: 601.
44. Kunishige K, Horie Y, Sawaoka AB (1992) in *Shock Wave and High-Strain-Rate Phenomena in Materials*, eds. Meyers MA, Murr LE, Staudhammer KP, Marcel Dekker.

Chapter 8
A New Processing for the Self-propagating High-Temperature Synthesis (SHS) Combined with Shock Compression Technique

Y. Gordopolov and A. Merzhanov

1. Introduction

Self-propagating high-temperature synthesis (SHS) is an efficient method of obtaining a wide range of materials and represents strongly exothermic interaction of reactants in condensed medium occurring in a combustion mode [1,2]. Thermal wave arising in a mixture of reactants is spreading spontaneously over matter and transforms it into reaction products showing valuable properties. The synthesis wave velocity makes a value of $10^{-3} \sim 10^{-2}$ m/s. High temperatures at the reaction zone (about 10^3K) promote diffusion and provide purification of products from contaminations.

The process taking place at the synthesis wave may be schematically represented as shown in Fig. 1 which shows the direction of the synthesis wave propagation (arrow), temperature profile, heat release rate, extent of conversion into reaction products. Several zone may be distinguished in the scheme. The first of them is the zone of starting substance. The second one is the zone of heating where chemical reactions has not yet been initiated. The third one is the zone of heat release in the course of chemical transformation, it determines the velocity of the synthesis wave propagation. What follows are the zones of after-burning, structuring and formation of final products, which already have no influence on the synthesis wave propagation velocity, but are of great importance, since they determine structure and properties of final reaction products.

The SHS products mostly are obtained in the form of powders or porous blocks which are then ground, and the articles of them are obtained by sintering using the conventional procedures of powder metallurgy. To obtain the low-porosity articles, the porous blanks are subject to the action of high pressures by using either presses, or the well known methods of shock wave compaction [3,4]. The latter method shows some certain advantages. It needs no expensive and bulky installation, since the force-inducing part of it is replaced by an explosive which is not too expensive. Moreover, explosion provides the conditions of equal loading over the entire surface of intricated articles which could not be attained in pressed condition. This makes it possible, for instance, to obtain the long-sized rods and hollow cylinders of high degree of homogeneity. Such a method of explosive treatment of SHS systems is of the utmost simplicity. It will be analyzed in the following section of the present communication mainly on the example of obtaining the high-density high-temperature superconductors (HTSC) from the powders synthesized by the SHS method [5-8]. In short, the data are also analyzed which are obtained in the studies on the shock wave usage for deposition of SHS powders of TiN, TiC and TiB_2 onto steel articles [9,10], as well as pre-treatment of compacted powders of Si_3N_4 for subsequent sintering [11].

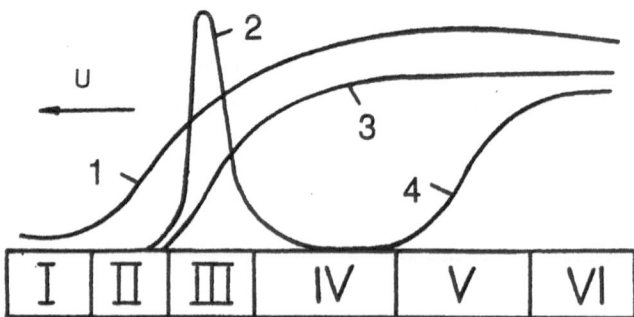

Fig. 1 Schematic representation of the synthesis wave: temperature profile, 1-rate of heat release, 2-extent of conversion, 3-concentration of final product, 4-U is combustion velocity.

The action of high dynamic pressures induced by detonation on the initial SHS compositions and SHS processes seems to be considerably hopeful. These methods of the shock-wave treatment of SHS systems are discussed in the third and fourth sections, respectively.

Of especial interest is the concomitant occurrence of SHS process and pressing. By the high pressure action on hot reaction products, properties of final products may be improved, since the common action of high temperatures and pressures behind the combustion wave front essentially influences the material structure. Besides, the intermediate stage of grinding is no more necessary and articles now become to be more compacted, since at high temperatures material is in the state of ductility in the course of pressing. Pressure may have both the static and dynamic character. The static pressure is reportedly giving good results in synthesizing hard alloys [2]. An interest to dynamic pressure is conditioned by the fact that by using explosives, very high pressures up to 10 GPa could readily be obtained. Besides, the shock wave velocities make a value of about 1 km/s, i.e., several orders of magnitude higher than the combustion wave velocities. This implies that the time of interaction is very small which provides a possibility of selective action of high pressure on any stage of synthesis. If the characteristic width of the post-effect zone at the combustion wave (the IV-VI zones in Fig. 1) is larger than SHS charge dimension, the latter will perceive them as the consequent stages of SHS process. In such a case, by varying the detonation initiation time, one may obtain (from the same initial SHS composition) relatively homogeneous but of different structure, and hence properties, final specimen. If the width of the post-effect zone is comparable or smaller than initial SHS charge, the shock-wave loading will result in obtaining specimen with a gradient of structure and physico-mechanical parameters. In addition, a possibility has to be mentioned of the shock-wave initiation of combustion in SHS charges, which has a quasi-volumetric nature, as well as suppressing the locally-initiated chemical reaction (suppression of the combustion wave). Versatile consequences of the shock-wave action on SHS systems makes them to be very promising in obtaining materials of modified parameters, as well as in the studies on the SHS processes.

An idea of concomitant performance of SHS and explosive compaction of hot reaction products has been born in the 70s [12], but realized only recently [13-26]. The fact is that up to the beginning of 70s several cells have already been suggested for saving condensed compounds formed upon the single and multistage shock-wave loading, as well as upon the dynamic isentropic compression, and a number of physico-chemical transformations have been brought about, including synthesis of diamond and diamond-like modifications of boron nitride described afterwards in reviews [27-28]. These cells had to be only slightly modified: to provide electric ignition of a reactive mixture; to remove gases formed upon combustion in a cell; to isolate explosive from hot synthesis products. Already then [12], the basic ideas have been formulated together with the most promising ways of investigating the combination of SHS with shock-wave loading, high density carbides and borides of IV-VI group metals have been obtained. Later, the method of combustion wave arresting by means of the shock-wave loading has been suggested and realized [15,17]. The shock wave effects on SHS in the complicated system Ti-C-Ni-Cr were studied [13-16], as well as in the hybrid system [17] Ti-N-O and in the binary system [18-24] TiC, TiB_2 and Hf-C. By using the method of dynamic pressing of heated reaction products, the composites $TiC-Al_2O_3$ were obtained [25] Combining combustion in the system $Y_2O_3-BaO_2-Cu$ with dynamic compaction [26] resulted in synthesizing the high-density HTSC.

The SHS process was also suggested [12] for use as a pulsed heater (chemical furnace) for attaining favorable temperature conditions for the shock-wave loading of matter placed in the bulk of reactive mixture. This method was then used [29-31] for promoting the shock-wave compaction of poorly deformable ceramics SiC.

In the early 70s, the feasibility of synthesizing various compounds by the shock-wave loading of initial SHS compositions has also been mentioned [12].This process was brought about [32-36] in the exothermic system Ni-Al to obtain intermetallide Ni_3Al, and in the system Ti-B to obtain [37] boride TiB_2. The feasibility of obtaining various solid coatings upon the shock-wave loading of powdered mixtures Ni-Al, Fe-Si, Cr-B, Cr-C, Cr-SiC, Cr-B4C, etc. onto metal base has been demonstrated [38]. Studied was the possibility of obtaining HTSC articles by the shock-wave pressing of the exothermic system of powders $Y_2O_3-BaO_2-Cu$ with subsequent high-temperature treatment in a furnace [39]. However, it should be outlined that such a shock wave treatment of SHS systems (starting exothermic compositions) is not directly related to the SHS problem, since in all the above mentioned case chemical reactions do not have a character of the self-sustaining layer-to-layer combustion. They occur either in the course of compression at a shock-wave (the shock-wave synthesis), or within the entire volume as a post-effect (the shock-wave initiation). Having no intention to present the state-of-the-art in such an extensive field with its own history as the shock wave synthesis of materials [40], in the present communication we are giving only an illustrative consideration of some examples to perform more complete analysis of all the possible shock wave treatments of exothermic compositions used for SHS.

2. Explosive Treatment of Final SHS products

As a rule, final SHS products are porous ceramics. Practice, problems, and potentialities of dynamic pressing of ceramic powders are widely known [41]. In view of this, the explosive treatment of various SHS materials may be expected to give good results. Let us consider some available examples.

Impressive advances were made in the field of high-temperature superconductivity during last years [42-44], stimulated by extensive studies in all the aspects of the problem. In technological aspect, two branches may be distinguished: (1) development of saving processings for manufacturing HTSC powders (raw material) and (2) development of processings for manufacturing HTSC finished articles. The SHS method turned out [45-48] to be productive in making HTSC powders (oxide ceramics) and more economical than the furnace technologies. All the known HTSC may be obtained by the SHS method, including yttrium, bismuth, and thallium ceramics. For instance, to synthesize the yttrium ceramics (Y-Ba-Cu-O, 1-2-3), the SHS system comprising the mixture of copper, yttrium oxide and barium peroxide powders is used. Reaction is brought about in oxygen flow, the synthesis wave velocity makes a value of 0.4-1.0 mm/s. In the absence of high pressure, the synthesis products are obtained in the form of brittle porous blocks not convenient for practical use. But they could be ground, and thus obtained powder may be compacted by the methods of explosive pressing [5-7]. Explosive pressing combines the shock-wave compaction of synthesis products and shaping of articles. In view of this, it was not surprising that the first studies on the explosive pressing of HTSC powders appeared just after discovery of HTSC phenomenon [48-50].

HTSC ceramics obtained by the SHS method show a number of particular features different from those of ceramics obtained by other methods. In particular, it is fine-grained with about 1 μm particle size and contains large amount of slightly bound oxygen which is probably due to the non-equilibrium character of synthesis. This feature may have an influence on compaction process, on kinetics of phase transformations, and on chemical reactions at grain surface under the action of shock waves.

The problem, which was being solved in Refs. [5,6], was to obtain the high-density uniform SHS HTSC ceramics in the form of a finished article of desired shape and electro-physical parameters (depending on its designation). Uniformity of HTSC is an essential condition, and without its fulfillment the service parameters of articles could not usually be achieved. The problems which are to be solved here are conventional for the problems of explosive pressing of powders. For instance, the requirement of uniformity is normal in fabricating the long-sized articles. The block diagram of explosive pressing setup is presented in Fig. 2. This scheme is being used for more than two decades (with some variations) both in a laboratory and industrial scales [3,4]. Starting powder is placed in a metallic cylindrical container which in turn is surrounded by a layer of explosive. Upon detonation onset at the upper end of a charge, a detonation wave is generated which slides alongside a cylinder element. Detonation products, pressure of which makes a value of dozens and even tens of GPa, are compressing the container at a large rate, and as a result an implosive conical shock wave arises which presses powder. To obtain uniform pressing, the regular reflection of shock waves should be organized within the central portion of specimen. This is one of the central problems of the given modification of explosive treatment of SHS products.

The experiments [5] were carried out with the yttrium ceramics Y_{123} prepared by the SHS method, preliminarily ground down to size below 100 μm and compacted to density 3.6 g/cm^3. The use was made of the cylindrical explosive press schematically presented in Fig. 2. A cell was mainly made of copper and a few stainless steel for comparison. The ratio of cell diameter to wall thickness was maintained to be constant which made it possible to avoid in energies of shell deformation. Cell diameter and wall thickness were 15, 20 and 25, and 1.5, 2.0 and 2.5 mm, respectively. One of the most important parameters, pressure at the detonation wave front, could be varied over the range 1-10 GPa by appropriate choice of explosive: ammonite 6 ZV and its mixtures with barium saltpeter, trinitrotoluene, RDX and their mixtures of varied composition and density. Pressure at the detonation wave front is determined by the detonation velocity; $P=0.25 \, x\rho D^2$, where P is maximal pressure (in GPa), ρ is density of explosive (in g/cm^3), D is detonation velocity (in m/s). The detonation velocity was monitored with electronic pickups.

Pressure in explosion products is not only parameter determining quality of pressing. An essential role is also played by: "history" of loading (i.e., pressure as a function of time); physico-mechanical parameters of powder grains, its initial porosity; material of a cell, its wall thickness, etc. However, most important of them are maximal pressure of explosion products and time of its action which may be characterized in an indirect way by the ratio of masses of explosive and powder to be compacted at fixed other conditions of this multiparametric process. Varying these two parameters, the authors [5] determined the curve of optimal modes of explosive compaction which is shown in Fig. 3(A) The curve is drawn by connecting the experimental points obtained under conditions of regular reflection of shock waves. The region above this curve corresponds to the conditions of overpressing, and below this curve to underpressing of specimen. Fig. 3(B) shows densities of homogeneously compacted specimen corresponding to the curve of optimal compaction modes. As is seen, maximal relative density of cross-sectionally uniform specimen, which may be attained within the frameworks of the given experimental setup, makes a value of about 90% of theoretically predicted density of yttrium ceramics. The obtained specimen showed the Meissner effect at 77K, but the transport current was not observed without sebsequent thermal treatment. T_o recover current superconductivity, an additional annealing is needed. Annealing was carried out in air at 905°C for 10h, then specimens were cooled down to 300°C in a furnace, kept at this temperature for 2 h, and then cooled down to 150°C in a furnace.

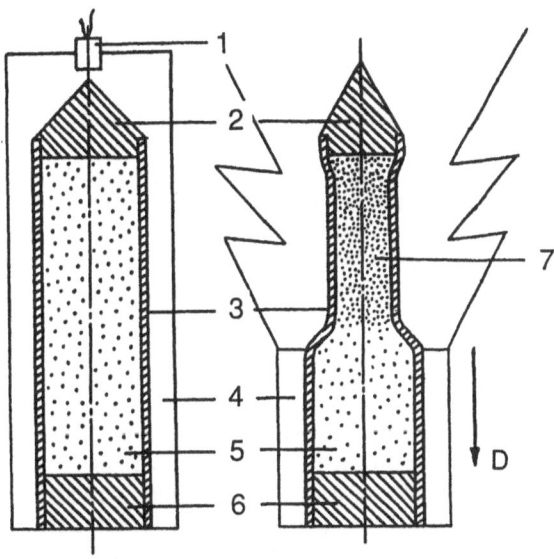

Fig. 2 Pressing of powders with cylindrical explosive piston: 1-electric detonator, 2-upper end-cap, 3-metallic ampoule, 4-explosive, 5-starting powder, 6-bottom end-cap, 7-high density product; D is detonation velocity.

Another scheme of explosive loading of the yttrium SHS HTSC ceramics was checked in Ref. 6. It was differing from that shown in Fig. 2 by the presence of metallic rod coaxially placed at the center of a cell. In this case, higher density in compacted specimen was achieved up to 97% of theoretically predicted density of yttrium ceramics. Homogeneity of pressed specimen was achieved over wider range of amplitudes and widths of pressure pulses. In experiments, the following explosive were used: RDX, mixtures of trinitrotoluene with RDX in ratios 50/50 and 30/70. Such parameters as pulse amplitude and its width were controlled by the explosive identity and its mass. Cell size was kept constant: 27 mm in diameter and 2 mm wall thickness. Cells were made of steel. Thickness of ceramic powder layer was 4 mm, density after preliminary pressing 3.6 g/cm³. Maximal pressure could be varied over the large range of 1~16 GPa, but the transport current in pressed ceramics has not been obtained at any explosive treatment, and subsequent annealing was needed to recover current superconductivity. To perform this procedure, pressed specimen had to be taken out of metallic shell (deformed container). This operation turned out to be very difficult to perform , since brittle ceramics were usually damaged during this procedure.

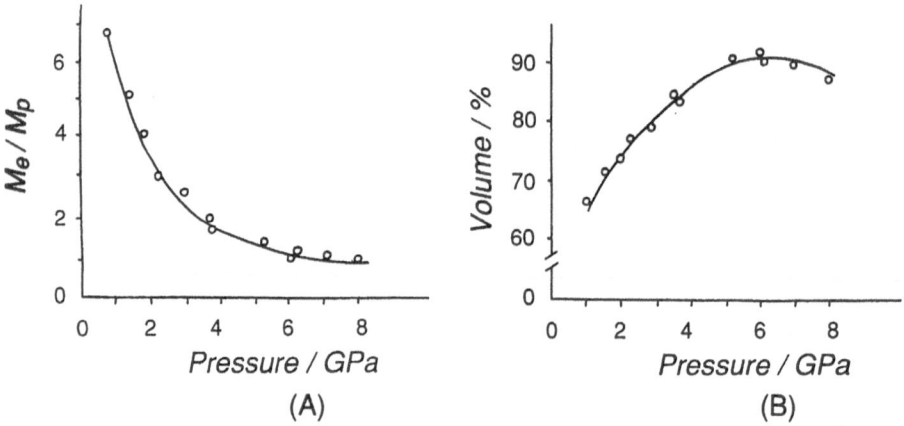

Fig. 3 Curves of optimal pressing for HTSC ceramics with cylindrical explosive press: m_e/m_p is the explosive/powder mass ratio, and V is relative density in % of theoretically predicted value.

This problem may be facilitated by using the loading scheme shown in Fig. 4. Powder of HTSC ceramics is placed in a thin-walled metallic tube, and loading is transferred via transmitting medium such as water, oil glycerol. To remove HTSC specimen from a thin-walled shell is much more easy to do. A series of experiments was carried out [6] within such a scheme. A steel cell 25 mm in diameter and 0.5 mm wall thickness was closed with two end-caps. Density of preliminary compression of SHS HTSC ceramics was 3.6 g/cm³. The cell was coaxially placed in a steel container 52 mm in diameter and 2 mm wall thickness containing either water, or glycerol. Ammonite 6 ZV and its mixtures with barium saltpeter were used as explosives in ratios 50/50 and 30/70 of varied thickness. Pressure could be varied within the range 2.7-7 GPa. Specimen of high homogeneity were obtained with relative density up to 95% of the theoretically predicted density of yttrium ceramics. The experimental data on pressing conditions for this scheme are presented in Table 1. In particular, such a scheme was used in making screens against magnetic fields. Specimens obtained were annealed in air under conditions described above. Specimens for studies on screening ability were machined from thus obtained material and represented hollow cylinders of 19 mm outer diameter, 2 mm wall thickness and 50 mm in length. Screening factors were measured by using induction pickups placed in the center of specimen. The following results were obtained: in the fields 0-120 0e at frequencies 20-150 Hz, maximal screening factors (ratio of fields outside and inside) made a value of 10^4 for longitudinal magnetic field and 116 for transverse magnetic field. Screening factors dropped by as much as twice upon frequency lowering from 100 down to 20 Hz.

Other schemes of explosive loading of HTSC powders were tried which extended the range of shock-wave parameters toward higher maximal pressures. Phase composition of yttrium ceramics was found to show no visible changes over the pressure range up to 20 GPa, though sometimes appearance of non-superconducting phase was observed in the form of films at grain interface. Pressure growth up to 25-30 GPa results in considerable increase in the amount of non-superconducting phase both at grain interface and in the bulk, which is probably due to oxygen losses during heating, to phase transformation and chemical interactions of ceramics with gases in pores. At low peak pressures 4~10 GPa, small amount of intergrain links is formed, predominantly only mechanical contacts between grains are formed which do not conduct transport current. At pressures 20~25 GPa, the number of intergrain links was found to increase. Increase in pulse width also enhances the amount of intergrain links.

The methods of obtaining composite materials and articles of the type SHS HTSC ceramics/metal were developed by using containers of especially designed shape and different loading schemes [7]. They are based on the fact that a deformed container is transformed into metallic matrix protecting article from damage, while HTSC layer provides the desired electro-physical parameters. In such a way, finished articles and blanks were obtained from yttrium HTSC ceramics for different purposes (Fig. 5). Typical parameters were as follows: the temperature of transition to the superconducting state 93-95K; transition width 1.2-1.5K; density of critical current reached a value of 2×10^3 A/cm^2 at 77 K in the absence of magnetic field for the best specimen; density of HTSC ceramics 90-97% of theoretically predicted value. Cylindrical and plane blanks may be subject to turning, milling, and other types of machining with no losses in electrophysical parameters which provides a possibility of fabricating articles of more complicated configuration.

As mentioned above, the SHS method enables synthesizing not only yttrium ceramics, but also other known HTSC. For example, SHS HTSC based on erbium (Er-Ba-Cu-O, 1-2-3) is readily pressed by explosion [8], By using the loading schemes similar to those described above and the methods described in Refs. 48-50, current-carrying specimen of Er_{123} ceramics deposited onto metal base of copper, stainless steel and titanium have also been obtained to date [8].

In a similar way, the wear resistant coatings of other SHS ceramics may be cladded onto metal bases. For instance, the SHS powders of TiN, TiC and TiB$_2$ in combination with metallic binder and without were cladded to steel articles of plane and cylindrical configuration [9,10]. Powder compaction was performed directly on a metal base by oblique shock wave induced by gliding detonation. Ammonite 6 ZV was used as an explosive.

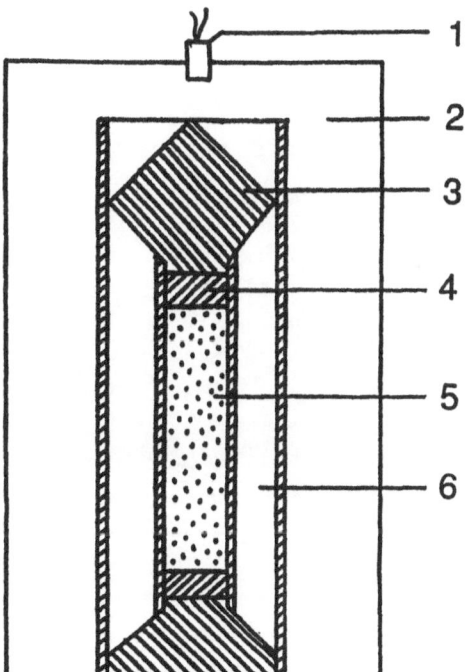

Fig. 4 HTSC pressing via transmitting medium: 1-electric detonator, 2-explosive, 3-aligning end-cap, 4-end-cap, 5-HTSC powder, 6-transmitting medium.

Table 1 Experimental data on explosive pressing of HTSC powders by using transmitting media

Explosive	Charge thickness, mm	Peak pressureat detonation wave, GPa	m_e/m_p* rel.units	Experimental results
AM 6 ZV Density of 0.8 g/cm^3	17.0	5	2.0	Overpressing
	22.0	6	2.8	Overpressing
	27.0	7	3.7	Regular reflection mode
AM 6 ZV / Ba(NO$_3$)$_2$ 50 / 50 Density of 1.0 g/cm^3	17.0	3	2.5	Underpressing
	19.5	3	3.1	Regular reflection mode
	24.5	4.5	4.1	Overpressing
	29.5	6	5.3	Overpressing
AM 6 ZV/Ba(NO$_3$)$_2$ 30 / 70 Density of 1.1 g/cm^3	19.5	2.5	3.4	Underpressing
	22.0	2.5	3.9	Underpressing
	27.0	4	5.2	Regular reflection mode
	32.0	5	6.5	Overpressing

*m_e / m_p = the explosive / powder mass ratio.

Fig. 5 Articles of HTSC ceramics prepared by shock-wave compaction of SHS products.

Extensive studies are now carried out on explosive pressing of large specimen of various ceramics [41]. Ceramics Si_3N_4 seems to be one of the most promising in this respect [51,52] SHS ceramics may also be used in this process. However, to date one cannot avoid cracks in the articles obtained by this method. Sintering is another way of fabricating ceramic articles. Interesting results were obtained by combining explosive pressing of SHS ceramics Si_3N_4 by weak shock waves up to density 70-80% (of theoretically predicted) with subsequent sintering by using catalysts of sintering [11]. In a number of experiments, sintering enhancement was observed. But no ceramics activation by shock waves was found in these experiments, and in view of this the reason for sintering enhancement remains to be unclear.

3. Shock Wave Effects in Starting SHS Compositions

In 70s it was stated that many SHS prepared compounds may be synthesized with no preliminary ignition of reactive mixture and no combustion [12]. This may be achieved by shock compression of reactive mixture which may be put into saving cells. Shock wave propagation in these systems results in strong shear deformations leading to disintegration and hence formation of various defects, chemical bond ruptures, non-equilibrium temperatures at contact sites between reactants, i.e., to some highly energetic non-equilibrium state. Initiation time at shock compression is thought [12] to have a value of 1 µs over entire volume of mixture at "hot" points including gas in pores. Reaction may be completed as a post-effect, or, which is of utmost interest, in the course of compression. It is manifested, e.g., by the Hugoniot curve profile for the mixture of copper and aluminum on which there is a portion confirming heat release in the course of compression [53]. As already mentioned [12], carbides, borides, silicides, intermetallides, solid solutions, nitrides and hydrides may be synthesized under conditions of shock compression. A number of these compounds have indeed been obtained by this method [27,28,54]. Feasibility of occurrence of multistage chemical reaction has also been outlined. For instance, upon single shock compression of the mixtures of titanium and paraffin or polyethylene, the mixtures of titanium carbide and titanium hydride were formed [27]. Disadvantage of these syntheses is in incompleteness of reaction and complexity of reaction product composition, which includes new unidentified phases.

Feasibility of obtaining novel useful materials by the shock-wave treatment of exothermic powders is nowadays extensively studied [40]. As an example, a modeling studies for the shock-wave initiation of chemical reaction in the system Ni-Al should be mentioned [33-36], which result in the formation of intermetallide Ni_3Al. This materials may be coated by utilizing explosion energy onto a surface of parts operating at high temperature in corrosive media as a protective coating [32]. An apparatus for the shock-wave synthesis of TiB_2 by explosive loading Ti-B powders up to 29.5 GPa has been described [37]. The shock-wave action was shown to initiate chemical reaction, with after-burning upon pressure release, which results in pore formation in final product. The process was modeled, and the respective data are also presented [37].

Of interest are the data obtained in the studies on shock-wave treatment of exothermic compositions on metallic bases [38], since in some cases this results in the formation of solid coatings. It was found [38] that there exist regions of formation of solid and liquid coating for every powder composition, transition from one of them to other occurs at some certain critical pressures. In the case of solid coatings, no new compounds and phases are formed, as well as no indications on the occurrence of chemical reaction were observed. In the case of liquid coatings, fast crystallization (quenching from liquid state), amorphization, and numerous physico-chemical transformations were found to occur. With powders Ni-Al, Fe-Si, Cr-B, Cr-C, Cr-SiC, Cr-B_4C, etc. taken as examples, feasibility of strong chemical interaction and synthesis reaction between powdered reactants in solid-liquid and liquid phases has been demonstrated [38]. As a result, continuous and layered coatings of carbides, borides, silicides, intermetallides, carboborides, etc. are formed at metallic surfaces, their metastable states being capable of stabilizing. The above-mentioned coatings are characterized by high microhardness.

Attempts were undertaken [39] to obtain articles of HTSC by shock-wave pressing of exothermic mixture of powders Y_2O_3-BaO_2-Cu. Loading scheme of Fig. 2 and its modifications were used to obtain articles of complicated configuration. Superconducting phase Y_{123} is formed in the form of separate inclusions in densely pressed mass of unreacted substance. Blanks of the 90-93% density were then subject to thermal treatment. In view of high density, chemical transformation in the SHS mode did not occur. Thermal treatment in furnace gave positive results [39]. Temperature slowly increased up to ignition temperature, and chemical reaction was of "quasi-volumetric" nature. It is important that blank shape retained its initial configuration in the course of thermal treatment. Thus obtained HTSC articles had a density of 60-70% of theoretically predicted value and standard electrophysical parameters.

4. Concomitant Occurrence of SHS and Explosive Pressing

As mentioned above, the utmost interest represent the direct shock-wave action on the SHS process. The simplest experimental setup for these studies is shown in Fig. 6. In the first case (Fig. 6 (A)), explosion products behind the gliding detonation wave compress the thin-walled cylindrical container. In the second one (Fig. 6 (B)), the use is made of the normally incident detonation wave to accelerate a metal piston compressing reaction products in a massive metallic container. In both cases, reaction is initiated by heated helix at the bottom of a container. At the top of container, pickups were installed for monitoring combustion wave propagation. Detonation was initiated at some moments after wave propagation. Some difficulties were found to arise due to necessity of thermal insulation of explosive, since the temperature of reaction products behind combustion wave is much greater than ignition temperature for the most of explosives. However, these difficulties were overcome by intrications in a structure of container. In a similar way, the problem of removing evolved gases outside container may also be resolved. All the experimental setups for the shock-wave loading of hot reaction products represent some modifications of the arrangements shown in Fig. 6.

It was outlined in the early studies [12] that concomitant occurrence of SHS and dynamic loading is more convenient to use for the mixtures of solid combustible and solid oxidizer, combustion of which is practically gasless. Attempts were undertaken to influence the various stages of SHS of titanium carbide and boride. The mixture was placed in a cylindrical cell 10 cm long 1 cm inside diameter. Shock compression was applied at the moment when the heating zone reached the opposite end of a cylindrical specimen. Under these conditions, the shock compression was assumed [12] to influence all the stages of SHS. Maximal pressures achieved in these experiments were in the range of 7-50 GPa. Reaction products were analyzed by the x-ray methods. Judging from appearance of new unidentified lines in the x-ray patterns, reaction products in the zones of heating and reaction were of compound composition. The products of after-burning and structuring represented the densely packed rods of titanium carbide or boride. No new phases were found. These data may be interpreted as follows. Heat of shock compression is added to heat released in chemical reaction. As a result, specimen may be heated up to temperatures above 2,000 K. Specimen is cooling down at atmospheric pressure according to the heat conduction laws. This results in annealing of new phases or the pressure-induced phases (if they were formed). For this reason, the further attempts were oriented on applying dynamic pressures at the last stages of SHS to obtain densely packed articles. The experiments were carried out [12] on shock compression of IV-VI group metal carbides and borides preheated up to 1,500-3,000 K. Rods, tubes and disks were obtained of the 97% density (of theoretically predicted value). The articles showed the finely dispersed structure with grain size of 1-3 μm. Analysis showed that grains of titanium carbide were of rounded shape and surrounded with the nickel-molybdenum binder.

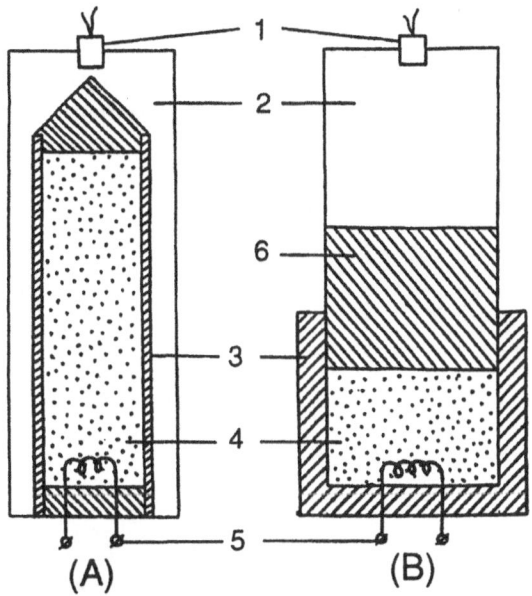

Fig. 6 Experimental setup for explosive loading of reaction products: 1-electric detonator, 2-explosive, 3-metallic container, 4-starting mixture, 5-igniting wire, 6-massive piston.

The schemes of the shock-wave compression were then suggested [12] for the gas evolving systems. Densely packed specimen of $TaC-Al_2O_3$ and $TiC-Al_2O_3$ were prepared under these condition from the mixtures of tantalum and titanium oxides with aluminum and soot. Porosity was of about 2%. Carbide grains were of 2-3 μm in size and uniformly distributed over oxide matrix. The articles showed enhanced resistance against heated oxygen and sulfuric acid.

These ideas received their further development in the further studies in the field [13-26]. The shock wave compaction of hot SHS products was studied [18-24] for the binary systems Ti-C, Ti-B and Hf-C. This technique was used to produce TiC and TiB_2 at greater than 98% of theoretical density and microhardness values which are equal to or greater than commercially available hot-pressed materials [18-21]. It was shown that the microstructures of the SHS materials do differ from the hot-pressed materials in significant ways leading to the possibility that new and unique structures can be fabricated. The effects of stoichiometry and the addition of third components like Cu, Fe, Mo, W, Al_2O_3 and ZrO_2 on the product microstructures and microhardnesses were investigated. It was shown [20] that the addition of metal or oxide additives can alter not only the microstructure of the formed product but can also modify the performance characteristics. Analysis of the TiC indicated [21] that density and microhardness increase as a function of the C/Ti ratio, with maximum values at the ratio of 1.0. A dynamic, finite-difference, heat flow model that predicts SHS reaction temperatures and propagation velocities, effects of material and process conditions on SHS, heat flow patterns during SHS compact cooling in post-densification fixtures was developed to control the processing of compacts made by simultaneous SHS and explosive consolidation [22,23]. Effects of explosive charge mass, powder compact containment design, and time delay between the SHS reaction and explosive detonation, were observed to critically affect the density , hardness, and microstructure of the final product (TiC). The explosive consolidation technique developed to fabricate combustion synthesized titanium carbide and titanium diboride was applied to hafnium carbide and binary HfC-TiC composites [24]. This technique was also applied [25] with the aim of obtaining the high-density dispersed-phase composites of TiC-Al$_2$O$_3$. In all these cases, modifications of the shock-wave loading shown in Fig. 6b were used in the experiments.

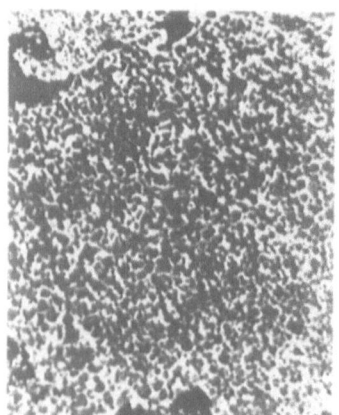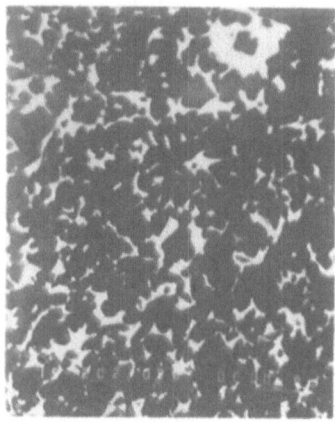

10 μm

Fig. 7 Structure of specimen obtained in a combustion mode from the mixture of powdered titanium and soot doped with nickel and chromium by the action of high pressure on reaction products (section metallographic specimen COMPO): (A)-dynamic pressure, (B)-static pressure.

The effects of the shock-wave loading were studied [13-16] in the system containing powdered mixtures of titanium and carbon (soot) doped with nickel and chromium. Both the schemes of shock-wave compaction of Fig. 6 were used in these experiments. Besides SHS charge composition, the delay time of detonation onset, size of explosive charge and type of explosive were varied. Microstructure and phase composition of the final product were studied by using the x-ray microanalyzers. Structure and properties of the obtained materials were found to depend markedly on the above-mentioned factors and to differ from those for the materials obtained under conditions of static pressure. Fig. 7(A) presents a microphotograph of a specimen obtained under conditions of dynamic pressure. Fig. 7(B) gives the same for a specimen obtained under conditions of static pressure. The explosively made materials were found to have smaller and equal-axis grains of titanium chromium carbide (dark regions) and more uniform distribution of the ductile nickel binder (bright region). Microphotograph shows the range of possible variation of grain size upon variation of loading parameters and detonation delay times. Density was of 97-99% of theoretically predicted value, hardness 90-92 H$_{RA}$ units. The feasibility of obtaining of ultrafine-grained ceramics (0.2-0.3 μm) was demonstrated [16] for the same system (TiC-Ni-Cr) which opens a possibility to develop ceramic materials undergoing transition into the ductile state at relatively low temperatures.

The interesting results are obtained in the case when detonation is initiated at the moment when combustion wave has not yet reached the specimen end. In such a case, in one experiment and at the same parameters of explosive loading, one may observe the effects of shock wave on all the stages of SHS (i.e., all the zones of the synthesis wave of Fig. 1, including the zone of initial matter).

The action of shock waves on the starting SHS compositions was shown to give results depending on the amplitude of loading, scheme of loading, and conditions of heat conduction. At other equal conditions, pressure growth initially leads to compaction of the SHS charge and deformation of grains with the formation of metallographic texture, chemical reactions being not still initiated. Upon further pressure increase, activation of reactants does occur: the large amount of defects is generated which is accompanied disintegration, mixing and heating of titanium and soot particles. The surface layers of metal undergo chemical reaction with nearest soot particles giving titanium carbide, which is arrested at the rarefaction wave due to sharp cooling of the system. The typical structure formed at these conditions is shown in Fig. 8. Thickness of titanium carbide film (gray region) makes a value of 2-3 μm, it separates titanium (bright region) and soot (dark region); the regions of molten titanium are absent, the reaction takes place probably in solid state. Upon further pressure growth, the extent of conversion and temperature of reactants at shock wave become to be sufficient to compensate cooling at the rarefaction wave. The "quasi-volumetric" initiation of exothermal mixture is brought about with subsequent after-burning in the "thermal explosion" mode.

At some certain conditions of the shock wave loading of the locally initiated SHS systems, the arresting (stopping) of the synthesis wave may be observed. Arresting of chemical reaction may be probably explained by sharp increase in thermal conductivity of a medium due to its compaction, leading as a result to larger heat removal from the system. Such an analysis may be of help in the studies on the mechanism of SHS processes (with account for the action of high pressures).

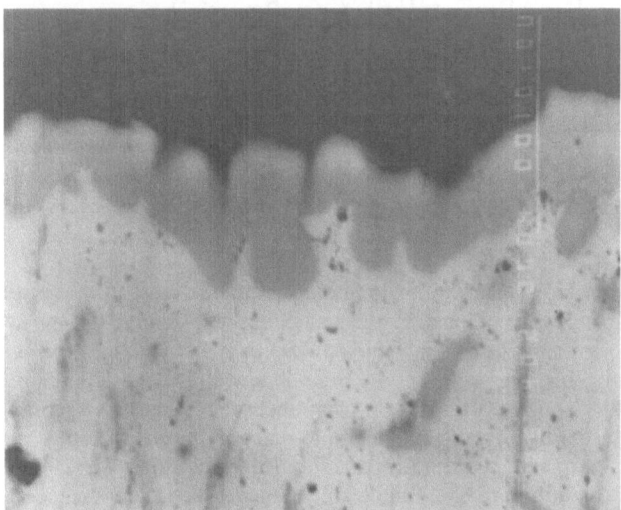

5 μm

Fig. 8 Carbide film formed upon interaction of titanium grain with soot particle under conditions of shock-wave loading (section metallographic specimen COMPO).

    The particular features of structuring under conditions of concomitant occurrence of SHS
in the hybrid system (e.g., metal-gas) and explosive loading were also reported [17]. Such an
approach is promising in resolving some technological problems: fabricating poreless
materials in the system gas-solid, layered materials, etc. The experimental setup for these
studies is presented in Fig. 9. Preliminarily pressed from metal powder, a specimen was
placed in a specially designed ampoule of conservation. Since a free access of air had to be
provided into the combustion zone, loading was performed with a thrown cylindrical striker.
The striker was thrown by a charge of explosive (mixture of ammonite 6 ZV with barium
saltpeter, 30/70). Combustion was initiated by tungsten helix from the bottom side. Blasting
cap was exploded at the moment of combustion wave passing over the desired portion of
specimen (monitored with thermocouples).

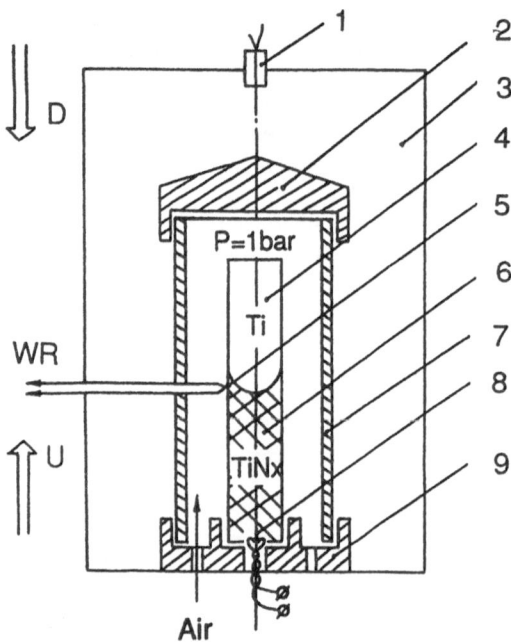

Fig. 9 Experimental setup for explosive action on the metal-gas system in the course of combustion: 1-electric
detonator, 2-upper end-cap, 3-explosive, 4-starting powder, 5- thermocouple, 6-synthesis products, 7-metallic
liner, 8-tungsten helix, 9-bottom end-cap. U is combustion velocity, and D is detonation velocity.

    The conditions were adjusted so that pressing should be close to the regular one. Under
the action of explosion, the combustion wave stopping was observed in the system Ti-N-O.
The cooling rate was found to have a value not less than 10^4 K/s (during combustion wave
stopping). The following interesting results were obtained [17]: (1) titanium interaction with
air in a combustion mode is occurring in two subsequent stages, the limiting stage being
interaction titanium-nitrogen; oxygen enters into reaction only at the stage of after-burning
(Fig. 10); (2) the "primary" structure responsible for combustion wave propagation is the
formation of the 2-3 μm film of Ti_2N in the kinetic mode, the leading temperature of process
being larger than the titanium melting point; (3) at the second stage, the process is conditioned
by cracking of the primary nitride film and by interaction of molten titanium with nitrogen,
including dissolution of the "primary" structures giving finely dispersed (grain size of 1 μm)

phase of a composition TiN_{1-x} (x=0.1-0.2); (4) the phenomenon of microstructural irregularities of the fine-grained islet-like zones type was discovered in the formation of the TiN_{1-x} phase due to structural inhomogeneity of heterogeneous medium; (5) material $TiN_{0.8}$ was found to have a density of 90~97% of theoretically predicted one.

Attempts were undertaken [26] to combine SHS and dynamic pressing to obtain high-density HTSC. The loading scheme was as shown in Fig. 6a. A cell was designed to provide oxygen filtration through the reaction products of a composition Y_2O_3-BaO_2-Cu. The systems were also tried with internal chemical courses of oxygen. Electro-physical parameters for thus obtained HTSC were the same as those for the conventional system Y_{123}. Further advance in this direction will allow to obtain HTSC with a value of critical current up to 10^4 A/cm^2.

Another application of SHS combined with dynamic compressing was proposed [12] to be the usage of the SHS process as a pulsed heater. This may turn out to be useful in the high-pressure physics, e.g., in synthesizing diamond, where there is a need in heating under pressure. Various compounds and their mixtures were placed [12] (in tubes of refractory metal) inside a reactive mixture which in turn was placed inside a reactive mixture which in turn was placed inside the ampoule. The thermocouple measurements showed that in the course of SHS the compound under investigation may be rapidly (in seconds) heated up to very high temperatures. Dynamic compression of SHS products and heated compound may be done at any stage : during temperature rise, at maximal heating, or at some moment during cooling down. It was shown that the pulsed SHS heating may be used for some specific purposes, such as, for instance, melting or evaporation of one or several components before compression, thermal activation of refractory component, etc., which promote the occurrence of those or other physico-chemical processes. As was stated [12], the cubic modification of boron nitride is formed under these conditions, but the wurtzite-like one as is the case at normal shock compression.

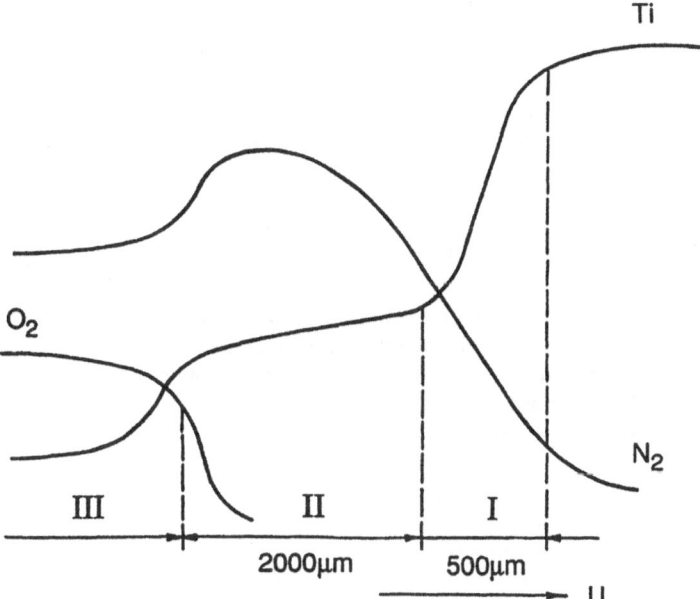

Fig. 10 Overall distribution of Ti, N, O content along the direction of combustion wave propagation; U is combustion velocity.

The method of pulsed heating was also utilized in Refs. 29-31. The SHS process in the Ti-C system was used as heat source for enlightening the shock-wave compaction of SiC ceramics.

In conclusion it may be stated that combination of SHS process with shock compression enables achieving inaccessible regions within the plane pressure-temperature both at high and low pressures. But there exist some limitations in the experiments with conservation of matter: very high residual temperature upon pressure release may result in transforming new phases into normal ones. These limitations may be eliminated [12] by performing fast quenching of synthesis products, or providing conditions of their cooling down at elevated residual pressures.

5. Conclusions

The most important advantage of the SHS method seem to be the absence of need in the external energy consumption, since the process is brought about at the expense of internal energy of system. Since high pressures at explosive detonation may also be attained without expensive installation, the combination of these two methods of creating extreme conditions (high temperatures and high pressures) is of undoubted interest for practice. The concomitant occurrence of SHS and shock-wave treatment of synthesis products (SHS/SWT) may be considered as a new efficient method of preparing various materials which cannot be reduced to single compaction and forming. The available data show that the action of high dynamic pressure may be used for controlling structure of materials synthesized in a combustion mode, though further studies are needed for research and development of the SHS/SWT method.

References

1. Merzhanov AG, Borovinskaya IP (1972) *Dokl. Akad. Nauk SSSR 204*: 366 (Russ.)
2. Merzhanov AG (1988) In: Self-Propagating High-Temperature Synthesis: Twenty Years of Search and Findings, ISMAN (preprint), Chernogolovka, (Russ. & Eng.)
3. Rinehart JS, Pearson J (1963) In: Explosive Working of Metals, A Pergamon Press Book: New York
4. Prümmer R (1987) In: Explosivverdichtung pulvriger Substanzen. Grundlagen, Verfahren, Ergebnisse: Springer-Verlag, Berlin, Heidelberg, New York, London, Paris, Tokyo (Germ.)
5. Fedorov VM, Gordopolov YuA (1990) In: The Study of Regimes for Explosive Pressing of HTSC SHS Ceramics, ISMAN (preprint), Chernogolovka, (Russ.)
6. Fedorov VM, Gordopolov YuA (1990) In: Shock-Wave Compaction of HTSC SHS Ceramics, ISMAN (perprint), Chernogolovka, (Russ.)
7. Gordopolov YuA, Fedorov VM (1990) In: HTSC Ceramics/Metal Composite Materials Fabricated by Explosive Pressing, VDNH SSSR (publicity), (Russ.)
8. Tabadze GF, Private Communication
9. Zotov NA, Private Communication
10. Sterser AA, Private Communication
11. Sharivker YuS, Fedorov VM, Private Communication
12. Adadurov GA, Borovinskaya IP, Merzhanov AG, Private Communication (1972-1987)
13. Gordopolov YuA, Shikhverdiev RM, Molokov IV, Bogatov YuV, Borovinskaya IP, Merzhanov AG (1988) In: The Study of Shock Wave Loading of Heated Reaction

Products at Synthesis of Refractory Alloys in Combution Waves, ISMAN (perprint), Chernogolovka (Russ.)

14. Gordopolov YuA, Shikhverdiev RM, Molokov IV, Bogatov YuV, Borovinskaya IP, Merzhanov AG (1988) *Proc. 7th Intern. Symp, Use of Explosive Energy in Manufacturing Metallic Materials of New Properties, Pardubice 2*: Oct., 324

15. Gordopolov YuA, Fedorov VM, Molokov IV, Shikhverdiev RM, Merzhanov AG (1989) *Proc. 10th Intern. Conf. on High Energy Rate Fabrication*, Ljubljana, Sept., 144

16. Gordopolov YuA, Molokov IV, Shikhverdiev RM, Pityulin AN, Efimov OYu, Zaripov NG, Petrova LV (1989) *Proc. 16th All-Union Conf. on Powder Metallurgy, Sverdlovsk*, May, 54, (Russ.)

17. Molokov IV, Mukasyan AS (1990) In: The Stopping of Combustion Wave by Explosive Effect on Ti-N-O System, ISMAN (preprint), Chernogolovka, (Russ.)

18. Niiler A, Kecskes LJ, Kottke T, Netherwood PH, Jr., Benck RF (1988) In: Ballistic Research Laboratory Report BRL-TR-2951, Aberdeen Proving Cround, Dec.

19. Niiler A, Kecskes LJ, Kottke T (1990) In: Combustion and Plasma Synthesis of High-Temperature Materials, edited by Munir ZA, Holt JB, VCH Publishers, Inc., New York, Weinheim, Basel, Cambrige, 309-314

20. Niiler A, Kecskes LJ, Kottke T (1990) *Proc . 1st U.S.-Japanese Workshop on Combustion Synthesis,* Japan, Jan. Paper T-5

21. Kecskes LJ, Kottke T, Niiler A (1990) *J. Am. Ceram. Soc. 73*: 1274

22. Greve HA, Advani A, Thadhani NN, Kottke T (1991) Presented et the TMS Symposium on Reaction Synthesis of Materials, New Orleans, Feb. for publication in Metallurgical Transactions

23. Advani AH, Thadhani NN, Greve HA, Heaps R, Coffin C, Kottke T (1991) Presented for publication in Scripta Metallurgica at Materialia, 31, Jan.

24. Kecskes LJ, Benck RF, Netherwood PH,Jr. (1990) *J. Am. Ceram. Soc. 73*: 383

25. Rabin BH, Korth GE, Williamson RL, (1990) *J. Am. Ceram. Soc. 73*: 2156

26. Fedorov VM, Shikhverdiev RM, Private Communication

27. Adadurov GA, Goldansky VI (1981) Uspekhi Khim. 50, Issue 10, 1810 (Russ.)

28. Adadurov GA (1986) *Uspekhi Khim. 55*: Issue 4, 555 (Russ.)

29. Akashi T, Sawaoka AB (1988) *Advan. Ceram. Mater. 3*: 288-290

30. Akashi T, Sawaoka AB (1988) *Kogyo Kayaku 49*: 278-284, (Jap.)

31. Akashi T, Sawaoka AB (1987) U.S. Patent, No.4, 655, 830, April 7

32. Mazein SA, Shmakov AM, Private Communication

33. Taylor PA, Boslough M, Horie Y (1987) Shock Waves in Condensed Matter, 395

34. Horie Y, Kipp ME (1987) *Shock Waves in Condensed Matter*, 387

35. Horie Y, Kipp ME (1988) *J. Appl. Phys. 63*: 5718

36. Boslough MR (1989) *Chem. Phys. Lett. 160*: 618

37. Maiden DE, Bianchini G, Holt B, Horning H, Kingman D (1985) *Proc. DARPA/ARMY SHS Symp.*, Daytona Beach, Florida, Oct. 359

38. Kaunov AM, Private Communication

39. Molokov IV, Private Communication

40. Graham RA, Morosin B, Venturini EL, Carr MJ (1986) *Ann. Rev. Mater. Sci. 16*: 315

41. Gourdin WH (1984) *Proc. Symp. Mater. Research. Soc. 24*, 307

42. Bednorz JG, Muller KA (1986) *Z. Phys. B-Condensed Matter 64*: 189

43. Chu CW et al. (1987) *Phys. Rev. Lett. 58*: 405

44. Wu MK et al. (1987) *Phys. Rev. Lett. 58*: 908

45. Nersesyan MD, Merzhanov AG (1990) "SHS in the High-Temperature Superconductivity Problem", Analytical Review 1969-1989: No.5111, Moscow (Russ.)

46. Merzhanov AG, Peresada AG, Nersesyan MD, Borovinskaya IP et al. (1988) *Pismo v ZETF 47*: Issue 11, 604 (Russ.)
47. Merzhanov AG, Lisikov SV, Nersesyan MD, Borovinskaya IP et ál. (1988) *Pismo v ZTF 14*: 1770 (Russ.)
48. Murr LE, Hare AW, Eror NG (1987) *Nature 329*: Sept. 37
49. Mur LE et al. (1988) *J. of Superconductivity 1*: 3
50. Murr LE, Monson T, Javadpour J, Strasik M, Sudarsan U, Eror NG, Hare AW, Brasher DG, Butler DJ (1988) *J. of Metals 40*: 19
51. Akashi T, Sawaoka AB (1987) *J. of Mat. Sci. 22*: 1031
52. Kamiya K, Ikazaki F, Uchida K, Goto A, Kawamura M, Tanaka K, Fujiwara S (1987) *Yogyo-Kyokai-Shi 95*: 480 (Jap.)
53. Dremin AN, Breusov ON (1978) *Nauka i Zizn No.2*: 28 (Russ.)
54. Batsanov SS (1986) *Uspekhi Khim. 55*: Issue 4, 579 (Russ.)

Chapter 9
Shock Wave Interaction in Solid Materials

K. Nagayama

1. Introduction

Most of the experimental studies on shock wave propagation in condensed media has been directed toward the study of material science by using the extreme conditions realized by shock compression [1-3]. In this sense, the shock wave is regarded as a useful tool of studying materials. Unlike gases or liquids, solid materials have a wide variety of properties, which is usually hard to discuss in a universal manner. The so-called law of corresponding states known for gas and liquid systems does not apply to solids at least in the same sense. One of the exceptional example of universalities in solid materials is the empirical Hugoniot relationship between shock velocity and particle velocity [4]. This fact seems rather curious, considering the wide variety of materials measured. At least within author's knowledge, the deep physical meaning of this universal law is still unknown. This law naturally comes from the form of equation of state, or from that of constitutive relations for the materials. It also includes much ambiguities, and is one of the major objectives of shock compression experiments.

Notwithstanding these situations, shock wave loading has been regarded as one of the established means of processing materials [5,6]. For the precise control of the process, one needs to know the details of the loading histories of the specimen. Complex assemblies of the material processing practice require the numerical simulation [7]. Knowledge obtained from these calculations, however, has limited precision mainly because of the uncertainty of the equation of state of the materials.

In the dynamic material processing, the shock propagation is not one-dimensional, but complex flow field is realized [7]. For example, the dynamic treatment process of materials by using cylindrically symmetrical assemblies using high explosive inevitably includes the phenomena of cylindrically converging shock wave propagation. Even in the case of using a plane wave generator, recent numerical simulation [8] has revealed that converging wave may propagate in the powdered thin specimen layer, placed perpendicular to the incident plane shock direction.

No extensive and systematic studies on complex flow problems in solids, however, has been continued so far. Naturally, one of the reasons of such trends is the lack of sufficient knowledge of the equation of state for solid media, through which shock wave propagates.

Problems of shock wave collision and convergence are exceptions. These problems have been studied concerning the realization of extremely high energy density states [9]. As is mentioned above, both these phenomena can be seen and therefore important in the material processing. As one of the unique application of these phenomena, the author [10] has used the cylindrically converging shock waves to compress the magnetic field.

Extremely high pressure over ten Mbar has been reported to be reached by the Mach reflection phenomena of two plane shock waves [11] in heavy metals. Al'tschuler et al, the same group has made a series of experiments on Mach reflection by using various metals. Krehl et al [12] observed Mach reflections in liquids by flash x-ray method. Syono et al [13] has found that Mach stem grows in the cylindrical assembly, and observed the effects of very high pressure over Mbar in the material processing by using a powder gun. Neal [14] has taken an X-ray shadowgraph of Mach reflection in aluminum, and the obtained data agrees well with those of Al'tschuler et al, and with numerical simulations. In the sense of gas dynamics, the strength of shock waves in solids seems weak in many situations, if one defines a shock Mach number. The above-mentioned experimental data are found to be not explained by the simple three shock model. This means the so-called von Neumann paradox [15] in solid state version. The generality of the result is unknown.

Most of the data that are mentioned has been obtained by using high explosive as a shock driver. High explosive method may be a convenient or a unique possibility for obtaining the desired flow conditions. The method, however, includes several disadvantages, i.e., : (1) the decay of pressure behind shock front, and (2) the shock collision takes place at not a line but a point, except for the cylindrical geometry, etc. These initial and/or boundary conditions make the phenomena complicated.

Our final goal is to know how and to what degree the fundamental features of the simple shock interactions occurring in solids is *universal*. For this purpose, one needs to know what material parameter is important, and how it determines the flow features. In other words, our standpoint is to know the material properties by observing the flow pattern. Within our knowledge of the equation of state for solids at high pressures and temperatures compatible with available Hugoniot data, no many parameters are required to express the phenomena theoretically. In this sense, we hope that the experimental simulations of the simple shock interaction phenomena together with the numerical simulation might be one of the important checking ground for the accuracy of the assumed equation of state model.

In this report, shock wave collision and convergence has been discussed both theoretically and experimentally. In section 2, new experimental procedures of generating converging shock wave and/or shock collision has been proposed, based on the impact of a projectile driven by a gas gun instead of using high explosives. In section 3, symmetrical convergence of shock waves in condensed media is considered theoretically giving insight on the effect of material properties to the converging process. Then, the generation of high magnetic field has been discussed as an example of the converging shock wave application. In section 4, the observation of the collision of two shock waves has been discussed with theoretical analysis. In the same section, the conically converging shock wave and resultant Mach disk formation in polyethylene has been observed by high-speed streak camera. The results has also been discussed based on the measured Hugoniots of polyethylene.

2. Gas Gun Based Methods of Realizing Wave Interaction

We have intended to apply the precise optical techniques developed in other research areas [16] to the impact shock study of solids. Although pulsed laser has been introduced to the area of high-speed phenomena already several decades ago, no extensive use of it has made to record the shock wave phenomena in condensed media as an image. Opaque property of many solid substances, necessity of multiple exposure to obtain velocity or acceleration information, might be the cause of the limited use of the laser method.

In this section, the development of an optical observation system of shock wave propagation in solids has been described [17]. The facility includes the high-pressure gas gun for shock wave generation, a compact but high-precision streak camera, and a flashlamp-pumped long pulse dye laser for the illumination of the phenomena. The present system has been designed mainly for the optical observation of shock wave interaction problems in condensed media.

2.1 Shock Wave Registration System

2.1.1 High-Pressure Gas Gun

A high-pressure gas gun is designed and constructed for the generation of a plane shock wave in condensed media precisely. Photograph of it is shown in Fig. 1. The specification of the gun is summarized in Table 1. Since we have intended that the experimental procedure of registrating shock wave parameters are optical, the gun was designed to match the requirements. The observation chamber has six 200 mm dia. windows for observation, and three flanges capable of connecting launch tube. As Fig. 1 indicates, two launch tubes are attached at the same time, one for the horizontal direction, and the other vertical. Due to this mechanism, we have two guns in a relatively narrow space, although two guns cannot be used at the same time.   Mainly, we use horizontal gun, because of the bigger bore diameter of 40 mm, suitable for wave interaction studies. The maximum available projectile velocity is about 500 m/s in the case of about 25-27 g projectile with 20 bar helium gas pressure. The most important point is that the reproducibility of the projectile velocity for the same experimental conditions is proved to be within about 2 %. This good reproducibility of velocity assures the precise control of trigger pulse for the optical measurement system. The details of the gun performance will be published elsewhere [17].

Fig. 1 Photograph of the high-pressure gas gun.

Table 1   Specification of the high-pressure gas gun

		horizontal gun	vertical gun
bore diameter	(mm)	40	20
launch tube length	(m)	2	1
high-pressure gas chamber volume	(m^3)	16 x 10^{-3}	4 x 10^{-3}
observation chamber volume	(m^3)	320 x 10^{-3}	
maximum gas pressure	(bar)	20	80
projectile mass	(g)	21-40	5-15

2.1.2 Compact High-Speed Camera

We have made a high-speed streak camera for optical registration of shock waves in condensed media. It is rather small but enough high streak velocity for the present purpose. The camera is of the optomechanical type, utilizing a rotating mirror [18]. Fig. 2 indicates the schematic of front and side view of the streak camera. The camera body is made of aluminum having the diameter of about 300 mm, and is 85 mm thick. The rotating mirror is a quadrilateral column of 10 x 10 x 23 mm. An air turbine spindle has been used for driving the mirror, whose maximum frequency is 9×10^4 rpm. The attainable writing speed is, therefore, estimated to be about 3 mm/μs. We use a double slit, whose opening is 10 μm x 10 mm each. A small aperture is located at some position as a spatial filter to cut the spurious light from the phenomena. The purpose of the compactness of the camera is multi-fold. Firstly, compactness reduces the construction cost of the camera body. Equipment, installation and alignment of all the optics has been done by ourselves. Second reason is the portability of the camera.

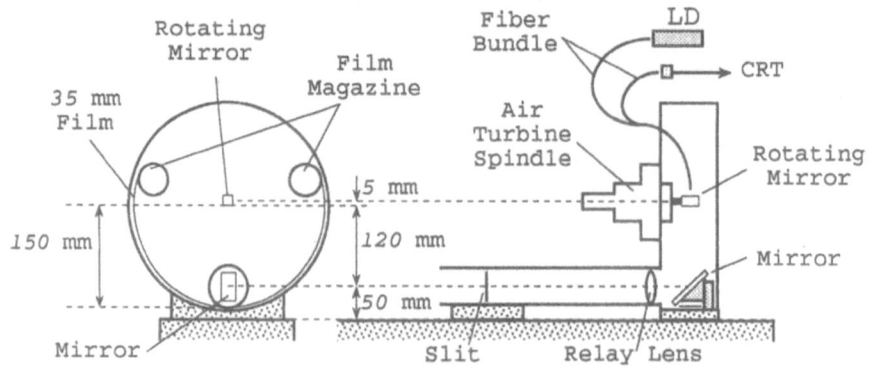

Fig. 2 Schematic of the streak camera.

Although the writing speed of the camera seems not so high, the use of the high-resolution film rather compensate the slower recording. This is possible, because of the relatively strong light intensity due to the smallness of the camera, and also of the relatively small F number of the relay lens system. In our case, Neopan F of the Fuji Film Co. of ASA 25 can be used, whose maximum resolution is 660 lines per mm.

2.1.3 Long Pulse Dye Laser

High-speed optical measurement requires enough intense light source. High-power flash lamp can be used for this purpose. Spurious flashes or lights, however, have been observed at and after the high-speed impact phenomena. In this sense, the illumination of the object should be stronger in intensity than the unnecessary flashes.

In order to overcome these problems, we have developed a flash-lamp pumped dye laser [19]. The most important requirement for the laser is the pulse duration, which must cover the characteristic time of the phenomena. To ensure the reliable synchronization of the onset of lasing to the phenomena, the pulse duration is desired to be much longer. The present laser consists of the two linear flash lamps in a double elliptical lamp cavity, to which the high electric currents are supplied from pairs of condensers, charging up to 128 J for each. Typical trace of the laser intensity monitor is shown in Fig. 3 together with the corresponding streak photograph for comparison. This trace and photograph was obtained before shot as a check for the camera system. In this photograph, the image of a double slit scans the film from left to right. The effective pulse duration is found to be around 25-30 μs. As seen from the intensity trace, laser power is not constant with time, but has a sharp peak in the early stage of duration, then decreases gradually. This undulation, however, is not so appreciable in the film density.

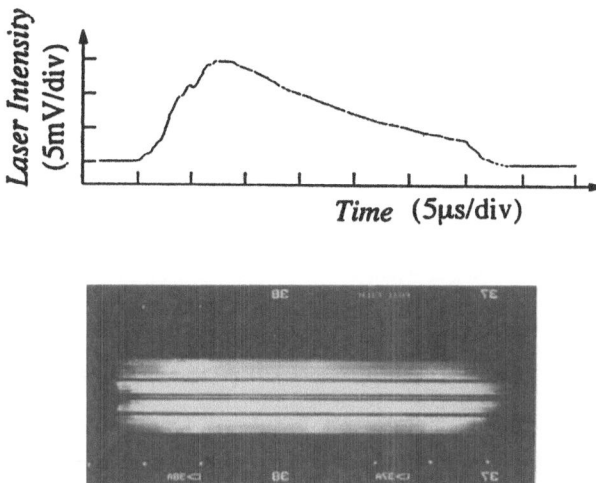

Fig. 3 Typical trace of the output laser pulse with the corresponding streak photograph.

The laser pulse duration does not seem to be long enough for shock study. The triggering of the laser should be very accurate within less than 10 μs. The trigger pulse is obtained from the projectile measurement system of the high-pressure gas gun, then it is sent to the delay pulse generator, then to the laser trigger circuit. A definite delay time must be calculated and set in advance of the experiments. The reproducibility of the projectile velocity within 2 % for the same initial conditions assures the accurate timing.

Part of the experimental results described in this report has been obtained in the High-Explosive Experimental Facility in the High Energy Rate Laboratory of Kumamoto University, when the author was a staff member of the facility.

2.2 New Procedure of Generating Shock Convergence or Collision

One of the most important and noticeable characteristics of shock propagation in solids is the stable interface of two materials. The combination of the materials and their shape of the interface might affect the shock wave front structure.

Based on this principle, it is possible to deflect the propagation of shock waves to the desired direction. The resultant wave front structure is determined by two factors, i.e., the shock velocity u_s and the shock impedance Z of both materials. The principal idea of this shock wave refraction is depicted schematically in Fig. 4. If a plane shock wave is driven to the interface from left, the refracted shock is transmitted. The refraction angle is determined by the difference in shock velocity and shock impedance of both materials. The estimation of the angle, however, is not very easy, since the transmitted shock velocity depends on the shock impedances of both materials.

It should also be noted that no materials can be regarded as an infinitely rigid against any substance. This situation is unique to the case of condensed media, and is quite different from that of gases. This will somewhat limit the possible shape of interfaces for the desired flow pattern.

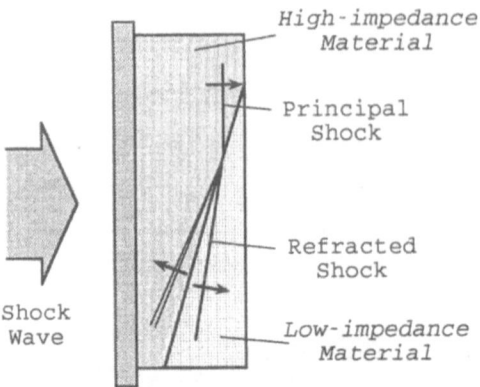

Fig. 4 Shock refraction at material interface (from [20] with permission).

According to this concept, we will propose a new experimental procedure of generating colliding shock waves, starting from projectile impact [20]. Fig. 5 shows a few examples of target assembly ideas. These are simple examples, and there will be a number of modifications. The figures indicate the cross section of the assemblies, and they can be understood as a two dimensional structure or a cylindrically symmetric structure. For cylindrical assemblies, lower impedance materials have the shape of a cone or a cylinder corresponding to Fig. 5(A) and (B). In this case, a cylindrically converging shock wave is generated, and it collides itself at the region of cylindrical axis.

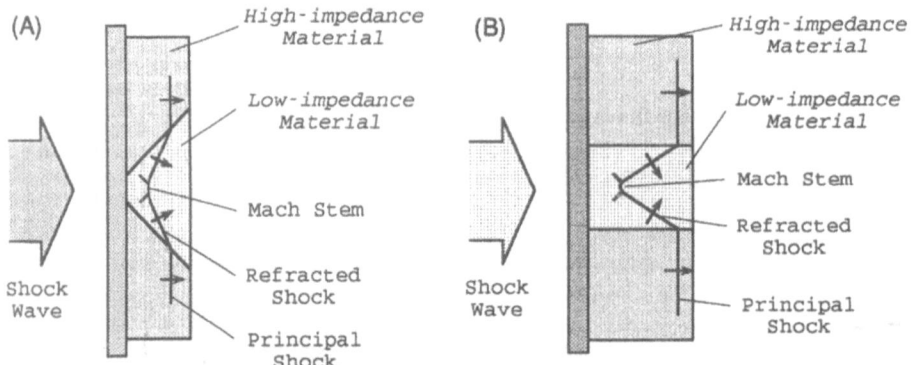

Fig. 5 Conceptual design of generating shock collision or convergence in solids (from [20] with permission).

It is emphasized again that the present method enables for the shock convergence or collision to be realized even starting from flat plate impact between the target assembly and the projectile driven by a gas gun. Therefore, it is possible to generate a well-controlled and constant-pressure shock wave in the assembly. This is one of the advantages of the present method, which is hard to realize by the conventional explosive methods. There might be some ambiguity of the initial conditions at the sharp top of lower impedance material, if the explosive method is used as a shock driver. This problem will also be overcome by the use of gas guns, although there will be another problem of machining the assemblies within tolerable precision.

Although the above-mentioned procedure of controlling wave propagation is suitable for many kinds of experiments, one of the disadvantages of the method is that only a part of the shock wave energy is transferred to the transmitted shock wave because of the difference in propagation direction. To obtain higher pressure shock wave, we have tried to use a projectile having special shapes. Examples of such projectiles and target assemblies are depicted in Fig. 6. These methods are suitable only for the cylindrically symmetric phenomena. We have used this method to observe Mach reflection in conically converging waves. In order to assure the precise and simultaneous impact over conical surface, a projectile with a long sabot is used. Use of the keyed gun is required for the two-dimensional experiments. Collision of two strong plane shock waves can be realized by such schemes.

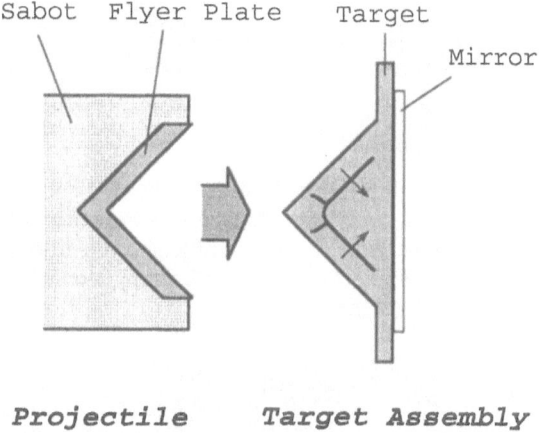

Fig. 6 A Design of projectile and target for generating shock collision or convergence in solids.

3. Symmetrically Converging Cylindrical Shock Waves in Solids

3.1 Approximate Theory of Converging Shock Waves in Condensed Media

A converging shock wave seems promising for the generation of ultra-high pressure states in condensed media, and is encountered commonly in many cylindrical material processing explosive assemblies. The realization of converging shock wave of precise cylindrical symmetry, however, is no easy task. It is still important to know the behavior of converging shock wave with ideal symmetry as a fundamental information on the effect of convergence. Our particular interest here contains the dependence of material properties on the converging process.

In this section, an approximate theory [21] of converging shock growth is described. Starting from the exact expression on the shock growth, we will obtain the approximate differential equation on the shock growth. The exact solution of the approximate equation is found, written by simple mathematical functions [22]. This expression enables us to see how material parameters affect the shock convergence trajectory. Such information will also be useful for the design of the explosive processing assemblies.

We will discuss briefly the behavior of a converging shock wave in condensed media, based on the fluid approximation. Consider a converging shock wave propagating at velocity $U_S(t)$ through a medium of initial density ρ_0. In the following, we will pay our attention only on the physical quantities just behind the shock front, which are uniquely determined by the shock strength at each instant. Observing the shock wave from the coordinate system moving at $U_S(t)$, three conservation equations of fluid dynamics for ρ, u, p can be reduced to the form

$$\frac{\delta\rho_f}{\delta t} - [U_s - u_f]\left(\frac{\partial\rho}{\partial r}\right)_f + \rho_f\left(\frac{\partial u}{\partial r}\right)_f + \frac{\alpha\rho_f u_f}{R(t)} = 0 \tag{1}$$

$$\rho_f\frac{\delta u_f}{\delta t} + u_f\frac{\delta\rho_f}{\delta t} - u_f[U_s - u_f]\left(\frac{\partial\rho}{\partial r}\right)_f + \rho_f[2u_f - U_s]\left(\frac{\partial u}{\partial r}\right)_f + \left(\frac{\partial p}{\partial r}\right)_f + \frac{\alpha\rho_f u_f^2}{R(t)} = 0 \tag{2}$$

$$\frac{\delta p_f}{\delta t} - [U_s - u_f]\left(\frac{\partial p}{\partial r}\right)_f = a_f^2\left[\frac{\delta\rho_f}{\delta t} - [U_s - u_f]\left(\frac{\partial\rho}{\partial r}\right)_f\right] \tag{3}$$

$$\frac{\delta}{\delta t} = \frac{\partial}{\partial t} - U_s(t)\left(\frac{\partial}{\partial r}\right)_f \tag{4}$$

where a denotes the local speed of sound and suffix f represents the values of quantities just behind the shock front, and α is a constant depending on the geometry being 0, 1, 2 for plane, cylindrical and spherical symmetry, respectively. Instead of the usual energy equation, the entropy equation is used, which expresses that the entropy along a fluid particle is kept constant. Combination of Eq.(1)-(3) yields

$$\frac{\delta\rho_f}{\delta t} = - \frac{\dfrac{\alpha\rho_f u_f[U_s - u_f]}{R(t)} + \left\{a_f^2 - [U_s - u_f]^2\right\}\left(\dfrac{\partial\rho}{\partial r}\right)_f}{\rho_f\dfrac{du_f}{d\rho_f} + U_s - u_f + \dfrac{1}{U_s - u_f}\left[\dfrac{dp_f}{d\rho_f} - a_f^2\right]} \tag{5}$$

where $du_f/d\rho_f$, $dp_f/d\rho_f$ denote the derivative along the Hugoniot curves. The sign of shock velocity and the particle velocity should be negative for converging shock waves.

From equation (5), it is revealed that the time variation of density ρ_f is determined by
(1) the shock strength at that instant, say ρ_f,
(2) the position of shock front, $R(t)$,
(3) the density gradient just behind the shock front, $(\partial\rho/\partial r)_f$.
The first term in the numerator of Eq.(5) gives the contribution from the area convergence. While the second term corresponds to the effect of the profile of shock wave on the propagation. If ρ_f is given, the denominator can be estimated by means of the equation of state for the fluid.

It is also noticeable that the values of gradients of other quantities, $(\partial u/\partial r)_f$, $(\partial p/\partial r)_f$, etc. cannot be arbitrary, but rather restricted by the definite relationship [21], which provides a constraint on the form of the shock wave. These gradients are also influenced by the initial profile of physical variables and cannot be uniquely determined by Eqs.(5)-(7). On the contrary, if the density gradient $(\partial\rho/\partial r)_f$ is evaluated at each instant, Eq.(5) can be integrated over t directly and one can know the density growth with time. It can be shown for the case of a weak shock wave the density gradient term in Eq.(5) is a higher order quantity and can be neglected compared with the first term. In this case, we have

$$\frac{\delta\rho_f}{\delta t} = -\frac{\dfrac{\alpha\rho_f u_f [U_s - u_f]}{R(t)}}{\rho_f\dfrac{du_f}{d\rho_f} + U_s - u_f + \dfrac{1}{U_s - u_f}\left[\dfrac{dp_f}{d\rho_f} - a_f^2\right]} \tag{6}$$

This equation has been proved to be meaningful especially for the shock pressure growth in cylindrical converging shock wave in solid media [21]. It is also emphasized that this approximation cannot be applicable for flows under any initial conditions, but suitable for flows under the almost constant-pressure support from outside. This might correspond to the situation that the material is surrounded by a very thick cylindrical explosive pad, which is not always the case of experimental setup. Therefore, the solution of Eq.(6) gives a kind of asymptotic behavior of converging waves.

In the following, it is shown that the above approximate differential equation has an exact solution under several assumptions for the equation of state. That can be written in elementary mathematical functions [21]. Again we will start with the exact expression, Eq.(5), which can be rewritten in terms of velocity variables, $U_s(t)$ and u_f as

$$\frac{\delta u_f}{\delta R} = -\frac{(U_s - u_f)\left[\dfrac{\alpha u_f}{R} + \dfrac{u_f}{U_s}\dfrac{dU_s}{du_f}\dfrac{(U_s - u_f)^2}{U_s - u_f\frac{dU_s}{du_f}}\left(2 - \dfrac{1}{\rho_0}\left(\dfrac{\partial p}{\partial \varepsilon}\right)_v\dfrac{u_f}{U_s}\Big|\dfrac{1}{\rho_0}\left(\dfrac{\partial \rho}{\partial r}\right)_f\right)\right]}{2U_s - u_f\dfrac{dU_s}{du_f} + \dfrac{1}{\rho_0}\left(\dfrac{\partial p}{\partial \varepsilon}\right)_v\dfrac{u_f^2}{U_s}\dfrac{dU_s}{du_f}} \tag{7}$$

where the relation $U_s(t) = dR(t)/dt$ is used. Several other convenient Hugoniot functions are used to derive this formula. If the second term of the numerator is neglected, we have a simple differential equation as

$$\frac{\delta u_f}{\delta R} = -\frac{(U_s - u_f)\dfrac{\alpha u_f}{R}}{2U_s - u_f\dfrac{dU_s}{du_f} + \dfrac{1}{\rho_0}\left(\dfrac{\partial p}{\partial \varepsilon}\right)_v\dfrac{u_f^2}{U_s}\dfrac{dU_s}{du_f}} \tag{8}$$

which naturally corresponds to Eq.(6), and can be solved under appropriate Hugoniot data and equation of state models. In this expression, the last term in the denominator of Eq.(8) are derived from that of Eq.(5), or (6), and it corresponds exactly to the contribution by the entropy increase.

To obtain the explicit expression for the shock growth, the following assumptions have been made here.
(1) Equation of state for condensed media obeys the Grüneisen assumption, namely,

$$(\partial p/\partial \varepsilon)_v = \rho\gamma, \tag{9}$$

where γ denotes the Grüneisen parameter.

(2) The behavior of the Grüneisen parameter is assumed to be

$$\rho\gamma = \rho_0\gamma_0 = \text{constant},\tag{10}$$

(3) The shock velocity U_S is related to the particle velocity u_f by the empirical linear relationship,

$$|U_s| = A + B|u_f|\tag{11}$$

where A and B are material constants. The sign of the absolute value is used to show that the two kinds of velocities have a negative sign.

By using these assumptions, Eq.(8) can be easily integrated over R. Introducing the degree of compression by means of the specific volume v,

$$\eta = 1 - \frac{v}{v_0}\tag{12}$$

the solution of Eq.(8) can be given by [21]

$$R^{-\alpha} = \text{const} \cdot \eta^2[1 - \eta]^{\frac{\gamma_0 B - B + 2}{B-1}}[1 - B\eta]^{\frac{-B+\gamma_0}{B-1}}\tag{13}$$

which includes the integration constant.

From this expression, one can derive very important conclusions concerning the effect of material parameters to the shock convergence behavior. It is seen that the material parameters meaningful for the shock growth are found to be only two parameters, that is, the non-linearity parameter B, in the Hugoniot, and the Grüneisen parameter γ_0. The effect of the Grüneisen parameter stems from the last term in the denominator of Eq.(5), or (6), and it corresponds exactly to the contribution of the increase in entropy.

Figure 7 shows the calculated converging shock growth by using the material parameters for aluminum. Shock growth is calculated with the initial shock strength in terms of the degree of compression, $h_i = 0.23$ at the reduced radius $R/R_0 = 1$. For comparison, shock growth for the case of $\gamma_0 = 0$ is also calculated and shown. Since the term containing the Grüneisen parameter stems from the increase in entropy, the change in the degree of compression for the case of no entropy increase is faster than that for the case of entropy increase. This can be explained by the temperature increase associated with the entropy change. Fig. 8 shows the similar plot of shock growth with varying only the non-linearity parameter, B. For aluminum, the value of B is known to be 1.338. It is shown that the value of B gives appreciable effects on the convergence process. The change in η with radius becomes smaller with increasing B. In other words, large non-linearity suppresses the density increase. To see this effect in terms of shock pressure, we have Fig. 9(A). Since the calculation starts with the same initial value of η, but it does not mean the same value in initial pressure. For comparison, Fig. 9(B) shows the similar plot with the same initial shock pressure. Tendency of shock pressure growth with radius looks opposite with B, and is quite different feature with that in Fig. 9(A).

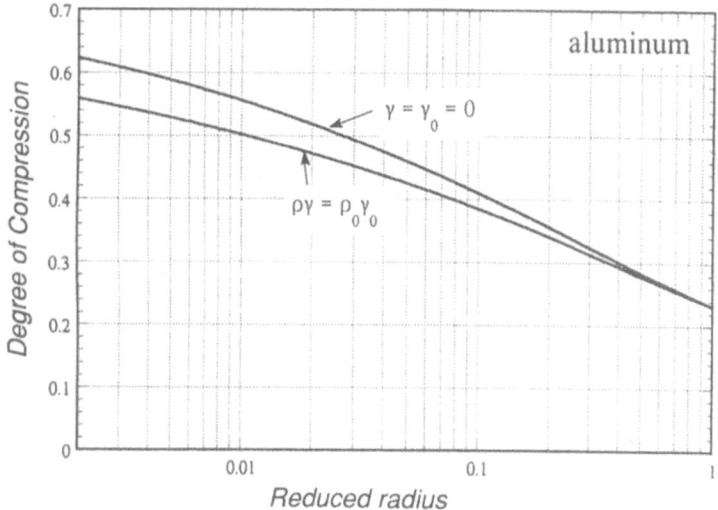

Fig. 7 Calculated converging shock growth in aluminum : effect of Grüneisen parameter.

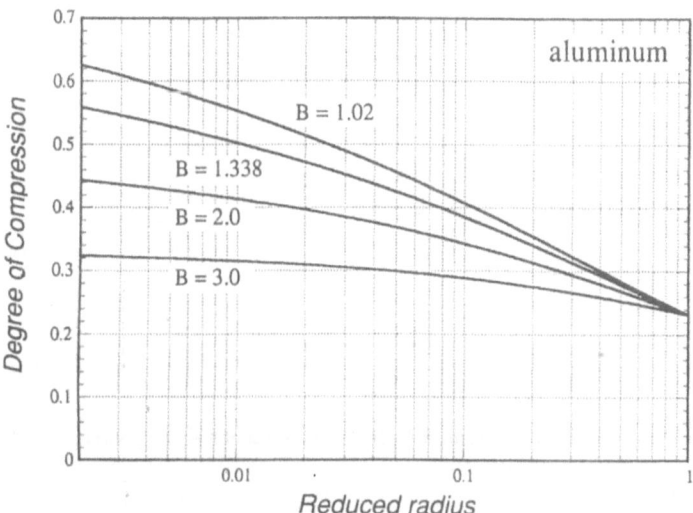

Fig. 8 Calculated converging shock growth in aluminum: effect of non-linearity parameter B.

Since the non-linearity parameter B reflects several physical properties of solids, namely, the non-linearity of the atomic configuration potential force, the thermal contribution to the sound velocity at high pressures and temperatures, it is not simple to figure out the effects of each factor on the flow field. Although these calculations are merely approximate, these simple results should be examined by comparing the exact numerical calculations in more detail in order to have more physical insight of the meaning of the parameter B.

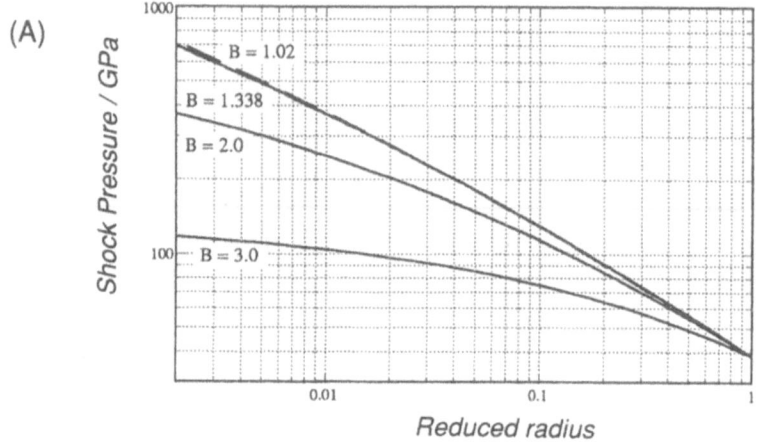

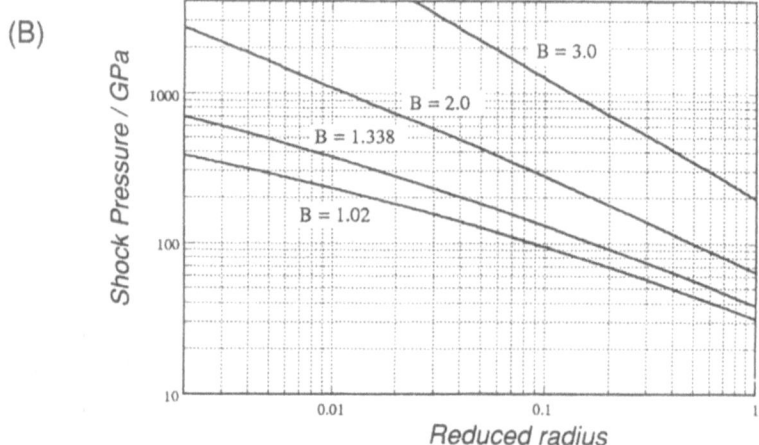

Fig. 9   Calculated converging shock pressure growth in aluminum : effect of non-linearity parameter B. Initial condition : the same value of η : (A), and p_H : (B)

3.2 Converging Shock Wave : A Unique Application

As an application of a converging shock wave in condensed media, a new procedure of magnetic flux cumulation has been introduced by the author [21, 23-25] and the group of Bichenkov [26-28], independently.   In this method, a high-pressure shock wave in powdered materials drives magnetic flux into a small region by cylindrical convergence as depicted schematically in Fig. 10.   By this process, the ultra-high magnetic field strength can be reached.   The ultra-high magnetic field itself provides a new vector for the material processing possibilities, while ultra-high pressure generated inside the small metal cylinder in the field can be applicable for the compression of highly compressible materials. Hawke et al have been demonstrated experiments for the almost isentropic compression of solid hydrogen [29]. The present method was proposed as an alternative of the conventional metal liner implosion method [30], which has an inherent disadvantage of magnetohydrodynamic instabilities at the metal free surface during the liner collapse process [31]. On the contrary, no free surface exists in the present method. The only boundary is the converging shock front in this case. Although the stability of the converging wave is not known for solids, the present method will be promising to provide better reproducibility of symmetrical convergence.

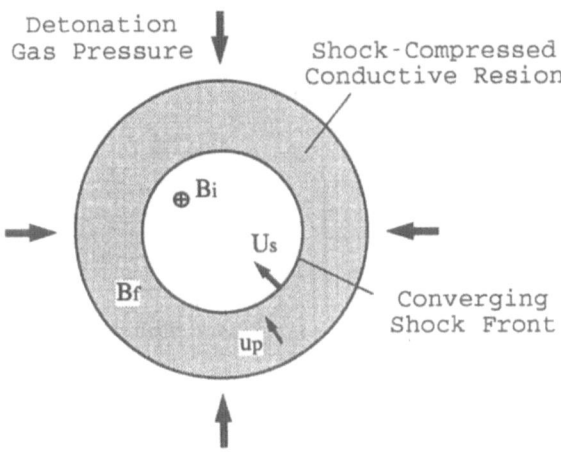

Fig. 10 Concept of converging-shock-driven magnetic flux compression (from [23] with permission).

The original idea of the present method is based on the metallic transition of semiconductors at very high pressures reached by the strong shock compression [10]. The only necessary condition for the magnetic flux to compress is that the electric conductivity should jump to very high values at the shock wave front. If a high-pressure shock wave with conductivity jump at wave front converges cylindrically to the axis of symmetry, high-conductivity region is created successively by the convergence of the shock wave. This conductive region can drive the magnetic flux to the region of lower conductivity, i.e., to the region ahead of the shock front. The mechanism of this flux motion can be discussed by means of magnetohydrodynamics [10].

It is found to be the case not only for semiconducting materials like silicon or germanium, but also for several powdered materials. The mechanism of conductivity jump for silicon powders may be attributed to the phase transformation to the metallic liquid at high pressures and temperatures.

It is demonstrated that even metal powders can be used instead of semiconducting materials, since the bulk electrical conductivity of metal powders is normally very low because of the very loose contact of metal grains. Rather as is shown later, the flux compression ratio is found to be the highest for light metal powders than any other materials tested.

In this section, the experimental results of the magnetic flux cumulation has been briefly reviewed for one of the typical examples of the application of converging shock wave to reach high energy density states. In this method, the material properties, mainly the Hugoniot curve of solids determine the magnetic flux cumulation process.

The mechanism of the flux cumulation can be described by the equation [10],

$$\frac{dB}{B} = -2\frac{u_p}{u_s}\frac{dR}{R} \ ,$$

(14)

where B denotes the magnetic flux density, and u_p denotes the particle velocity just behind the front, i.e. $u_p = u_f$. In this equation, the magnetic flux is driven by cylindrically converging shock waves. From this equation, it is said that the flux compression efficiency depends upon the ratio of the particle velocity to the shock velocity,

$$\alpha \equiv 2\frac{u_p}{u_s}$$

(15)

Equation (14) suggests that highly compressible materials are suitable in the present situation. This indicates that flux compression ratio depends on the material to be used as a wave transmitting medium, and on the process of shock convergence. In other words, the flux compression signal contains information on the shock convergence process. The obtained signal has been processed to show that the flux compression process is really governed by Eq.(14).

Since the signal is obtained as a function of time, we will develop an approximate formula written in terms of time t. If the ratio α can be regarded as almost constant during shock convergence process, Eq.(14) can easily be integrated to be

$$\frac{B}{B_0} = \left(\frac{R}{R_0}\right)^{-\alpha}$$

(16)

where the suffix 0 denotes the initial value. Let us start with the expression for the shock velocity

$$u_s = \overline{u_s} + a\left[t - \frac{\tau}{2}\right]$$

(17)

where $\overline{u_s}$ denotes the mean shock velocity, and the parameter a fulfills the condition, $a\tau \gg \overline{u_s}$. And τ denotes the shock convergence time defined as

$$\tau \equiv \frac{R_0}{u_s}$$

(18)

By noting that the shock front position R at time t is expressed as

$$R(t) = R + \int_0^t u_s dt\tau$$

(19)

Under these simple assumptions, the time evolution of flux density can be represented as

$$\ln\left[\frac{B}{B_0}\right] = -\alpha \left\{\ln\left[1 - \frac{t}{\tau}\right] + \ln\left[1 + \frac{at}{2\overline{u_s}}\right]\right\}$$

(20)

At least in the late stage of convergence, i.e., as time t approaches to τ, the second term of Eq.(20) can be neglected compared with the first term. This means that the two variables $\ln[B/B_0]$ and $\ln[1-t/\tau]$ are linearly related. We will see later how this relation is realized by the obtained signals.

We have performed a series of flux compression experiments by varying material and also porosity, i.e. the initial density. For most of the materials tested, flux compression signal were detected with different amplitude depending on the material. The dependence of the flux multiplication ratio upon the initial density of the material is plotted in Fig. 11. Details of the results are published elsewhere [23-25]. Fig. 12 shows a schematic of the explosive assembly for flux compression tests. These experiments have been performed at High Energy Rate Laboratory of Kumamoto University. In this assembly, six detonators are fired simultaneously to produce an approximately symmetric converging shock wave in powder medium. Magnetic flux compression signal was detected by a small inductive coil until it is destroyed by the converging wave.

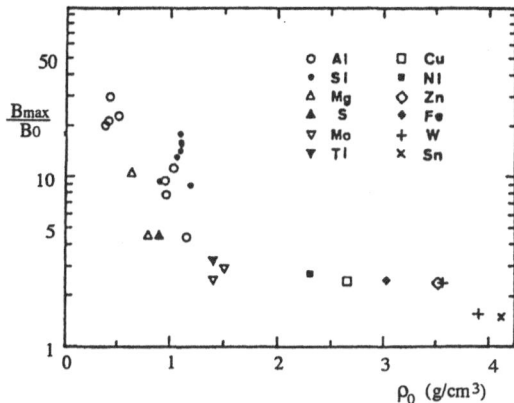

Fig. 11 Dependence of magnetic flux compression ratio on material density.

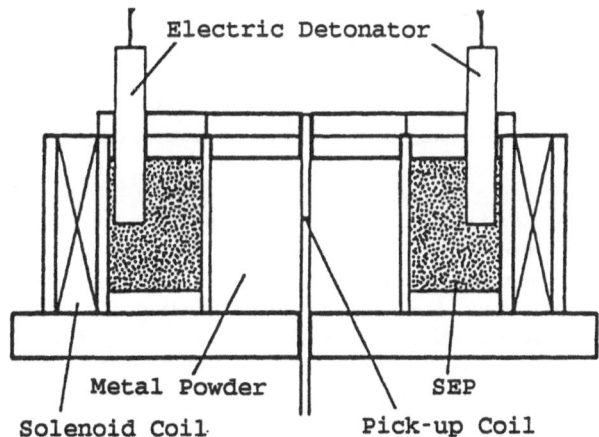

Fig. 12 Explosive assembly of magnetic flux compression test: multi-detonator assembly.

In order to discuss the experimental results in more detail, the normalized flux density history $B(t)/B_0$ is plotted against $[1-t/\tau]$, as inferred from Eq.(20). Fig. 13 shows these plots especially for light metal powders. The experimental results of silicon powders of different initial densities are also included in Fig. 13. The implosion time τ has been estimated by the initial diameter of the powder and the duration of the flux compression time. It is found in Fig. 13 that in the case of light metal powders as well as of silicon powder, flux compression process can be expressed as almost a straight line in this plot. This result may be interpreted as follows : the slope of these lines may correspond directly to the ratio $\alpha = 2u_p/u_s$ as suggested by the discussion of this section. Furthermore, it is plausible that the change in shock velocity during converging process is not so large.

The value of this slope can be compared with the estimated value from the available Hugoniot data for powders. In the results of flux compression experiments using silicon powder, whose initial density is 1.06 (g/cm^3), the value of the above-mentioned slope is measured to be 1.01 from Fig. 13. It is demonstrated from the measured Hugoniot data [25] for Si powder of similar density that the value of a is estimated to be around 1.0 in the region $u_p>0.5$ (km/s). This value agrees very well with that of slope. This fact indicates strongly that at least for semiconductor powders like silicon, flux compression process in the present experiments can really be represented by Eq.(14).

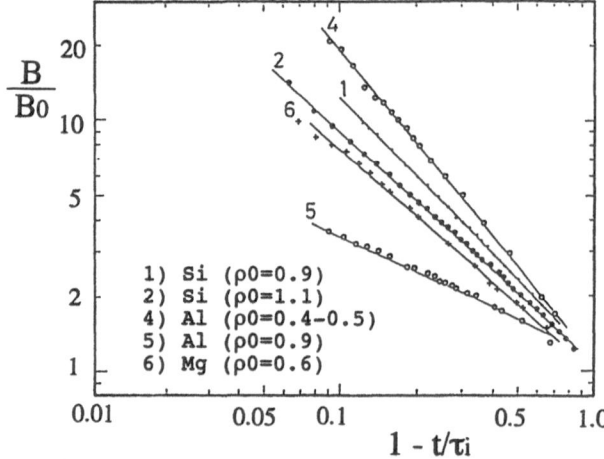

Fig. 13 Log(B/B$_0$)-log(1-t/τ) plots for magnetic flux compression process by using metal powders and silicon powder.

The same comparison for light metal powders, aluminum and magnesium, however, is found to give slightly different results. Hugoniot compression curve has been calculated by Wagner et al for aluminum powders of very high porosity, the value of which is very near that used in the present experiment. From their results, the p-v Hugoniot for the present powder is shown to be anomalous. The value of a for Al powder of corresponding porosity estimated from the Hugoniot is around 1.5-1.6 in the wide region of shock strength. While the value of slope of the present log-log plot for the case of aluminum powder of about 0.3 g/cm^3, is about 1.2. Although this value is the largest one among powders used in the present experiment, it is somewhat smaller than that estimated from Hugoniot. It is plausible in the case of metal powders that flux loss due to the prevention of sweeping out the flux from metal grains during compression process is the main cause of the difference in the above two values. This is explained by the skin effect of metal surface layer. The skin depth for the present condition calculated by using the time scale of flux compression time is about 1 mm, while mean particle diameter of aluminum powder is about 10 μm, which seems to be sufficiently smaller than skin depth. The flux sweep-out process from each metal grains during the progress of shock front is supposed to occur in much shorter times. In this sense, some amount of flux trapped within the grains has kept unswept from them. Very simple conclusion may be obtained from these considerations that it is favorable to use metal powders of finer and finer grain size for the efficient flux compression.

The value of α for the case of magnesium powder is found to be the second largest, while solid density of it is the smallest among materials tested here. This is simply explained by the value of initial density of magnesium powder, which is the smallest except for aluminum powder. It is suggested, however, from these considerations that magnesium powder may have the larger possibility than aluminum powder, if we can obtain highly porous powder of magnesium as aluminum.

The same plot for the experimental results of heavy metal powders is found to be quite different tendency compared with light metal powders. In such plot, flux compression may seem strongly suppressed at some intermediate stage of shock convergence. Since the grain size of these powders is not same, it is shown that the effect of flux loss due to the trap of it in grains depends on the kind of powders. Therefore, this flux loss cannot explain these results. The slopes of these plots for the case of various powders in the early stage of flux compression are almost the same or slightly smaller than those of light metal powders, while the magnitude of slope decreases drastically to very small values. The reason of this phenomena can be explained as follows. Grains of light metal powders like Al or Mg may have a thin oxide layers of a good insulator, although no special treatment of insulation of the powders is made before shot. If this may be the case, it is plausible for premature breakdown to occur in the case of heavy metal powders, which prevents strongly from compressing magnetic flux into the region ahead of shock front. It is desirable to give the insulating-treatment for metal powders before use because of the above reason.

It is demonstrated here that metal powders as well as semiconductor powders are promising for the efficient flux compression in the new method of generating pulsed ultrahigh magnetic field. Within the present experimental conditions, it is found that very light metal powders like aluminum or magnesium are promising.

4. Collision of Plane Shock Waves and Mach Stem Produced by Conical Convergence

As mentioned in the Introduction, there have been many chances to encounter the wave collision in the dynamic material processing practice. Elementary processes of such wave collisions can be figured out to the following two cases. That is, the plane wave collision of unequal strength, which can take place in the case, as depicted in Fig. 14. In this case, the weaker wave is generated by the shock velocity difference of neighbouring substances. Irregular reflection of shock waves can occur even in such cases. If this is the case, it is supposed to appear the high temperature region in the wave interaction region, because of the slip line (or layer) behind the stem. This high-temperature region expands as the wave proceeds. This non-uniform temperature distribution might influence strongly the properties of the processed materials, in some cases. Since at least one of the shock wave is weak, it is plausible not to have a well-defined Mach stem, but to have a somewhat blurred and rounded wave structure. We have demonstrated this behavior afterwards by experiments in a slightly different situation.

Another case is naturally the collision of two plane shock waves of equal strength. This problem has been pursued by several authors as explained in the Introduction. Although this is, of course, the special case of the former situation, it is regarded as important because of its simplicity for analysis and interpretation of the experimental results. Furthermore, conically converging wave, which will be of particular importance in the material processing, containing two aspects of wave collision and convergence, has some connection to this problem.

This section treats with the above two topics by presenting the results of our experiments with a simple theoretical analysis.

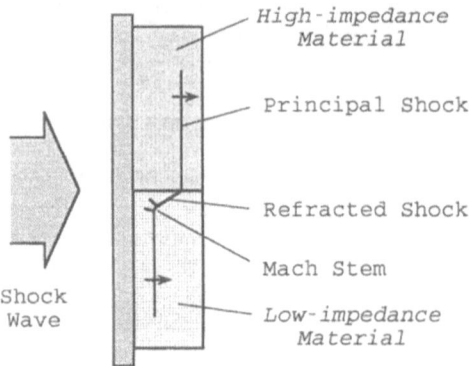

Fig. 14 Collision of shock waves of unequal strength.

4.1 Regular and Irregular Reflection

Regular reflection of two shock waves is possible only for the limited collision angle, 2α defined in Fig. 15. Because of the non-linearity properties of materials, the reflected shock waves intersects at the collision point with an angle different from 2α. Irregular reflection is defined simply when the regular reflection cannot be the case, i.e., the collision angle exceeds some critical angle $2\alpha_{cr}$. We would like to know what material properties will affect the critical angle and how. For this purpose, very simple theoretical consideration of regular reflection of two plane shock waves of equal strength is made here.

Let us start with the situation as shown in Fig. 15. The shock and particle velocity of the incident shock is denoted as u_s and u_p. Two shock waves of the same parameter collides at an angle 2α, and reflected at an angle 2β defined in the figure. The problem is to know the reflected angle β and the reflected shock strength by giving the incident shock strength and collision angle α.

Since the shock front contains no jump of the velocity component parallel to the shock wave front, particle velocity behind the incident shock has the direction to the shock propagation. While D_r denotes the reflected shock velocity observing from the laboratory system of reference. If this flow configuration does not change in time, we can choose a coordinate system moving at the same velocity as that of the collision point C of Fig. 15.

Since the collision point moves at the constant velocity, we have

$$u_s \sin \beta = D_r \sin \alpha \qquad (21)$$

If the flow field is observed from this frame of reference, the fluid particle motion is tracked as arrows in the figure. The flow direction is deflected twice at the incident shock front, and also at the reflected shock front. Fig. 15 contains two particle trajectory lines. The distance between these two lines is directly proportional to the specific volume in that area. Trajectory deflects always to the direction to the symmetry plane at the incident shock front, which assures the density increase at the front. The direction of deflection of the trajectory is opposite at the reflected wave front.

We will also define the parameter of the reflected shock wave, u_{sr} and u_{pr}, which denote the shock velocity and particle velocity for reflected shock waves relative to the incident shock wave. After a somewhat complicated but straightforward calculation, we obtain for u_{sr}

$$u_{sr} = u_s \frac{\sin \beta}{\sin \alpha} + u_p \cos (\alpha+\beta) \qquad (22)$$

while we arrive at the expression for u_{pr} in the same way as

$$u_{pr} = u_p \frac{\cos \alpha}{\cos \beta} \qquad (23)$$

The unknown variables contained in Eq.(22) and (23) to be determined are the following three, i.e., u_{sr}, u_{pr} and β. To close the system we need another relationship between these variables, which should be the reflected Hugoniot compression curve between u_{sr} and u_{pr}. This will be discussed below.

The regular and irregular reflection criterion for the material depends upon the shock Hugoniot compression curve, i.e., principal Hugoniot and the reflected or secondary Hugoniot curve centering at a state on the principal Hugoniot. The form of these curves is determined by the equation of state for the material. Since the shock wave strength realizing in solids can be regarded as weak in the sense of the shock Mach number, the strength of the reflected shock is still said to be weak. In order to have an insight on the non-linearity of the material properties to have effects on the Hugoniot curves, we will try to derive an approximate formula for the reflected shock Hugoniot in terms of the shock velocity or particle velocity for the states on the principal Hugoniot.

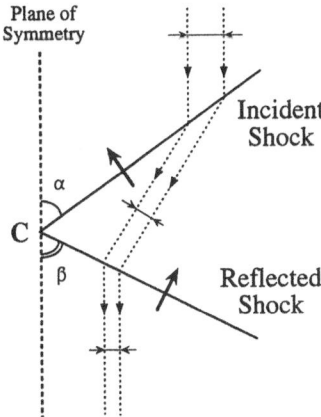

Fig. 15 Schematic of regular reflection of two plane shock waves.

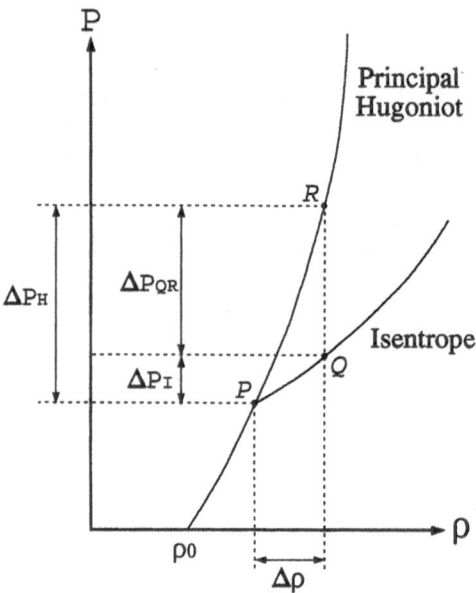

Fig. 16 Pressure-density diagram : Principal Hugoniot and an isentrope.

The schematic pressure-density diagram is shown in Fig. 16 to explain how to derive the reflected Hugoniot. The shock Hugoniot starting from the state P on the principal Hugoniot can be regarded almost as an isentrope, if it has a form

$$u_{sr} = A_2 + B_2 u_{pr} \tag{24}$$

where A_2 and B_2 are the material constants. It is shown to be the case, the higher order terms of the reflected Hugoniot in shock-particle velocity plane seems to have small effect on the form of it in a relatively wide range of pressures. The contribution of these terms can be estimated from the examples to within about 1 %, in the range of over 1 km/s of particle velocity, u_{pr}.

The first term of Eq.(24) is shown to be the speed of sound at the state P on the principal Hugoniot,

$$A_2 = \left(\frac{\partial p}{\partial \rho}\right)^{\frac{1}{2}}_{S_A} \tag{25}$$

while the second coefficient B_2 is given by

$$B_2 = \frac{1}{2} + \frac{1}{4}\frac{u_s}{u_s - u_p}\frac{\rho_0\left(\frac{\partial^2 p}{\partial \rho^2}\right)_S}{\left(\frac{\partial p}{\partial \rho}\right)_S} \tag{26}$$

These two coefficients are derived as a function of u_s and u_p by equating the pressure change Δp_H along the path from point P to R, equal to that along the path from point P to Q, Δp_I plus that along the path from point Q to R, Δp_{QR}. These are written in terms of the change in particle velocity, Δu_p instead of density or specific volume up to its second order.

Finally, we have two equations determining two parameters, A_2 and B_2. They are written as

$$\left(\frac{\partial p}{\partial \rho}\right)_S = \frac{(u_s - u_p)^2}{u_s - u_p \frac{du_s}{du_p}}\left[u_s + u_p\frac{du_s}{du_p} - \frac{1}{\rho_0}\left(\frac{\partial p}{\partial \varepsilon}\right)_v \frac{u_p^2}{u_s}\frac{du_s}{du_p}\right] \tag{27}$$

$$\rho_0\left(\frac{\partial^2 p}{\partial \rho^2}\right)_S \frac{\left(u_s - u_p\frac{du_s}{du_p}\right)^2}{(u_s - u_p)^4} = \frac{du_s}{du_p} + \frac{1}{2}u_p\frac{d^2 u_s}{du_p^2}$$

$$+ \left(\frac{\partial p}{\partial \rho}\right)_S \frac{\frac{du_s}{du_p} - 1}{(u_s - u_p)^3}\left[u_s - u_p\frac{du_s}{du_p}\right] + \frac{1}{2}\left(\frac{\partial p}{\partial \rho}\right)_S \frac{u_p\frac{d^2 u_s}{du_p^2}}{(u_s - u_p)^2}$$

$$- \frac{1}{\rho_0}\left(\frac{\partial p}{\partial \varepsilon}\right)_v \frac{u_p^2}{u_s}\frac{du_s}{du_p}\left[1 - \frac{1}{2}\frac{du_s}{du_p} + \frac{u_s - u_p\frac{du_s}{du_p}}{u_s}\left\{\frac{1}{2\rho_0}\left(\frac{\partial p}{\partial \varepsilon}\right)_v - \frac{\frac{d}{dv}\left(\frac{\partial p}{\partial \varepsilon}\right)_v}{\rho_0\left(\frac{\partial p}{\partial \varepsilon}\right)_v}\right\}\right] \tag{28}$$

It is emphasized here that at least up to this point we have assumed only on the form of the equation of state to be the Grüneisen type. The derivative, $(\partial p/\partial \varepsilon)_v$, contained in Eq.(27) and Eq.(28) corresponds to $\rho\gamma$ as in Eq.(9). Details of these derivation will be reported elsewhere.

If two assumptions on the equation of state and Hugoniot have been made as in Eq.(10), (11), the above two equations are written to be

$$A_2 = (u_s - u_p)\sqrt{\frac{u_s + Bu_p - \gamma_0 B\frac{u_p^2}{u_s}}{A}} \tag{29}$$

$$B_2 = \frac{1}{2}\left[1 + (B-1)\frac{u_s}{A} + B\frac{u_s - u_p}{A}\frac{1 - \gamma_0\frac{u_p^2}{u_s^2}\left(1 - \frac{B}{2} + \gamma_0\frac{A}{2u_s}\right)}{1 + B\frac{u_p}{u_s} - \gamma_0 B\frac{u_p^2}{u_s^2}}\right] \tag{30}$$

If these equations are combined with Eqs.(22),(23), the whole flow field of the regular reflection, and/or the Mach scheme of the simple three shock configuration can be calculated. From these discussions, it is noticeable that the phenomena can be conveniently described by the shock Mach number, just as in the case of ideal gases. This situation suggests that the von

Neumann paradox stated in the area of gas shock reflection will be applied to the area of shock reflection in solids. We will discuss this point again later.

Based on the measured Hugoniot of polyethylene, the regular reflection characteristics has been calculated by using several of the above equations, (22)-(24), (29), and (30). The reflected shock pressure is calculated against the collision angle, and is shown in Fig. 17. The incident shock velocity in polyethylene is assumed to be 3.0 km/s. The critical incident angle is calculated to be 37 degrees. There has also been calculated the dependence of the critical angle upon the initial shock strength, and is shown in Fig. 18. The critical angle decreases very drastically with increasing incident shock strength. This result may be attributed to the very strong non-linearity of polyethylene in this pressure region.

4.2 Experimental Procedures

In order to study the above-mentioned problems, three kinds of experimental assemblies are designed, which can be used in the gun-based experiments. These assemblies have been used to observe the following phenomena.
(1) collision of two plane shock waves of equal strength,
(2) collision of two plane shock waves of unequal strength,
(3) wave structure comprising by a conically converging shock wave.

Figure 19 shows these three different setups. The target assembly of Fig. 19(A) is used to see the phenomena, (1). Therefore, this figure has to be considered to have two-dimensional structure. Fig. 19(B) can be used to observe the two shock wave collision of unequal strength, (2). In this assembly, the material boundary is not parallel to the direction of shock wave propagation. This assembly is designed in order to see part of the phenomena produced by the assembly of Fig. 19(A). As explained later, an assembly, in which the material boundary is parallel to shock direction is difficult to use in the present experiment using mirror destruction method to detect the shock wave arrival. The conically converging shock waves have been produced by using the assembly of Fig. 19(C).

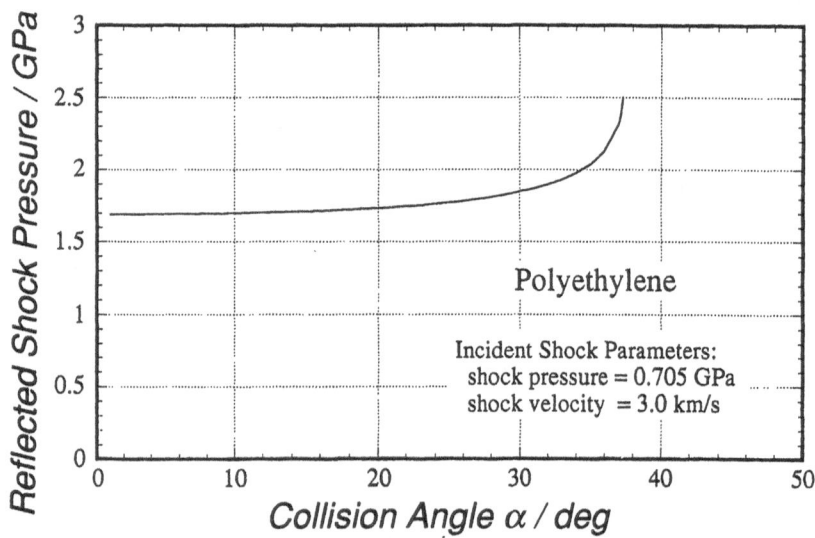

Fig. 17 Calculated reflected shock pressure for polyethylene against the collision angle.

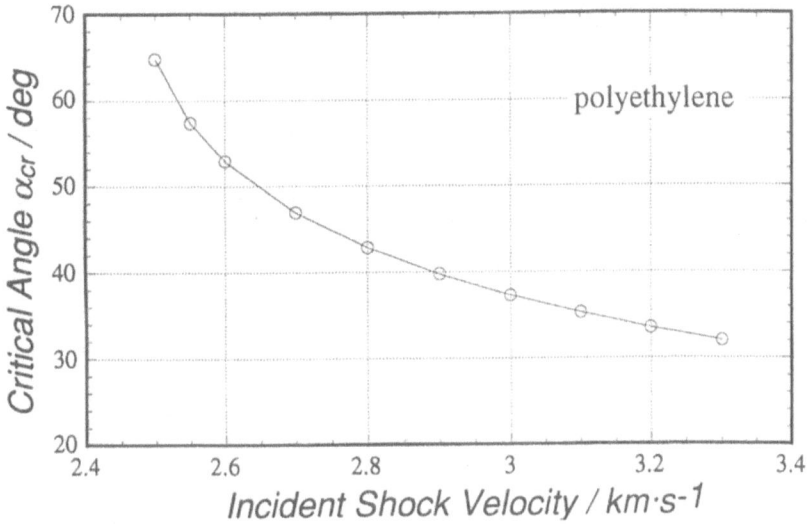

Fig. 18 Calculated critical angle against the initial shock strength.

All the assemblies consist of two pieces of materials, aluminum and polyethylene(PE). This material combination is chosen by the following reason. The attainable projectile velocity by the high-pressure gas gun is limited up to about 500 m/s. Polymeric material is chosen as the lower impedance material, which has enough non-linearity even in relatively lower pressure region, which can be attained by the present gun facility. Wave interactions has been observed always in polyethylene. The initial density of polyethylene is measured to be 0.94 g/cm^3.

The specimens are carefully polished, and combined to keep smooth contact at the material interface. In case of the assembly shown in Fig. 19(C), due to the difficulties of machining, the shape of the polyethylene piece is not exactly conical, but a very small portion of the cone top is ground away to adjust the aluminum piece. Surface mirror is placed over the free surface of the composite assembly. Smear camera catches the history of destruction process of the mirror, by recording the light intensity reflected from it illuminated by the pulse laser.

In order to discuss the experimental results of the wave interaction, shock wave parameters of polyethylene has been measured based on the impedance mismatch method with a mirror destruction technique by a high-speed smear camera. These measurements have been performed to get Hugoniot data of polyethylene specimen for the pressure range, which is realized by the wave interaction experiments, since the available Hugoniot data for polyethylene is mostly limited to the very high pressures [32,33]. One of the streak photographs of impedance mismatch experiments is shown in Fig. 20. As seen in the photograph, shock velocities of aluminum and polyethylene are measured simultaneously. Photograph shows the elastic and plastic shock front in aluminum, while only one wave front is detected in polyethylene as in other polymer like PMMA, although weak shock wave front cannot be detected in this experimental procedure. In this experiment, the shock velocity of polyethylene is measured to be 3.4 km/s.

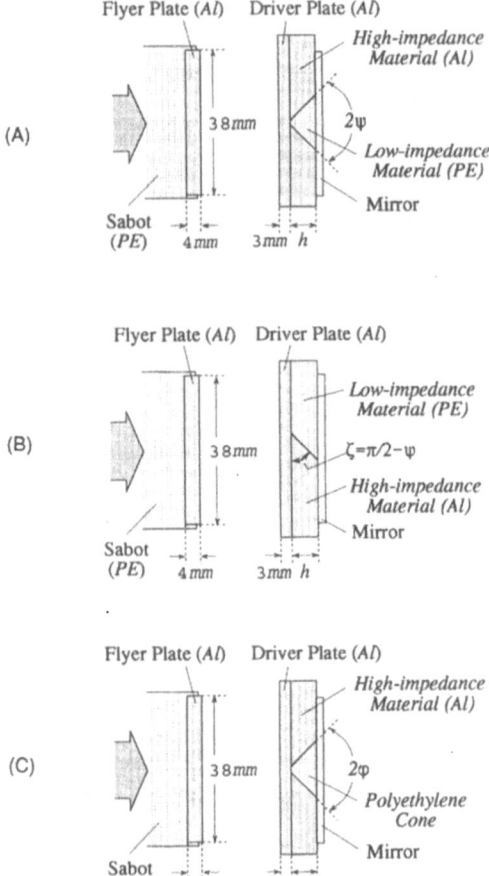

Fig. 19 Three kinds target assembly for shock collision and convergence : generation of (A) collision of two plane shock waves of equal strength ; (B) collision of two plane shock waves of unequal strength ; (C) conically converging shock waves.

We have also measured the ultrasonic longitudinal velocity of the specimen by using a pulse transmission technique. The sound velocity is measured to be 2.4 km/s, for the ultrasonic frequency of 5 MHz. These values depend strongly upon the manufacturing process of the polyethylene, and the specimen used in the present experiment seems to have relatively high density indicating highly crystallized structure. The measured Hugoniot for polyethylene was found to be consistent with this sound velocity. The estimated slope B of shock-particle velocity relationship is 2.4. This value is much larger than that of very high-pressure Hugoniot. This result indicates the strong non-linearity of polyethylene in the measured low-pressure region. If the shock wave in polyethylene has a relaxation region or a double structure due to elastic-plastic behavior as in the case of PMMA, it is necessary to know the details of the particle velocity histories as well as to have simple u_S-u_p Hugoniot point. Much more sensitive technique of Hugoniot measurement like VISAR [34] might be required to know the details of the behavior of polyethylene Hugoniot in such pressure region.

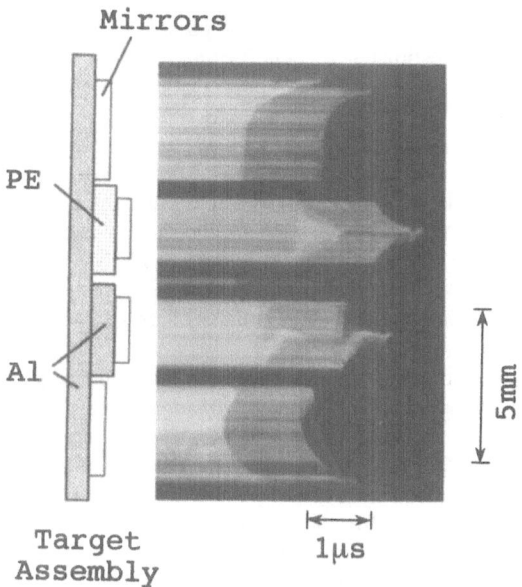

Fig. 20 Streak photograph of shock wave velocity measurement for polyethylene and aluminum by the impedance mismatch method.

4.3 Results and Discussion

4.3.1 Shock Wave Collision of Equal Strength

A reproduction of typical streak photographs of this kind of experiments is shown in Fig. 21. Note that the streak photographs do not directly correspond to the structure of the converging shock and Mach stem region. They show the instant when the mirror of each point is destructed. The obstructed and curved part of the photograph indicates the relatively later arrival of shock to the surface, i.e., the region corresponds to that of polyethylene. In this experimental condition, the shock wave in aluminum should have double wave structure, i.e., the elastic precursor and the succeeding plastic wave. It is seen from the photographs that the propagation of these waves is recorded in the region of aluminum near the aluminum-polyethylene boundary. These two wave fronts are seen as the boundary of light intensity changes. Consequently, two-step change can be observed even in polyethylene region near aluminum boundary, which corresponds to the two successive shock waves induced by the transmission of elastic-plastic waves of aluminum. In this sense, these two wave fronts does not mean the elastic-plastic waves of polyethylene. Photograph clearly shows the catchup process of the second shock to the first one in polyethylene. After catchup, only one wave front was recorded. This result is also supported by the experimental data of other experiments which are performed to study the Hugoniot compression curve of polyethylene for the similar pressure range. The present photographs are discussed later by using the Hugoniot data.

As is seen from the figure, the region of wave collision has no well-defined boundary, but is smoothly curved. This kind of wave structure has been observed commonly in any experiments of our experimental conditions. The wave structure has a linear part of plane waves, which is smoothly coupled to the curved region. The planar part of the structure should correspond to be the refracted wave front by the passage of the material interface of aluminum and polyethylene. The estimated wave deflection relative to the material interface is around 22 degrees. It should be mentioned that these analysis is valid only for the locally steady flow field approximation. The propagation velocity of the plane wave portion is found to be 3.0 km/s. Since the value of this shock velocity is slightly larger than the measured sound velocity of the specimen, the observed phenomena corresponds to the shock collision of weak shock waves. The result shows that in this case it is still possible to realize irregular shock reflection even in the case of weak shock waves. The shock velocity of the central part of the wave interaction region, i.e., the Mach stem propagates with the velocity equal to about 3.5 km/s. This value is not very fast, but is still large value compared with those of incident wave velocity. In this case, pressure enhancement remains slightly larger than two times, due to the large collision angle.

Although it is not easy to define the boundary between the colliding plane wave portion and wave interaction region, we have estimated the growth angle of Mach stem. It is about 26 degrees. This value is extremely large compared with the previously reported values for the different materials and initial conditions. As noted from the expression for reflected Hugoniot coefficients, Eq.(29), (30), the fundamental material parameters are also limited to B and γ_0, if one defines the shock Mach number for the description of shock strength. Hugoniot measurement for polyethylene showed that at least the non-linearity parameter B has a very large value. The present result can be partially understood by considering the above Hugoniot characteristics. There still requires the more precise time-resolved measurement of shock Hugoniot for polyethylene in this pressure region to have the definite conclusion on this flow problem.

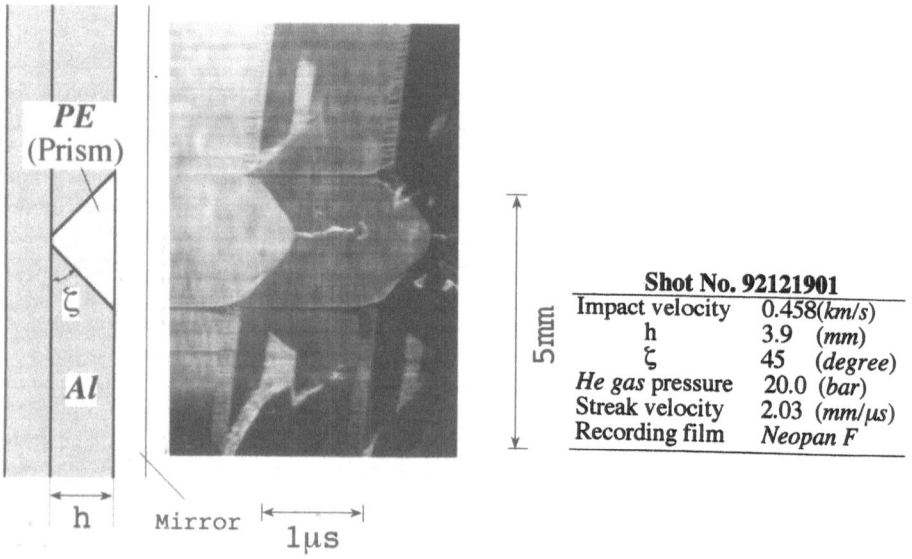

Fig. 21 Streak photograph of collision of two plane shock waves of equal strength.

4.3.2 Shock Wave Collision of Unequal Strength
Since the shock wave collision of equal strength might be important from the standpoint of the fundamental scientific research, that of unequal strength will find much importance in the

practice of material processing. We have measured the wave shape for the collision by using the assembly of Fig. 19(b). An example of the streak photographs is shown in Fig. 22. In the photograph, it is possible to identify the shock arrival time, but it contains unknown and spurious flashes after the shock arrival at the surface of the specimen.

In this case, a weaker wave portion of the photograph will be similar to the results of the previous section. Since a weaker shock (almost the same strength with the previous experiments) collides with a stronger wave, the interaction region is expected to be broader. Another shock in polyethylene is determined by the simple shock impedance mismatch from the aluminum driver plate. Shock velocity is estimated to be about 3.2 km/s. This value is not far from the central part of the wave interaction region by the collision of two shocks of equal strength. These two shock waves of different shock velocity meets in polyethylene at an angle of approximately 160 degrees. As is seen in Fig. 22, wave interaction region does not seem to be wider than the previous experiments. Since the interaction region is smoothly continued to the planar wave front, it is difficult to conclude how spread the interaction region. The difference of the initial condition of the present case from that of collision of same strength shocks is that the collision angle is larger in the present case, and one of the shock strength is also larger in this case. These conditions are supposed to be the easier situation for the irregular reflection.

The wave front envelope is more smoothly connected to the wave interaction region with smaller curvature than those in the case of same-strength shock collision. It is not known until now that the growth angle is peculiar to the initial condition of shock collision, or it can be attributed to the special material properties. The author feels that the latter explanation might be the case, i.e., by using the two material parameters, B and Y_0.

The study of these problems will be an important suggestion on how to design the material processing assemblies. That is, the impedance mismatch of the specimen and casing will have much effects on the loading and temperature history and distribution in the specimen.

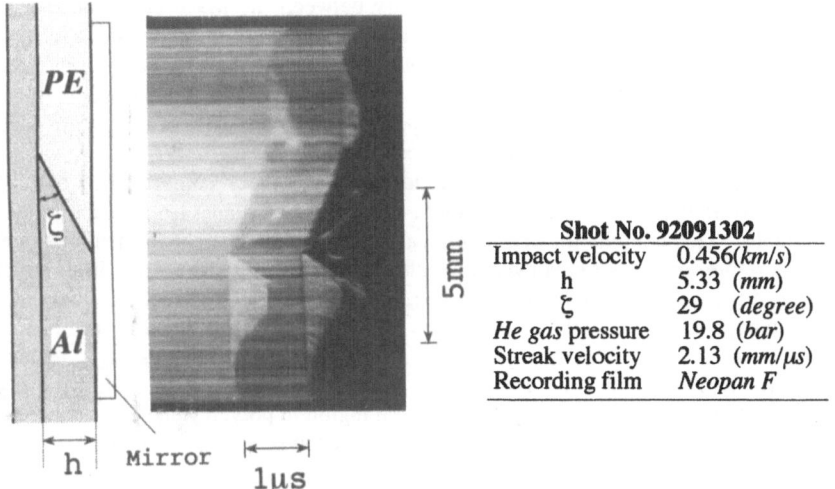

Shot No. 92091302	
Impact velocity	0.456(*km/s*)
h	5.33 (*mm*)
ζ	29 (*degree*)
He gas pressure	19.8 (*bar*)
Streak velocity	2.13 (*mm/μs*)
Recording film	*Neopan F*

Fig. 22 Streak photograph of collision of two plane shock waves of unequal strength.

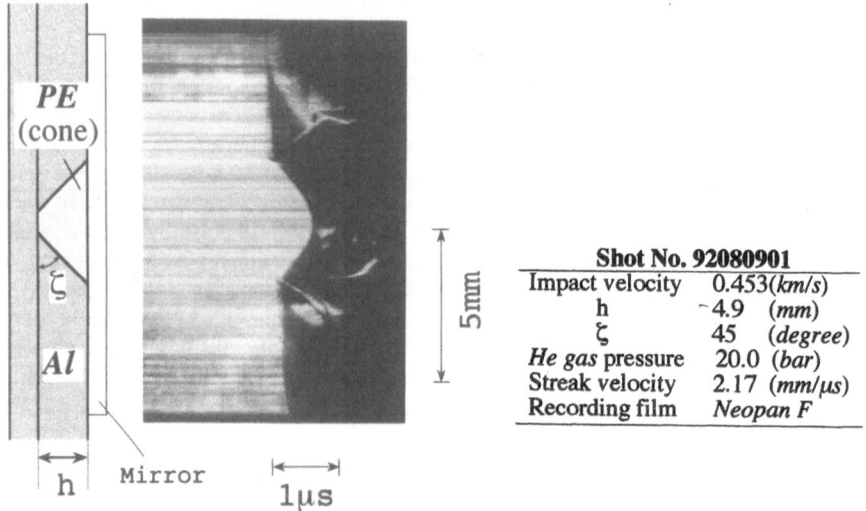

Fig. 23 Streak photograph of conically converging shock waves produced by flat plate impact.

4.3.3 Conically Converging Shock Wave Structure

We have performed a series of experiments by varying the shape of interface and propagation distance. The reproduction of a typical streak photograph is shown in Fig. 23. As shown from the photograph, the conically converging front is seen to be continuously curved, and it is very difficult to discriminate the boundary of the incident converging wave with the Mach stem, which is expected to develop at some definite angle. Estimating the shock velocity in aluminum region, the streak record is processed to represent the spatial profile of the wave at some particular instant. Although the boundary between the Mach stem and the incident converging wave cannot be identified clearly, the progression of the phenomena seems almost in a self-similar fashion, although the resultant wave structure does not seem to interpret very easily. The estimated collision angle of these two experiments is about 70 degrees, some 25 degrees larger than the top half angle of the polyethylene cone. These results are apparently different from those of plane shock collision experiments.

By comparing the present result with those of plane shock collision, it is found that in case of converging geometry, there cannot be recognized no planar front region in the streak photographs. This exhibits the effect of area convergence. The above-mentioned collision angle, therefore, includes the convergence effect. In this sense, the wave front structure is determined by two factors, the convergence effect and the wave interaction. It is difficult in this photograph to figure out which part of the front structure is attributed to each factor.

The incident shock velocity in polyethylene can be calculated by this angle, assuming the shock velocity in aluminum is constant. It is about 3.3 km/s even in the peripheral region of the cone. While the shock velocity at the central region in polyethylene is estimated to range 3.6 km/s. The shock pressure enhancement is supposed to be not very appreciable.

The most striking result of this experiment is that the continuously curved wave front suggests the strong non-linearity of the mechanical properties of polyethylene as inferred from other experiments. According to the published Hugoniot data for polyethylene in very high pressure region, the behavior of polyethylene exhibits no strong non-linearity. In other words, the effect of wave convergence and the effect of Mach reflection might attribute in a combined manner to the curved wave structure in polyethylene. The present results together with the Hugoniot measurements of polyethylene strongly suggest that the non-linear property of shock Hugoniot might be the key to understand these phenomena.

5. Concluding Remarks

This report describes two fundamental shock wave flow phenomena, shock wave collision and convergence with constant interest to the tie to material processing. The author tries to discuss the problem theoretically or rather analytically, although most of the treatment is an approximate one. These analytical approaches supported by the experimental data are, in some sense, much more instructive to understand the phenomena than the numerical calculations. Because the analytical expression gives clear insight to what is the key of the phenomena. We are trying to have simple understanding of these two problems by a few of material parameters.

The author has proposed to use the shock Mach number to describe these flow problems just like gases. Use of it will provide promising possibility of rather universal description. In the case of Mach reflection, most of the experimental results are categorized to weak shock collision, and they are piled to the many examples of von Neumann paradox. The necessary condition of this criterion is that the shock is weak. The author, however, have a hope to observe a real Mach reflection in condensed media. This might be realized in powdered materials or mixtures, where the effective shock Mach number is relatively large, since the effective equilibrium sound speed is small.

References

1. McQueen RG, Marsh SP, Taylor JW, Fritz JM and Carter WJ (1970) *High Velocity Impact Phenomena* (ed. R. Kinslow), Academic Press, New York.
2. Murri WJ, Curran DR, Peterson CF and Crewdson RC (1974) Advances in High-Pressure Research Vol. 4, ed. R.H. Wentorf, Jr, Academic Press, London and New York, p.1.
3. Shock Compression of Condensed Matter (1990), Proc. APS Top. Conf. Albuquerque, 1989, ed. S.C. Schmidt, J.N. Johnson and L.W. Davison, North-Holland Publ. Co., Amsterdam.
4. Rodean HC (1977) J. Appl. Phys., 48, 2384.
5. Prummer R (1987) *Explosiveverdichtung Pulveriger Substanzen*, Springer-Verlag, Berlin.
6. Meyers MA and Pak HR (1985) J. Mater. Sci., 20, 2133.
7. Mader CL (1966) Los Alamos Scientific Laboratory Report No. LA-3578.
8. Reaugh JE (1987) J. Appl. Phys., 61, 962.
9. Keeler RN and Royce EB (1971) *Physics of High Energy Density*, ed. P. Cardirola and H. Knoepfel , Academic Press, New York and London.
10. Nagayama K (1981) Appl. Phys. Lett., 38, 109.
11. Al'tshuler LV, Kormer SB, Bakanova AA, Petrunin AP, Funtikov AI and Gubkin AA (1962) J.E.T.P. (English Translation), 14, 986.
12. Krehl P, Hornemann U and Heilig W (1977) Proc. 11th Int. Symp. Shock Tubes & Waves, ed. B. Ahlborn, A. Hertzberg and D. Russel, Univ. Washington Press, Seattle, 1977.

13. Syono Y, Goto T and Sato T (1982) J. Appl. Phys., 53, 7131.
14. Neal T (1975) J. Appl. Phys., 46, 2521.
15. Ben-Dor G (1992) *Shock Wave Reflection Phenomena*, Springer-Verlag, New York.
16. Collier RJ, Burckhardt CB and Lin LH (1971) *Optical Holography*, Chap.11, Academic Press, New York and London.
17. Nagayama K and Mori Y (1992) submitted to Proc. 20th Int. Cong. High-Speed Photo. Photonics, 21-25 Victoria, Canada 1992.
18. Dubovik AS (1968) *Photographic Recording of High-speed Processes*, [English Translation], Pergamon Press, Oxford.
19. Schäfer FP (1989) *Dye Lasers - Topics in Applied Physics Vol.1*, ed. F. P. Schäfer, Chap. 1, Springer-Verlag, Berlin.
20. Mori Y and Nagayama K (1992) submitted to Proc. 20th Int. Cong. High-Speed Photo. Photonics, 21-25 Victoria, Canada 1992.
21. Nagayama K and Murakami T (1976) J. Phys. Soc. Japan 41, 359.
22. Nagayama K (1992) submitted to J. Phys. Soc. Japan.
23. Nagayama K, Oka T and Mashimo T (1982) J. Appl. Phys. 53, 3029.
24. Nagayama K and Mashimo T (1982) in : *High Field Magnetism, Proc. Int. Symp. High Field Magnetism, Osaka, Japan* (North Holland, Amsterdam, 1983)
25. Nagayama K and Mashimo T (1984) in : *Proc. 3rd Int. Conf. Megagauss Magnetic Field Generation and Related Topics, Novosibirsk, USSR* (Nauka, Moscow, 1985)
26. Bichenkov EI, Gilev SD and Trubachev AM (1980) Zh. Prikl. Mekh. Tech. Fiz. 5, 125.
27. Gilev SD and Trubachev AM (1982) Picima b JTF, 8, 914.
28. Bichenkov EI, Gilev SD and Trubachev AM (1984) in : *Proc. 3rd Int. Conf. Megagauss Magnetic Field Generation and Related Topics, Novosibirsk, USSR* (Nauka, Moscow, 1985)
29. Hawke RS, Duerre DE, Heubell JG, Keeler RN and Klapper H (1972) Phys. Earth Planet. Inter., 6, 44.
30. Fowler CM, Garn WB and Caird RS (1960) J. Appl. Phys., 31, 588.
31. Somon JP (1969) J. Fluid Mech., 38, 769.
32. Nellis WJ, Ree FH, Trainor RJ, Mitchell AC and Boslough MB (1984) J. Chem. Phys. 80, 2789.
33. LASL Shock Hugoniot Data, ed. by S.P. Marsh, Univ. Calif., Berkeley (Berkeley, Los Angels, London, 1980).
34. Barker LM and Hollenbach RE (1970) J. Appl. Phys., 41, 4208.

Index